CONTENTS

Aims and Scope

Behavioral Group Therapy: An Annual Review is a multidisciplinary serial publication which is aimed at presenting significant research reports, clinical group studies, review articles, and other material related to the application of behavioral treatment procedures in group therapy settings. The primary purposes of this *Annual Review* are: (1) to serve as an up-to-date and central source of information about the important recent clinical work and research involving behavioral group therapy, (2) to indicate areas of needed research and to stimulate scientific inquiry in these areas, and (3) to provide a forum for communication among those doing clinical work and research of this type. In order to serve a desirable and needed training function, its articles include case material, detailed descriptions of therapeutic methods, and transcriptions of group therapy sessions in order to illustrate how target behaviors are identified, how treatment methods are selected and applied, how difficulties are handled, and how therapeutic progress is evaluated. Articles which critically analyze and discuss important theoretical, social, professional, and ethical issues involved in the practice of behavioral group therapy also are included.

This *Annual Review* has two major thrusts, one theoretical and one practical. The theoretical thrust is designed to answer the question: "Will research and case material using behavioral group approaches enable us to understand group psychotherapy in general as a lawful, predictable process?" The practical thrust is directed to the question: "Can analyzing group therapy from a learning point of view direct our attention to practical improvements in the practice of group therapy, whatever the theoretical persuasion of the therapist?" Even more specifically, this series addresses the question of which behavioral procedures have proven to be particularly amenable to implementation in group therapy situations and with what types of problematic behaviors and patient populations.

Included in this series will be articles describing behavioral group treatment in the following areas: (1) emotional and avoidance behaviors, (2) habitual and addictive behaviors, (3) children's behaviors and family interactions, (4) adolescent and delinquent behaviors, (5) adult social and sexual behaviors, (6) behaviors of institutionalized patients and inmates, (7) miscellaneous target behaviors and techniques (e.g., group treatment of stuttering, chronic pain, etc.), and (8) behavioral marital therapy. Although the scope of the *Annual Review* is broad enough to include a wide range of content and settings, two general limitations are observed. First, the *Annual Review*'s focus is on the application of behavioral procedures in group therapy settings, rather than on the more general area of behavior modification with a group of patients, such as through the use of hospital token economies or structured classroom programs. Second, although significant work is being done in the area of applying behavioral techniques in vocational counseling and training groups, this *Annual Review* concentrates on group work in which more "clinical" behavioral problems such as those enumerated above are the targets for change.

Contributors

Robert P. Archer

Adult Programs Department, Florida Mental Health Institute, Tampa, Florida

Hubert E. Armstrong, Jr.

Department of Psychiatry and Behavioral Sciences, University of Washington School of Medicine and Harborview Community Mental Health Center, Seattle, Washington

Cornelius B. Bakker

Department of Psychiatry and Behavioral Medicine, Peoria School of Medicine, Peoria, Illinois

Marianne K. Bakker-Rabdau

Department of Psychiatry and Behavioral Sciences, University of Washington School of Medicine, Seattle, Washington

Francisco X. Barrios

Department of Psychiatry, Baylor College of Medicine, Houston, Texas

Jeffrey R. Bedell

Adult Programs Department, Florida Mental Health Institute, Tampa, Florida

Judith S. Blau

Department of Psychology, Moravian College, Bethlehem, Pennsylvania

Curtis D. Booraem

Booraem, Flowers and Associates, Newport Beach, California

Thomas J. Coates

Department of Psychiatry and Behavioral Sciences, Johns Hopkins University School of Medicine, Baltimore, Maryland

Nancy B. Cohn

Department of Psychology, University of Utah, Salt Lake City, Utah

Eugene Couture

Department of Psychology, West Virginia University, Morgantown, West Virginia

Mary Cray

Department of Psychology, West Virginia University, Morgantown, West Virginia

Gay R. Davis

The Epilepsy Center, Harborview Medical Center and the University of Washington School of Medicine, Seattle, Washington

Pamela Dickens

Department of Psychology, West Virginia University, Morgantown, West Virginia

Barry A. Edelstein

Department of Psychology, West Virginia University, Morgantown, West Virginia

John V. Flowers

Booraem, Flowers and Associates, Newport Beach, California, and Fairview State Hospital, Costa Mesa, California

Katherine E. Goethe

Department of Psychology, Virginia Commonwealth University, Richmond, Virginia

Karen A. Hartman

Department of Psychology, University of California at Riverside, Riverside, California

William M. Kallman

Department of Psychology, Virginia Commonwealth University, Richmond, Virginia

Barbara Kenney

Department of Psychology, University of California at Irvine, Irvine, California

Susan Kidder

Department of Psychology, University of California at Irvine, Irvine, California

Nancy Lusebrink

Department of Psychology, West Virginia University, Morgantown, West Virginia

Rochelle J. Mann

Department of Psychology, University of California at Irvine, Irvine, California

Herbert A. Marlowe, Jr.

Adult Programs Department, Florida Mental Health Institute, Tampa, Florida

Cheryl Perry

Heart Disease Prevention Program, Stanford University, Stanford, California

Henne R. Queisser

The Epilepsy Center, Harborview Medical Center and the University of Washington School of Medicine, Seattle, Washington

David Roth

Department of Psychiatry, Michigan State University, East Lansing, Michigan

Mary J. Rotheram

Department of Psychology, California State University, Los Angeles, California

Wayne R. Smith

The Epilepsy Center, Harborview Medical Center and the University of Washington School of Medicine, Seattle, Washington

Bruce Tapper

Santa Ana Psychiatric Hospital, Santa Ana, California

Johan Verhulst

Department of Psychiatry and Behavioral Sciences, University of Washington School of Medicine, Seattle, Washington

Garrie Wein

Department of Psychology, University of California at Irvine, Irvine, California

Brian L. West

Department of Psychology, Virginia Commonwealth University, Richmond, Virginia

Gary A. Williamson

Department of Psychiatry, University of Pittsburgh School of Medicine and Veterans Administration Medical Center, Pittsburgh, Pennsylvania

Contents of
Behavioral Group Therapy, 1979

Part One

Skills Training Approaches

The four chapters included in Part One reflect an important recent development in contemporary psychotherapy—the integration of purely educational procedures (such as structured curricula and lesson plans, didactic teaching, and formal exercises) and psychological techniques (such as modeling, behavioral rehearsal, feedback, and reinforcement) into psychoeducational skills training programs.

In Chapter 1 Jeffrey Bedell, Robert Archer, and Herbert A. Marlowe, Jr., review the literature supporting a skills deficit model of psychopathology and describe a problem solving skills training program for psychiatric hospital inpatients which is based on this model. Training program participants are taught to understand a problem solving model which has four major elements: (a) problem recognition, (b) problem definition, (c) generation of alternatives, and (d) evaluation of alternatives and decision. After learning this model through didactic lectures and structured exercises, participants work in groups to develop greater competence in employing the problem solving procedures. Of particular importance to this group work is the use of psychotherapeutic games, therapist-facilitated role-playing activities, client journals, and between-groups homework assignments to enhance problem solving skills.

Chapter 2 by Brian West, Katherine Goethe, and William Kallman describes the systematic use of cognitive training procedures and behavioral rehearsal in a group therapy program aimed at developing and maintaining heterosocial skills (i.e., skills relevant to social interactions with members of the opposite sex). Here again, as with other chapters in this section, a central role is given to coordinating therapeutic and educational procedures. Skills training is carried out in the context of a mixed-sex therapy group and is organized into progressive stages according to the difficulty of the skills, the complexity of skills integration, and the degree of contact among group members required for each stage. Predetermined behavioral goals for each stage are used to assess individual client development and to determine when to change the treatment emphasis.

Judith Blau, in Chapter 3, reports the results of a study designed to measure selected effects of participation in assertiveness training groups for married women. Particular attention is focused on the issues of: (a) whether assertiveness training groups actually help women to be more assertive, (b) what effects, if any, such training has on the satisfaction married couples experience in their relationships, and (c) whether groups which focus on general assertiveness skills and groups which focus on assertiveness skills specific to the marital situation have different effects. Group therapists and social skills trainers will also be interested in this chapter's discussion of the advantages of grouping clients on the basis of specific skills deficits rather than using more heterogeneous methods of grouping.

In Chapter 4 Barry Edelstein, Eugene Couture, Mary Cray, Pamela Dickens, and Nancy Lusebrink present evidence to support the view that chronic psychiatric patients tend to be deficient in day-to-day problem solving skills (particularly in identifying the main problem and selecting the best alternative solution), and they examine the effects of formalized skills training with this population. Their results indicate that chronic psychiatric patients can be taught (in a group setting) to improve their problem solving abilities, even by staff with very little or no training in problem solving or group therapy. In addition, this chapter discusses the advantages of a group format over individual instruction and pinpoints some potential problems which may be encountered in attempting to involve chronic patients in an active training program.

Chapter 1

A Description and Evaluation of a Problem Solving Skills Training Program

Jeffrey R. Bedell
Robert P. Archer
Herbert A. Marlowe, Jr.

Abstract

Skills training has become an accepted psychotherapeutic approach to alleviating behavioral dysfunctions and facilitating the growth of human potential. The finding that deficits in problem solving ability are significantly related to a wide range of psychopathologies indicates the need for training programs addressing the development of problem solving skills. This chapter describes an ongoing problem solving skills training program that develops cognitive and behavioral competency in a four-phase approach to problem solving. A controlled evaluation of the program indicated an increased awareness of effective problem solving skills and improved problem solving performance for a group of clients receiving training. This chapter suggests that problem solving may be an important basic coping skill that can be improved by a comprehensive skills training program.

The goal of skills training approaches in psychotherapeutic settings is to provide patients with those "physical, intellectual and emotional skills needed to live, learn and work in the community with the least possible amount of support from agents of the helping profession" (Anthony, 1977, p. 660). Gordon, Patterson, Bates, Weathers, Bedell, and Hatcher (1979) have subdivided the global concept of individual life skills into the following functional categories: survival skills (e.g., medication maintenance), daily living skills (e.g., personal hygiene), personal skills (e.g., anxiety management and relaxation), social skills (e.g., communication skills), vocation skills (e.g., job skills), leisure skills (e.g., recreational skills), and integrative skills (e.g., values clarification and personal goal setting). Each of these functional skill areas is viewed as a common core set of behaviors that are observable and measurable and which can be learned by the individual through the application of psychotherapeutic principles. Skills training programs are typically developed with strong attention given to the principles of behavior therapy, including sequential learning, immediate feedback, successive approximation, and stimulus and response generalization.

Within the area of skills training therapy, problem solving has generated a substantial degree of theoretical attention, but relatively little applied clinical evaluation (Mahoney, 1974). Thorndike (1898) provided a widely accepted definition of *problem* from his studies of problem solving in lower mammals. Thorndike stated, "A problem exists when the goal that is sought is not directly attainable by an act that is currently available in the animal's repertory; the solution calls for either a novel action or a new integration of available acts" (p. 26). Similarly, Davis (1973) has defined a problem as a "stimulus situation for which an organism does not have a ready response" (p. 12). Emphasizing situational components, D'Zurilla and Goldfried (1971) have used the term problematic situation, defined as a stimulus for which "no effective response alternative is immediately available" (p. 108). As defined by D'Zurilla and Goldfried, a problematic situation may include external stimuli, as well as participant produced stimuli, i.e., problematic stimuli originating in the attitudes, thoughts, or feelings of the individual. Problem solving has been defined by D'Zurilla and Goldfried as a "behavioral process, whether overt or covert in nature, which (a) makes available a variety of potentially effective response alternatives for dealing with the problematic situation, and (b) increases the probability of selecting the most effective response from among these various alternatives" (p. 108). The ability to solve problems, as defined above, is often considered to be a basic skill that is critical to the psychological well-being and social competence of the individual. Hence, the development of problem solving skills would be an important therapeutic endeavor, one worth consideration by behavior therapists. In their discussion of the relevance of behavior modification techniques to problem solving approaches, D'Zurilla and Goldfried (1971) have called for more systematic evaluations of the therapeutic effectiveness of the problem solving training programs now in use.

The purpose of this chapter is to describe and evaluate a problem solving training program. The chapter will begin with: (a) a review of the literature regarding the debilitating effects of behavioral deficits in living skills areas, including problem solving skills; (b) a discussion of the therapeutic effectiveness of skills training programs, including problem solving skills training; and (c) a presentation of a comprehensive problem solving model developed by D'Zurilla and Goldfried. Then the chapter will focus on a problem solving skills program that was developed in a residential treatment setting. This section will also present the results of an evaluation of the therapeutic outcome of the problem solving skills training program.

Living Skills Deficits and Behavioral Disorders

While there are many perspectives on behavioral disorders, the scope of this chapter will be limited to presentation of a skill deficit model of psychopathology. This model assumes that individuals who display abnormal behaviors frequently have impaired abilities. A person's behavior may be regarded as skill deficient if inappropriate or abnormal behavior develops as a result of marked limitations in the individual's capacity to respond to life situations with adaptive coping responses. The major assumptions of this perspective are that (a) many classes of psychiatric illnesses may be related to behavioral deficits in living skills areas, and (b) improvement or recovery may be facilitated as a result of effective skills training programs. We will begin by reviewing the literature on the first assumption.

A number of authors have pointed out that social isolation marked by a relative absence of social interactions constitutes the typical behavior pattern of hospitalized psychiatric patients (Bockhoven, 1963; Doty, 1975; Dunham & Weinberg, 1960; Ullmann, 1967). Also, this same behavior pattern has often been found to characterize the patients' prehospital adjustment (Barthell & Holmes, 1968). Further, the work of Platt and Spivack and their colleagues has consistently demonstrated that psychiatrically disturbed individuals are deficient in social skills when compared to normal populations (Intagliata, 1978; Platt, Scura, & Hannon, 1973; Platt, Siegel, & Spivack, 1975; Platt & Spivack, 1972b, 1974; Platt, Spivack, Altman, Altman, & Peizer, 1974).

Zigler and Phillips and their colleagues have also demonstrated in a series of studies that psychiatric patients have behavioral deficits in social-interpersonal maturity (Levine & Zigler, 1973; Phillips, Broverman, & Zigler, 1966; Phillips & Zigler, 1964; Zigler & Levine, 1973; Zigler & Phillips, 1960, 1961a, 1961b, 1962). Among their findings, Zigler and Phillips (1961a, 1961b) reported that patients who differed in levels of social competence exhibited different types of psychiatric disorders and that premorbid social competence was positively related to satisfactory posthospitalization life adjustment. Further, skills deficits have been identified as an important antecedent condition for depression (Libet & Lewinsohn, 1973), schizophrenia (Eisler, Blanchard, Fitts, & Williams, 1978), alcoholism (Marlatt, Kosturn, & Lang, 1975; Miller & Eisler, 1977), heterosocial anxiety (Curran, 1977), and sexual deviation (Abel, 1976). Thus, deficits in life skills have been associated with a wide range of psychopathologies and may constitute an important factor in these disorders.

Problem Solving Skills Deficits

Researchers have investigated behavioral deficits specifically related to problem solving skills in relation to a variety of mental health stressors and problems. Experimental studies based upon "normal" populations have compared the problem solving abilities of distressed and nondistressed marital dyads and reported that distressed couples engaged in significantly less problem solving behavior (Birchler, Weiss, & Vincent, 1975; Patterson, Hops, & Weiss, 1975; Patterson, Weiss, & Hops, 1976; Vincent, Weiss, & Birchler, 1975; Weiss, Hops, & Patterson, 1973).

Findings have also indicated problem solving deficits are of importance with regard to the child and adolescent psychiatric populations. Kempler and MacKenna (1975) conducted a case study on drug-dependent adolescents and reported the families of these children showed marked deficits in mutual problem solving. In an experimental investigation Ferreira and Winter (1965, 1966) found that families exhibiting marked pathology required longer periods to make decisions and derived less fulfillment from those decisions. Platt, Altman, and Altman (Note 1) and Platt et al. (1974) showed that adolescent psychiatric patients scored significantly lower on measures of interpersonal problem solving skills than normal adolescents. Several additional studies have indicated that suicidal adolescents have impaired problem solving abilities (Levinson & Neuringer, 1971; Rosenberg & Latimer, 1966; Shaw & Schelkun, 1965; Toolan, 1962; Tuckman & Cannon, 1962). Other research indicated deficiencies in problem solving on the part of disturbed latency-aged children (Shure & Spivack, 1972), delinquent girls (Frank & Quinlan, 1976), disadvantaged preschool children (Shure & Spivak, 1975; Shure, Spivack, & Jaeger, 1971; Spivack & Shure, 1974; Shure & Spivack, Note 2, Note 3, Note 4), hyperactive boys (Kuchta, 1975), and impulsive teenagers (Spivack & Levine, 1963).

In the area of adult substance abusers Marlatt et al. (1975), Miller and Eisler (1977), and Intagliata (1978) reported that alcoholics show significant deficiencies in interpersonal problem solving skills. Studies by Platt et al. (1973) and Platt (1975) found that compared to nonaddicts, heroin addicts demonstrated less ability to conceptualize means of reaching goals in real-life problem situations.

Among other areas of adult psychopathology, problem solving deficits have been noted. Self-destructive behavior has been linked to diminished problem solving capacity by a number of writers (Binswanger, 1958; Cavan, 1928; Dublin & Bunzel, 1933; Menninger, 1938; Neuringer, 1964;

Shneidman, 1957, 1961, 1969). Platt and Spivack (1972a, 1973) found that short-term psychiatric inpatients were deficient in a wide range of interpersonal problem solving skills, and Trower, Bryant, and Argyle (1978) reported a trend toward less problem solving skill among lower class psychiatric patients. Platt et al. (1975) examined the degree to which adult clients differed from normals on what constitutes a hierarchy of effective and socially appropriate means of solving interpersonal problems. Psychiatric subjects were found to be deficient in their ability to generate viable means of solving personal and interpersonal problems and in providing valid rationales for their solutions. The two groups did not differ, however, in their capability to recognize the relative effectiveness of means of resolving the problems when these were presented for inspection.

Overall, the literature strongly supports the conclusion that problem solving deficits, as a subset of the adaptive skills behavioral area, are significantly associated with a variety of child, adolescent, and adult behavioral problems and disorders.

Skills Training Programs

The skills deficit model of abnormal behavior is further supported by reports of the therapeutic effectiveness of skills-oriented therapeutic interventions. Margules and Anthony (1976) reviewed the skills training approach to the treatment of psychiatric patients and found substantial evidence for two conclusions. First, psychiatric patients suffering from a wide variety of psychopathologies can learn a variety of physical, emotional, and intellectual skills; second, when skills training is integrated into a comprehensive treatment program, patients' rehabilitation outcomes can be significantly improved. Various skills training programs have been shown to be effective with a wide variety of populations, including current and former psychiatric patients (Giebink, Stover, & Fahl, 1968; Goldsmith & McFall, 1975; Goldstein, 1974; Goldstein, Martens, Hubben, Van Belle, Schaaf, Wiersma, & Goedhart, 1973; Goldstein, Sprafkin, & Gershaw, 1976; Gutride, Goldstein, & Hunter, 1973), moderately distressed couples (Patterson, Hops, & Weiss, 1973), and withdrawn schizophrenics (Hersen, Edelstein, Turner, & Pinkston, 1975).

Despite the extensive literature on the importance of problem solving skills, especially for psychiatric populations, few treatment programs have been designed to specifically teach problem solving skills to psychiatric patients. Siegel and Spivack (1976) reported two pilot studies of a problem solving skills program conducted with chronic patients. The results of these studies, conducted for the purpose of determining whether the problem solving program could change the problem solving cognitions of chronic patients, were inconclusive. Coche and Flick (1975) also investigated problem solving training among residential psychiatric patients.

Results indicated significant gains in problem solving skills for the group exposed to problem solving training. In addition, the length of hospital stay was significantly lower for both the problem solving group and a placebo group as compared to a control group.

Other problem solving programs have been shown to be effective in studies with alcoholics (Intagliata, 1978), couples in the early stages of marital conflict (Patterson et al., 1973), hyperactive children (Kuchta, 1975), impulsive boys (Giebink et al., 1968), and college students anxious about vocational decisions (Mendonca & Siess, 1976). Finally, Sarason (1968) and Sarason and Ganzer (1969a, 1969b, 1973) have shown a skills training program based on modeling of appropriate behaviors to be effective in improving the problem solving skills of adolescents who had been adjudicated as delinquent.

In summary, problem solving training programs have shown promise in terms of effectively increasing participants' skill levels. The results of training programs with residential psychiatric patients, however, have not been conclusive. Certainly, the therapeutic outcomes have not been as impressive as expected, considering the central and basic role of problem solving skills in psychopathology. Part of this problem has resulted from the fact that treatment programs were often poorly conceptualized and shallow in content and procedures. The potential for the clinical utilization of a well-designed problem solving program would appear great. The most comprehensive and thoughtful model for problem solving training has been presented by D'Zurilla and Goldfried (1971).

A Problem Solving Training Model

The D'Zurilla and Goldfried (1971) problem solving model is based on a careful analysis of the problem solving process and empirical support for various critical components and assumptions. D'Zurilla and Goldfried (1971) identified five stages of the problem solving process: (a) general orientation, (b) problem definition and formulation, (c) generation of alternatives, (d) decision making, and (e) verification and evaluation. Each of these stages will be described in the following sections.

General Orientation

The first stage in the problem solving process, general orientation, refers to the mental set of the problem solver. As is generally recognized, a particular approach to a problem will result in a certain range of responses to the problem (Scheerer, 1963). Heppner (1978) formally defined problem set as an "inferred predisposition that influences a client to behave in a certain manner which can be facilitative or disruptive, depending on whether it moves the client toward or away from an effective procedure and solution" (p. 367).

A wide range of mental sets is possible, evinced by everyday colloquialisms ranging from "it's hopeless" to "no problem at all." D'Zurilla and Goldfried posited that the optimal mental set for problem solving should include the following: (a) a verbal acknowledgment that problematic situations constitute a normal aspect of everyday life, coupled with an acknowledgment that it is possible to cope with most of these situations in an effective manner; (b) a recognition or identification of problematic situations when they occur; and (c) an inhibition of the tendency to either respond to the first impulse or to avoid the situation in toto by doing nothing at all.

The importance of perceiving problems as "normal" and "solvable" has been empirically related to effective problem solving performance. In an early study Bloom and Broder (1950) found that successful problem solvers in an experimental task had greater confidence in their general ability to solve problems presented to them than their unsuccessful counterparts. Nash (1975) found children who were encouraged to view solutions as possible were significantly better in their capability to produce original solutions than children who were not given that set. Other studies have presented evidence that people who maintain positive expectancies of their ability to control environmental reinforcers in general tend to be better problem solvers (Elstein, Kagan, Shulman, Jason, & Loupe, 1972; Lefcourt, 1966; Rotter, 1966; Sprafka & Elstein, Note 5).

Much evidence also exists to support the hypothesis that good problem solvers are not impulsive but engage systematically in several problem solving behaviors over a period of time. Bloom and Broder (1950) found unsuccessful problem solvers tended to be random and unanalytic in their approaches to the problem and to select solutions impulsively on the basis of what initially "seems right." In theoretical articles Thurstone (1924) and Dollard and Miller (1950) pointed out that the first step in all but the most simple types of reasoning is to "stop and think." Their position was that immediate responses preclude sufficient time for selection of the most appropriate and effective course of action. One of the basic foundations of Osborn's (1963) creative problem solving process, supported by findings of Parnes and Meadow (1959, 1960), is the ability to inhibit the tendency to utilize the first idea made available.

Epstein, Hallahan, and Kaufman (1975) reviewed the literature on exceptional children and concluded that impulsive cognitions play an important role in the problem solving skills deficiencies found in this population. In support of this view Klein, Blockovich, Buchalter, and Huyghe (1976) found that reflective children made significantly fewer errors than impulsive children on a convergent problem solving task. In sum, these studies clearly indicate that inhibition of impulsivity is related to improved problem solving.

Problem Definition and Formulation

Getzels (1975) quoted Einstein as saying, "The formulation of a problem is often more essential than its solution" (p. 180). Following the individual's general orientation to the problem, the second stage of the problem solving process consists of the operational definition of central aspects of the problem and the formulation or classification of situational elements into subgroups such as (a) relevant or irrelevant, (b) primary or secondary goals, and (c) major subproblems, issues, or conflicts. The importance of specificity and comprehensiveness in this stage is heavily emphasized by Crutchfield (1969), D'Zurilla and Goldfried (1971), Heppner (1978), Osborn (1963), and Parnes, Noller, and Biondi (1977). Real-life problematic situations are often ambiguous, presenting only partial amounts of the information that would be helpful in solving the problem. The greater the initial ambiguity of the problem, the more important it becomes to clearly define and formulate the solution. The underlying principle of this problem solving stage is that the individual is forced by the formulation process to make relevant what may at first seem to be irrelevant (Crutchfield, 1969; D'Zurilla & Goldfried, 1971; Getzels, 1975).

Empirical support for the importance of problem definition and formulation comes from several sources. Bloom and Broder (1950) found successful problem solvers tended to translate difficult and unfamiliar terms into simpler, more concrete or familiar terms. In contrast, unsuccessful problem solvers tended to accept vague or unfamiliar concepts. DeGroot (1965) and Simon and Barenfield (1969) noted that successful problem solvers first gather all available information and facts. Reviews of the problem solving literature (Bourne, Ekstrand, & Dominowski, 1971; Davis, 1966; Davis, Train, & Manske, 1968; Duncan, 1963) consistently conclude that the more information a person possesses about a problem, the easier it is to solve the problem. Reed and Abramson (1976) found specification of problem subgoals facilitated the solution to larger and more complex problems. Elstein et al. (1972) found accuracy of medical diagnosis was associated with greater thoroughness of cue acquisition and with greater accuracy of cue interpretation. Similar findings are reported by Sprafka and Elstein (Note 5). Kleinmuntz (1968) found that the more experienced clinical neurologists focused on those initial patient interview questions which were most likely to maximize information yield, i.e., they knew what information to obtain and took adequate time to do so before final diagnosis.

In sum, substantial empirical evidence supports the importance of a problem definition and formulation stage to problem solving.

Generation of Alternatives

The purpose of this third stage, which occurs after the problem has been recognized and defined, is to generate possible solutions appropriate to the particular problematic situation. While generating solutions may appear to be an easy task, Heppner (1978) identified social inhibitory factors which may limit solution generation. The characteristics of the problem, such as limited clues about its nature, can restrict the problem solver's performance (Bourne et al., 1971; Burack & Moos, 1956; Burke, Maier, & Hoffman, 1966; Maier, 1930; Marks, 1951; Reid, 1951). The first response of the problem solver may also serve as a cue to emit a restricted range of behaviors, thereby excluding possibly useful alternatives (Janke, 1962; Scheerer, 1963). Not only can initial responses elicit an inappropriate response pattern, but past experiences of the problem solver may introduce incompatible or inhibitive response tendencies with relation to a new problem (Bugelski & Huff, 1962; Jacobus & Johnson, 1964; Safren, 1962; Schulz, 1960). Heppner (1978) cited several studies which indicate that emotionally laden stimuli associated with frustration or failure reduce the number of alternative problem solving patterns in animals, as well as human subjects (Everall, 1935; Feather, 1959; Hamilton, 1916; Hamilton & Krechevsky, 1933; Lantz, 1945; Patrick, 1934a, 1934b; Solley & Stagner, 1956).

One of the most widely accepted techniques for developing alternatives is that of brainstorming (Osborn, 1963). Brainstorming has four basic rules. Criticism of ideas is ruled out and judgment is withheld until the decision-making stage is reached. Free-wheeling discussion is welcome, and seemingly inappropriate, unrelated, or wild ideas are encouraged in the hope of stimulating other ideas. Emphasis is placed on quantity, based on the assumption that a larger list of ideas will result in a higher probability of an optimal solution. Finally, combination and improvement should be attempted after the list of possible solutions is exhausted. There have been a number of studies which have supported the efficacy of brainstorming as a technique for solution generation (Bouchard, 1971; McDonald, 1975; Parnes et al., 1977; Turner & Rains, 1965). In addition, D'Zurilla and Goldfried (1971) reviewed a number of studies which support each of the four rules of brainstorming.

In addition to brainstorming, a number of other techniques have been developed for creating alternative solutions. Among these are: creative problem solving (Parnes, 1972, 1977); synectics (Gordon, 1971; Gordon & Poze, 1972; Prince, 1968); general semantics (Noller, 1971); creative analysis (Upton & Samson, 1963); bionics (Papanek, 1972, 1977); sociodrama (Torrance, 1975; Torrance & Myers, 1970); idea checklists (Parnes, 1977); forced relationships (Parnes, 1977); attribute listing (Crawford, 1954; Parnes, 1977); morphological synthesis (Allen, 1962, 1966; Parnes, 1975; Zwicky, 1957); alteration (deBono, 1977); functional visualization (Eberle, 1977; Taylor, 1977; Walkup, 1967); space visualization (Eberle, 1977); and creative dramatics (Davis, Helfert, & Shapiro, 1973). Many of these techniques have empirical support, although none is as extensively utilized as the brainstorming methodology.

Decision Making

The fourth stage of the problem solving process is decision making. Decision making can be defined as the process of selecting one course of action from a number of alternatives (Bross, 1953). The D'Zurilla and Goldfried process entails several skills identified as important by Clark, Gelatt, and Levine (1965) and Krumboltz (1965), including: (a) assessing the probabilities of success of a particular solution, (b) assessing preferences for a solution, and (c) assessing the consequences of various alternatives. Additionally, Egner and Jackson (1978) have detailed a decision making model that encompasses a cyclical process of (a) probability determination, (b) consequences assessment, and (c) flexible choice leading again to probability determination.

The importance of decision making skills is evinced by the fact that consistent positive relationships have been established between decision making and career maturity (Crites, 1973; Dilley, 1965; Holland, Gottfredson, & Nafziger, 1975; Jepsen, 1974; Mathewson & Orton, 1963), socioeconomic status (Goldstein, 1974; Jepsen, 1974), and grade point average (Jepsen, 1974). Despite the importance of decision making, Egner and Jackson (1978) reported that a review of the literature indicates no comprehensive research studies related to teaching decision making skills. Egner and Jackson (1978), however, do report increased career maturity with a pilot program designed to teach career decision making skills.

Verification and Evaluation

The fifth and final problem solving stage involves (a) testing the plan of action selected in the decision making stage and (b) matching the outcome of the response to some standard. Heppner (1978) posited that the two skills required for successful completion of this stage include the ability to identify outcomes and the ability to compare these outcomes to a standard. The ability to detect unsuccessful outcomes can cue the individual to resume the problem solving process. Equally as important, the ability to identify a successful outcome signals the end of a particular problem solving process, thereby eliminating needless further efforts, and serves as a reinforcing event which increases the probability of the individual's engaging in future systematic problem solving attempts.

The D'Zurilla and Goldfried model of problem solving is comprehensive and supported by empirical studies. Many of these component skills have been utilized in training programs and have been shown to be associated with effective problem solving. It is, therefore, reasonable to expect that a comprehensive skills training program that addresses the major components of this five-phase model would be effective in increasing mental health clients' problem solving ability. In the next section a program based upon the D'Zurilla and Goldfried model of problem solving will be described and evaluated.

Problem Solving Skills Training Program

This section of the chapter describes an ongoing group therapy program designed to teach clients a sequential method of problem solving. This training program attempts not only to provide guidance and support regarding clients' current problems but also to provide a set of skills that can be employed in the solution of future problems as they inevitably arise. In utilizing the model for psycho-educational treatment programs proposed by Bedell and Weathers (1979), it was first necessary to define the skills to be trained and to define the general curriculum of the program. The skills that were the foundation of this program were adapted from the work of D'Zurilla and Goldfried (1971) and are listed in Table 1.1. As noted in this table, the program teaches skills in four major processes of problem solving, each of which is divided into subskill areas.

Table 1.1
Problem Solving Processes Trained in the Therapy Group

1. Problem Recognition
   A. Acknowledge the problem
   B. Identify the problem
   C. Accept the problem as your own responsibility
2. Problem Definition
   A. Break the problem down into workable units
   B. State the problem in such a way that it appears solvable
3. Alternatives Generation
   A. Brainstorm
   B. Change frame of reference
   C. Adapt a solution from a similar problem
   D. Make an idea check list
   E. Draw a problem diagram
4. Alternatives Evaluation and Decision
   A. Predict possible outcomes of alternative solutions
   B. Determine likelihood of each outcome
   C. Determine desirability of each outcome
   D. Rank order alternative solutions
   E. Make decision

Table 1.2 briefly describes the 12-session program, which is divided into two sections: (a) Awareness Training and (b) Skill Enhancement/Generalization Training. In the Awareness Training section clients are taught to understand the problem solving model, including the major phases of (a) problem recognition, (b) problem definition, (c) alternatives generation, and (d) alternatives evaluation and decision. This information is taught through the educational techniques of didactic lectures and learning exercises. In the Skill Enhancement/Generalization Training section the clients develop behavioral competence in using the problem solving procedures. First, they practice each of the four phases of problem solving separately; then they employ them sequentially as a complete process. Skill Enhancement and Generalization sessions use psychotherapeutic games (Bedell & Weathers, 1979; Weathers, Bedell, Marlowe, Gordon, Adams, & Reed, 1978), therapist-facilitated activities, client journals, and homework assignments to develop the individual's problem solving competencies. In order of sequence, clients first practice problem solving with hypothetical problems in a role-play format and later apply the procedures to their personal real-life problems.

As can be seen in Table 1.1, the processes used to develop skill in each of the four phases of problem solving are very similar. Each phase of problem solving (e.g., problem recognition) is the primary target of one Awareness Training session, one Skill Enhancement/Generalization session using a psychotherapeutic game, and one session that focuses on less structured exercises and the clients' personal problems as monitored in their journals. A detailed description of the therapeutic activities relating to any one of the four problem solving phases will, therefore, serve as an illustration for the other three parts of the program.[1] The Alternatives Generation phase of treatment will be illustrated here. These activities encompass the primary content of Sessions 3, 9, and 10 of the therapy program.

Awareness Training Section

The Awareness Training sessions attempt to provide the client with an understanding of important concepts, build the necessary vocabulary, and lay the groundwork for subsequent development of problem solving behaviors. With regard to the Alternatives Generation phase of the program, a didactic lecture describes the concept and five different methods

---

[1] The content and procedures of the entire program are described in the *Problem Solving Therapist's Manual* available from the first author, Jeffrey R. Bedell, Ph.D., Florida Mental Health Institute, 13301 N. 30th Street, Tampa, Florida 33612.

Table 1.2
A Description of the Problem Solving Skills Training Program

Awareness Training Section

Session 1. Introduction to Problem Solving
Problem Recognition
- Didactic lectures and exercises

Session 2. Problem Definition
- Didactic lectures and exercises

Session 3. Alternatives Generation
- Didactic lectures and exercises

Session 4. Alternatives Evaluation and Decision
- Didactic lectures and exercises

Skill Enhancement/Generalization Training Section

Session 5. Problem Recognition
- Structured role-play practice, with a psychotherapeutic game

Session 6. Problem Recognition
- Introduction of journal use
- Practice of problem recognition, with exercises, journals
- Homework: Completion of problem recognition exercises in journal

Session 7. Problem Definition
- Structured role-play practice, with a psychotherapeutic game

Session 8. Problem Definition
- Review of problem recognition homework
- Practice of problem definition, with exercises, journals
- Homework: Completion of problem recognition exercises in journal

Session 9. Alternatives Generation
- Structured role-play practice, with a psychotherapeutic game

Session 10. Alternatives Generation
- Review of homework
- Practice of alternatives generation, with exercises, journals
- Homework: Completion of alternatives generation exercises in journal

Session 11. Alternatives Evaluation and Decision
- Structured role-play practice, with a psychotherapeutic game
- Practice of four-phase problem solving

Session 12. Alternatives Evaluation and Decision
- Review of homework
- Practice of alternatives evaluation and decision making, with exercises, journals
- Practice of four-phase problem solving
- Wrap-up and review

for generating alternative problem solutions. Subsequent to the lecture, exercises that illustrate these five methods of generating alternatives are performed by the group, under the direction of the therapist.

Sample didactic lecture. The content and style of the awareness training sessions are well defined by the program but can be modified to fit the orientation of individual therapists and the needs of the client group. A summary of an example lecture will illustrate the Awareness Training part of the Alternatives Generation Program and will be presented here. This lecture, of course, is similar in style and purpose to those presented on Problem Recognition (Session 1), Problem Definition (Session 2), and Alternatives Evaluation and Decision (Session 4).

> The purpose of this lecture is to introduce you to the alternatives generation step of the problem solving model. In particular, it will be teaching you five different methods of generating alternative solutions to problems. You will also be given the opportunity to practice these methods by doing exercises as we have done in past groups.
>
> You have now learned to recognize your problem and define it so that it is in a more workable form. Now you are going to use your creativity and come up with some ideas about alternative solutions to your problem. This process of generating alternative solutions is Step 3 of our problem solving model. We are giving you five methods so that you can pick the best course of action for working with your own problem. However, if you have any suggestions of your own about how to generate alternative solutions, you can also bring these up in our group.
>
> It is important to have several different solutions to a given problem because the first solution that comes to your mind may not be the best. It is likely that when solving a problem you may either act too fast by taking the first solution that comes to your mind or you might be so disorganized in your approach that the more you try, the more confused you become. Hopefully, these methods will help you avoid these common mistakes in solving your own problems. Therefore, you can use either one or several of these alternative generating techniques.
>
> The first method we will talk about is *brainstorming.* Brainstorming is best done in a group, but it can be done individually. When you use brainstorming, you think up, say out loud, and then write down solutions or activities without stopping to think about how good or bad the ideas are. This approach is probably very different from the way you normally think of alternative solutions. Normally, a person would think of one solution, decide how good or bad it was, and then either use it or think of another solution. By doing this, you make two mistakes. You usually accept the alternative solution that you like without comparing it with other good solutions that you have not yet considered. Also, you tend to decide quickly how well you like an alternative, thereby sometimes rejecting solutions that might have worked well had you looked at them more closely. What brain-

storming does is allow you to bring up all your ideas so that you have many alternatives from which to choose. By getting all of your ideas out into the open first, you can then take the time to compare them and to decide how well you like them and how well they solve your problem. In summary, brainstorming has two ground rules: (a) to say any idea you think of and (b) not to decide how good or bad the idea is until after you come up with all the ideas and solutions of which you can think.

The second method of generating alternative solutions is *changing your frame of reference (or point of view)* and then attempting to think of a solution. Very often people get so bogged down with a problem that they cannot see the many possible alternatives that are open to them. Often you are so busy looking for a complicated solution that you miss a very simple one, or the first solution which comes to mind is not a very good one.

A third method for generating alternative solutions is *adapting a solution from a similar problem* to your problem. It is often the case that the solution selected by one person in one situation would not apply to another person in the same situation. However, this solution could be modified somewhat to meet the needs of the person and the situation. For example, suppose that you have been looking for a job and you have been given the runaround at the employment agency. You hear that a friend of yours has been having a similar problem in a different agency and that this friend found a solution. This person, who is somewhat aggressive in your opinion, demanded to see the director of the employment agency. Upon seeing the director, your friend insisted upon some direct answers about job leads. Realizing that you are not an aggressive person, could you use this person's solution for your problem or could you modify your friend's solution to fit your needs?

A fourth method of generating alternative solutions is *writing out a checklist of ideas* to consider as you think about possible solutions. This list could vary from person to person and from problem to problem. Basically, the list should resemble a shopping list to stimulate new ideas for solving your problems. For example, a sample list might contain these questions: (a) Can I change my behavior? (b) Can I change other people's behavior? (c) Can I get more money? (d) Can I change my living situation?

The last method of generating alternative solutions is *drawing a problem solving diagram*. The purpose of the problem diagram is to provide a schematic or concrete representation of the steps involved in solving problems. This procedure is analogous to the flow chart which is used by professionals to organize a step-by-step course of action for solving a work problem or implementing a plan. To use this method, you draw a diagram of the problem using symbols to represent different parts of the problem. You might use a stick figure to represent a person and a dollar sign to represent money. Arrows can be used to indicate the direction or sequence of events in solving your problem.

For each of these five methods for generating alternative solutions there is an exercise to help reinforce the didactic lecture. These exercises are performed during the same therapy session in which the didactic lectures are presented. For illustration, the brainstorming exercise from the Alternatives Generation session will be described.

Examples. In the brainstorming exercise the participants stay in a large group, and the therapist poses problems to them from a prepared list or makes use of other relevant situations. As each of the sample problems is presented to the group, they attempt to brainstorm solutions. The therapist writes the clients' suggestions on the blackboard, reinforcing clients who suggest solutions, no matter how impractical or farfetched. Clients are discouraged from judging or evaluating the suggested solutions, since this behavior is inconsistent with the brainstorming procedure. After a sufficient list of solutions is compiled, the therapist discusses with the group the process of brainstorming. Key points developed by the therapist are that creativity is facilitated by not screening ideas and that ideas which seem to be impractical help to produce other valuable and practical ones.

Four other alternatives generation exercises are conducted to illustrate the concepts of (a) changing one's frame of reference, (b) using a similar problem, (c) writing a question list, and (d) using a problem diagram. In each exercise clients discuss the process of the exercise and the value of the procedure as it relates to problem solving. For example, to introduce the concept of changing one's frame of reference, the therapist requests participants to connect nine dots in a figure, using only four straight lines without lifting their pencils from the paper. Most people fail to solve this puzzle because the normal frame of reference dictates that all the lines be drawn within the boundaries of the matrix of dots. Once the person is able to change the frame of reference and utilize the area outside of the matrix of dots, he can usually solve the puzzle.

Another method of generating alternative solutions is using a solution from a similar problem, as already mentioned. To introduce this concept, the therapist conducts an exercise in which a simulated problem is solved by altering a solution found to be effective by a friend. To illustrate the alternative generation process of writing a question list, the therapist groups participants in triads and gives a series of simulated problems. As they think about alternative solutions, they generate a series of questions that can be posed to each alternative. Each new question may in turn generate other questions and alternatives. For example, in response to an alternative, the question "Can I change my behavior?" may be suggested. This question in turn may stimulate another question, "Can I rearrange my living situation?" Other standard questions of this sort are involved to guide the alternatives generation process. Triads later compare and discuss their lists of questions.

The final method of generating alternatives is drawing a problem diagram. Each triad of clients is provided a problem situation (e.g., buying a new car) and a series of boxes arranged in random order which contain the names of various responses (e.g., bank, insurance company, registration center, car inspection center, etc.). Group members are requested to read the problem and discuss in their small groups how to arrange the boxes to represent the most successful problem solving sequence. Each triad is told to reach a consensual solution, and the various response sequences developed are then shared and discussed between groups.

The lectures and exercises, such as brainstorming, are short, personally involving, and interesting to the clients. Often, however, it may take some skilled prompting to get the client to use the technique illustrated. This is only natural, since these clients may never have used these techniques before. It has always been the case, however, that after some examples are modeled by the therapist at least one client will catch on to the idea, and this client can then help to explain the exercise to others and encourage their participation. Also, the exercises that are conducted during the first few therapy sessions require little in the way of self-disclosure and risk to the group members. The safety of these early exercises encourages responding and helps to establish cohesion and trust among group members so that personal problems can be dealt with in later sessions.

Skill Enhancement and Generalization Training Section

The Skill Enhancement section of the program is designed to facilitate the development of behavioral competency in solving problems. Generalization training procedures attempt to ensure that effective problem solving behavior will be performed by the clients in their natural community environments. The process employed to develop problem solving behavior and to promote generalization relies heavily on a psychotherapeutic game technology, in conjunction with structured therapist-facilitated exercises, open-ended group processing, and homework assignments. The context of the group session centers around role-play practice with hypothetical problem situations and also includes the clients' real-life problems. Much of the material used in the therapy group regarding real-life problems is taken from the clients' daily journals, which are kept to record progress occurring during the therapy group and in everyday life.

Behavioral skill in each of the four phases of problem solving is developed sequentially. Continuity between treatment sessions is maintained through the use of the clients' journals. For example, clients recognize four personal problems during Session 6 and enter them into their journals. In Session 8 these problems are submitted to the problem definition process, and the results are also written in the journal. Subsequently, in

Session 10 alternative solutions are developed for these problems, and in Session 12 the journal materials are used to conduct the Alternatives Evaluation and Decision phases of problem solving.

Homework assignments are the primary mechanism for generalization training in this therapy program. Homework that encourages the application of newly learned problem solving skills is assigned in Sessions 6, 8, and 10. These homework experiences are reviewed at the beginning of the following therapy session. Success is reported and help is given when difficulty has been experienced in applying the problem solving processes outside of the therapy group. Using homework in conjunction with the client journal has proved to be useful in facilitating generalization of skills. For example, a client might make a notation in her journal regarding how she had attempted to deal with a family problem that had occurred one weekend while visiting at home. This journal entry is subsequently brought up in the therapy group, and the four-phase problem solving approach is applied to its solution. After gaining new knowledge and skill by working on this problem in the group, the client moves to a homework assignment to deal with a problem during the next family visit and then must report the outcome of the homework assignment to the group. This process could be repeated a number of times until the client deals effectively with the identified problem situation.

Using the treatment sessions that deal with the Alternatives Generation phase of the program as an example, the next section of the chapter will illustrate the Skill Enhancement and Generalization area of the problem solving program. Alternative generation skills are practiced and developed in Sessions 9 and 10. They are also practiced in Sessions 11 and 12, when clients use all the problem solving procedures together in four-phase problem solving exercises. The processes of Sessions 9 and 10 will be described here.

Sample skill enhancement game. The psychotherapeutic game has the following components: (a) a game board that guides group interactions, (b) game cards to set the role-play situations, (c) a spinner to instruct the client as to which method of generating alternatives to use, and (d) vote cards that are used by fellow group members to provide performance feedback. The essence of the clients' training experiences during the alternatives generation game is as follows. Four clients are situated around a playing board. A player selects a problem card from the stack and reads the problem to the group. This sets the role-play situation about which the client will practice alternatives generation. Examples of problem cards that may be drawn are: "You are trying to find a job. You have gone through the phone book and called a few places but don't know what else to do," and "You want to visit your parents for the weekend, but your car is in the shop for repairs." Next, the client whirls a spinner located on the game board and performs the task to which the spinner points. It will be one of four tasks: (a) brainstorm, (b) change frame of reference, (c) make an idea checklist, or (d) make a problem diagram. The client's task is to elaborate on the problem posed by the problem card and to develop alternatives in the fashion indicated by the spin. Having done this, the player is given evaluation feedback by the other players. The number of points awarded to the player for performance (better performance yields more points according to an established criterion) determines the number of spaces to move around the game board. The next player proceeds as did the first, and the game continues for at least 45 minutes. The first player to go around the board is the winner of the game. As the game continues, each player has repeated trials with various ways of generating alternatives to problem solutions, is given feedback on performance, and is able to observe and provide feedback on the performance of others.

Therapeutic games are used in this program as an important therapeutic modality. Clients find them to be involving, novel, and nonthreatening. The peer orientation of the game facilitates the positive therapeutic effects that group members can have on one another. The more skilled clients are often able to help their gaming partners in a manner that is not always achievable by the mental health professional.

Sample exercises and journal use. Session 10 also centers on developing the behavioral components of alternative generation. This session, however, calls for much more intensive therapist involvement, is much less structured, and deals with more personalized problems. The therapist conducts two exercises during this therapy session, the first of which is a large group exercise. Everybody selects a problem from a check list which is part of the problem solving materials. Once everyone has a problem, the therapist facilitates the generation of problem solutions, using each of the alternative generation methods (brainstorming, etc.). The therapist guides the exercise by modeling and prompting as necessary, making sure that everyone receives an opportunity to practice and develop alternative generation skills.

The second exercise in this session uses the clients' journals. The group members break into triads and use the various alternative generation methods on as many as possible of the four personal problems each client has identified in the journals. This session builds upon work done in previous therapy sessions in which clients have broken problems into workable units. Each client shares a problem with the other members of the triad and generates alternative solutions. They receive help in doing so from other group members and the group leader. In turn, each client is able to help other members of the triad to develop alternative solutions for their problems. After each person has had an opportunity to work in this exercise (usually in short sessions totaling 1 hour), members re-form into a large group and discuss their experiences with the guidance of the therapist.

In addition to collecting case history data and clinical impressions regarding the effectiveness of the problem solving training program, we have made a systematic and ongoing effort at the Florida Mental Health Institute to empirically evaluate this program based upon objective data sources. The following section describes our preliminary findings in this regard.

EVALUATION OF THE PROBLEM SOLVING PROGRAM

A controlled evaluation of the problem solving treatment program was conducted to determine its effectiveness with severely disturbed patients. The problem solving therapy group described previously was presented to clients in the context of an active, multifaceted treatment program. Treatment outcome effects could, therefore, be potentially related to a number of therapeutic interventions experienced by the subjects. Because of this source of confounding, an experimental design that controlled for the effects of all treatment groups except the problem solving therapy was employed. The present outcome evaluation compared the degree of improvement of subjects who experienced the entire treatment program, with the exception of problem solving. The control group engaged in recreational activities for a time period equivalent to that spent in problem solving therapy by the experimental group. Differences in outcome between these two groups could, therefore, reasonably be seen as related to the effects of the problem solving therapy.

Because of the broad treatment experience provided to both the experimental and control groups, all subjects could be expected to make significant gains in general psychological and social functioning. Further, some of these general gains might be reflected in improvement in problem solving ability if measurement of treatment outcome was not uniquely sensitive to gains in the problem solving skills area. Therefore, a battery of problem solving assessments was developed to evaluate cognitive and behavioral learning that could be directly linked to problem solving training.

The raw data in the present study were improvement scores obtained by subtracting pretreatment test scores from posttreatment scores. The resulting scores indicated the amount of improvement on that measure. Results were evaluated with one-tailed t tests since it is reasonable to expect that neither the control nor experimental subjects would significantly decrease in problem solving ability as a result of treatment.

Subjects

Subjects were 20 adult psychiatric patients (10 experimental, 10 control) who participated in residential treatment. These subjects were selected from a pool of 26 subjects and matched as closely as possible on pretest scores obtained on the Problem Solving Knowledge and Information Test (PKIT) to ensure that groups were initially equivalent in problem solving knowledge before the first phase of treatment began.

A comparison of the experimental and control groups indicated an average age of 25.4 years and 30.2 years, respectively. There were six male subjects in the experimental group and five males in the control. The diagnoses of the subjects in the two groups were also comparable, with 60% being psychotic (mostly schizophrenic) and the rest being neurotic (mostly depressive), except for one experimental subject who was diagnosed as having a schizoid personality. All of the experimental and control subjects had less than 4 months of hospitalization prior to beginning this experiment, were involved in residential treatment, and received medical-psychiatric treatment in addition to the procedures described here.

Assessment

The three assessment devices that were developed for this study were specifically designed to measure (a) knowledge and information about problem solving, (b) self-reports of problem solving ability in everyday situations, and (c) behavioral performance in a simulated problem situation.

Knowledge and information test. The Problem Solving Knowledge and Information Test (PKIT) is a 32-item multiple choice instrument that measures the subject's knowledge level as it pertains to information about problem solving processes. A general information score is derived by adding the number of correct responses. Item content included problem recognition and definition, generation of alternate solutions to problems, and decision making. A large pool of test items was rationally determined, and preliminary versions of the scale were piloted on psychiatric patients. Items were selected for inclusion in the final version of the scale based upon appropriateness of reading level, clarity, and lack of ambiguity.

Self-evaluation test. The Problem Solving Self-Evaluation Test (PSET) assesses the individual's self-evaluation of typical problem solving abilities. The inventory consists of 40 items requiring subjects to rate the frequency with which they perform certain problem solving behaviors. Ratings are done on a four-point Likert Scale: almost never, sometimes, often, and almost always. Subjects rate their behavior in areas such as how they define problems, develop alternative solutions to problems, and make decisions in daily living. Other inventory items deal with frequency of hesitation before taking action to solve a problem, a consideration of alternative solutions, and frequency of solicitation of advice from significant others. The content of problem solving items included in this scale has been identified as important in the research literature (e.g., D'Zurilla & Goldfried, 1971).

Performance evaluation test. The Problem Solving Performance Evaluation Test (PPET) measures the subject's ability to solve a problem situation that might be typically experienced in everyday life. The assessment consists of two parallel scripts which describe, in a realistic fashion, a multifaceted social problem being experienced by an individual. The problem situation involves interpersonal, personal, and financial components which must be considered and resolved if the person is to solve the problem.

Two problem exercises are used, one for pretreatment testing and one for posttreatment. Two alternative forms were necessary so that subjects' posttreatment performance would not be unduly affected by their experience with the pretreatment exercise. The parallel forms were carefully constructed so that they present problems of equal difficulty which encompass the same or similar problem components.

After the subjects studied the script of the problem simulation, they were required to respond in five ways: (a) identify the problem in the story, (b) list basic elements of the problem, (c) indicate possible problem solutions, (d) rank the possible solutions in order of their feasibility regarding the solution of the main problem, and (e) provide reasons for selecting each of the possible problem solutions. The five responses were scored by a staff member who had been specially trained to reliably use specified criteria and who was blind to the respondent's group assignment and to whether the responses were pre- or posttreatment assessments.

Since the PPET is a measure of actual problem solving performance, it was not readily subject to the effects of faking or social desirability. The validity of using a simulation exercise as a performance evaluation is well accepted; many behavioral therapists use simulation exercises for evaluation.

Results

The improvement scores for the Problem Solving Therapy Group and the Control Group were analyzed for the PKIT, PSET, and PPET by using the t statistics. The Problem Solving Therapy Group and the Control Group, as indicated previously, were matched on the basis of pretreatment PKIT scores. A comparison of pretreatment scores of the experimental and control groups on each of the three outcome measures (PKIT, PSET, PPET) yielded no significant t tests. Consequently, it may be assumed that before treatment, the experimental and control groups were equivalent on the three outcome variables used in this research and that differences between the experimental and control groups on the improvement scores could be attributed to the effects of the Problem Solving Therapy Group.

Knowledge and information. The average improvement on the PKIT for the Problem Solving Therapy Group was 4.6 responses, as compared to 2.4 for the Control Group. The results of the analysis conducted on the individual PKIT change scores were significant ($t\ (18) = 1.84$, $p < .05$), indicating a greater degree of improvement associated with the Problem Solving Group.

Self-evaluation. The average improvement in self-evaluated problem solving, as measured by the PSET, was 1.47 points for the therapy group and 0.6 for the control group. Although these scores indicate greater change for the therapy group, the statistical comparison with the control group on the individual improvement scores was not significant (t (12) = 0.12, nonsignificant). The Problem Solving Group did not appear to result in significantly increased self-evaluation of ability.

Performance evaluation. Performance in the simulated problem situation (PPET) showed an average 2.7 point improvement for the Problem Solving Therapy Group and a slight decline (−0.2) in average performance for the controls. The comparison of individual performance changes was highly significant (t (18) = 3.92, $p < .01$), indicating greater behavioral improvement for the subjects receiving problem solving therapy.

Discussion and Summary

The skill deficit conceptualization of psychopathology and behavioral disorders is gaining increased acceptance as basic concepts become better developed. To alleviate the skill deficits underlying behavioral disorders, treatment programs are increasingly adopting a skills training orientation. In general, research suggests that this is an effective method of treating behavioral disorders.

Also, problem solving deficits have been shown to be significantly related to a variety of psychopathologies in child, adolescent, and adult populations. Despite the rather extensive literature dealing with problem solving skills, there have been few reports of therapeutic skills training programs. Evaluations of programs that have been reported have shown inconclusive evidence of their effectiveness.

The present chapter has reviewed the problem solving literature, described a comprehensive treatment and skills training program, and reported the results of a controlled evaluation. The problem solving process proposed by D'Zurilla and Goldfried (1971) appears to be the most comprehensive and best supported model in the literature. This model was cast into a psychoeducational format and defined the primary content of the skills training program. The results of the evaluation indicated significantly greater improvement as a function of problem solving skills training with regard to clients' knowledge and understanding of the process and the behavioral performance of problem solving activities. These results are consistent with the orientation of a psychoeducational program since it emphasizes both the acquisition of cognitive understanding and behavioral performance.

Of course, further evaluations of the effectiveness of skills training therapy groups are necessary. Long-term follow-up evaluations should examine the generalizability of skills acquisition when patients return to community settings. It will also be important to refine the problem solving program based upon empirical investigations to eliminate unnecessary parts and add components that would improve outcome. These efforts could lead to a shortening of the program, which would save staff and client time and facilitate involvement in other areas of treatment. Finally, it is crucial to determine in future research those client populations that can most benefit from skills training programs. In addition to evaluating variables such as client diagnosis, such studies should develop behavioral assessment measures of problem solving readiness skills. Applications of this type of program to outpatient clients and in growth and prevention training groups for nonmental health populations would seem appropriate and might prove useful. In any case, the results of the present report provide information suggesting that skills training programs can be operated to effectively increase the problem solving abilities of mental health clients.

Reference Notes

1. Platt, J. J., Altman, N., & Altman, D. S. *Dimensions of interpersonal problem solving thinking in adolescent psychiatric patients.* Paper presented at the Annual Meeting of the Eastern Psychological Association, Washington, D.C., May 1973.
2. Shure, M. B., & Spivack, G. *A preventive mental health program for young "inner city" children: The second (kindergarten) year.* Paper presented at the Annual Meeting of the American Psychological Association, Chicago, August 1975.
3. Shure, M. B., & Spivack, G. *Interpersonal cognitive problem solving intervention: The second (kindergarten) year.* Paper presented at the Annual Meeting of the American Psychological Association, Chicago, August 1975.
4. Shure, M. B., & Spivack, G. *Interpersonal problem solving intervention for mother and child.* Paper presented at the Annual Meeting of the Society for Research in Child Development, New Orleans, June 1977.
5. Sprafka, S. A., & Elstein, A. S. *What do physicians do? An analysis of diagnostic reasoning.* Paper presented at a Conference on Research in Mental Judgment, Michigan State University, East Lansing, May 1974.

References

Abel, G. G. Assessment of sexual deviation in the male. In M. Hersen & A. S. Bellack (Eds.), *Behavioral assessment: A practical handbook.* New York: Pergamon Press, 1976.

Allen, M. *Morphological creativity.* Englewood Cliffs, N.J.: Prentice-Hall, 1962.

Allen, M. *Psychodynamic synthesis.* West Nyack, N.Y.: Parker, 1966.
Anthony, W. A. Psychological rehabilitation—a concept in need of a method. *American Psychologist,* 1977, *32,* 658-662.
Barthell, C. N., & Holmes, O. S. High school yearbooks: A non-reactive measure of social isolation in graduates who later become schizophrenic. *Journal of Abnormal Psychology,* 1968, *73,* 313-317.
Bedell, J. R., & Weathers, L. R. A psycho-educational model of skill training: Therapist-facilitated and game-facilitated applications. In D. Upper & S. M. Ross (Eds.), *Behavioral group therapy, 1979: An annual review.* Champaign, Ill.: Research Press, 1979.
Binswanger, L. The case of Ellen West. In R. May, E. Angel, & H. F. Ellenberger (Eds.), *Existence.* New York: Basic Books, 1958.
Birchler, G. R., Weiss, R. L., & Vincent, J. P. Multi-method analysis of social reinforcement exchange between maritally distressed and non-distressed spouse and stranger dyads. *Journal of Personality and Social Psychology,* 1975, *31,* 349-362.
Bloom, B. S., & Broder, L. J. *Problem solving processes of college students.* Chicago: University of Chicago Press, 1950.
Bockhoven, J. S. *Moral treatment in American psychiatry.* New York: Springer, 1963.
Bouchard, T. J. Whatever happened to brainstorming? *Industry Week,* 1971, *22,* 3-7.
Bourne, L. E., Jr., Ekstrand, B. R., & Dominowski, R. L. *The psychology of thinking.* Englewood Cliffs, N.J.: Prentice-Hall, 1971.
Bross, I. D. *Design for decisions.* New York: Macmillan, 1953.
Bugelski, B. R., & Huff, E. M. A note on increasing the efficiency of Luchins' mental sets. *American Journal of Psychology,* 1962, *75,* 665-667.
Burack, B., & Moos, D. Effect of knowing the principle basic to solution of a problem. *Journal of Educational Research,* 1956, *50,* 30-33.
Burke, R. J., Maier, N. R. S., & Hoffman, L. R. Function of hints in individual problem solving. *Journal of American Psychology,* 1966, *79,* 389-399.
Cavan, R. S. *Suicide.* Chicago: University of Chicago Press, 1928.
Clark, R., Gelatt, H. B., & Levine, L. A decision-making paradigm for local guidance research. *Personnel and Guidance Journal,* 1965, *44,* 114-119.
Coche, E., & Flick, A. Problem-solving training groups for hospitalized psychiatric patients. *Journal of Psychology,* 1975, *91,* 19-29.
Crawford, C. *How to make training surveys.* Los Angeles: Author, 1954.
Crites, J. O. *Career maturity inventory: Research and theory.* Monterey, Calif.: CTB/McGraw-Hill, 1973.
Crutchfield, R. S. Nurturing the cognitive skills of productive thinking. In *Life skills in school and society.* Washington, D.C.: Association for Supervision and Curriculum Development, 1969.
Curran, J. P. Skills training as an approach to the treatment of heterosexual social anxiety. *Psychological Bulletin,* 1977, *84,* 140-157.

Davis, G. A. Current status of research and theory in human problem-solving. *Psychological Bulletin,* 1966, *66,* 36-54.
Davis, G. A. *Psychology of human problem solving: Theory and practice.* New York: Basic Books, 1973.
Davis, G. A., Helfert, C. J., & Shapiro, G. R. Let's be an ice-cream machine!: Creative dramatics. *Journal of Creative Dramatics,* 1973, *7,* 37-48.
Davis, G. A., Train, A. J., & Manske, M. E. Trial and error vs. "insightful" problem solving: Effects of distraction, additional response alternatives, and longer response chains. *Journal of Experimental Psychology,* 1968, *76,* 337-340.
deBono, E. Information processing and new ideas–lateral and vertical thinking. In S. J. Parnes, R. B. Noller, & A. M. Biondi (Eds.), *Guide to creative action.* New York: Scribner's, 1977.
DeGroot, A. D. *Thought and choice in chess.* The Hague, Netherlands: Mouton, 1965.
Dilley, J. S. Decision-making ability and vocational maturity. *Personnel and Guidance Journal,* 1965, *44,* 423-427.
Dollard, J., & Miller, N. E. *Personality and psychotherapy.* New York: McGraw-Hill, 1950.
Doty, D. W. Role-playing and incentives in the modification of the social interaction of chronic psychiatric patients. *Journal of Consulting and Clinical Psychology,* 1975, *43,* 676-682.
Dublin, L. I., & Bunzel, B. *To be or not to be: A study of suicides.* New York: Random, 1933.
Duncan, C. P. Effect of instruction and information on problem solving. *Journal of Experimental Psychology,* 1963, *65,* 41-45.
Dunham, H. W., & Weinberg, S. K. *The culture of the state mental hospital.* Detroit: Wayne State University Press, 1960.
D'Zurilla, T., & Goldfried, M. Problem solving and behavior modification. *Journal of Abnormal Psychology,* 1971, *78,* 107-126.
Eberle, R. F. Developing imagination through scamper. In S. J. Parnes, R. B. Noller, & A. M. Biondi (Eds.), *Guide to creative action.* New York: Scribner's, 1977.
Egner, J. R., & Jackson, D. J. Effectiveness of a counseling intervention program for teaching career decision-making skills. *Journal of Counseling Psychology,* 1978, *25,* 45-52.
Eisler, R. M., Blanchard, E. B., Fitts, H., & Williams, J. G. Social skill training with and without modeling for schizophrenic and non-psychotic hospitalized psychiatric patients. *Behavior Modification,* 1978, *2,* 147-172.
Elstein, A. S., Kagan, N., Shulman, L. S., Jason, H., & Loupe, M. J. Methods and theory in the study of medical inquiry. *Journal of Medical Education,* 1972, *47,* 85-92.
Epstein, M. H., Hallahan, D. P., & Kaufman, J. M. Implications of the reflectivity-impulsivity dimension for special education. *Journal of Special Education,* 1975, *9,* 11-25.

Everall, E. E. Perseveration in the rat. *Journal of Comparative Psychology*, 1935, *19*, 343-368.

Feather, N. T. Subjective probability and decision under uncertainty. *Psychological Review*, 1959, *66*, 310-315.

Ferreira, A. J., & Winter, W. D. Family interaction and decision-making. *Archives of General Psychiatry*, 1965, *13*, 214-223.

Ferreira, A. J., & Winter, W. D. Stability of interactional variables in family decision-making. *Archives of General Psychiatry*, 1966, *14*, 352-355.

Frank, S., & Quinlan, D. M. Ego development and female delinquency: A cognitive-developmental approach. *Journal of Abnormal Psychology*, 1976, *85*, 505-510.

Getzels, J. W. Problem finding and the inventiveness of solutions. *Journal of Creative Behavior*, 1975, *9*, 180-182.

Giebink, J. W., Stover, D. O., & Fahl, M. A. Teaching adaptive responses to frustration to emotionally disturbed boys. *Journal of Consulting and Clinical Psychology*, 1968, *32*, 366-368.

Goldsmith, J. B., & McFall, R. M. Development and evaluation of an interpersonal skills-training program for psychiatric inpatients. *Journal of Abnormal Psychology*, 1975, *84*, 51-58.

Goldstein, A. P., Martens, J., Hubben, J., Van Belle, H. A., Schaaf, W., Wiersma, H., & Goedhart, A. The use of modeling to increase independent behavior. *Behaviour Research and Therapy*, 1973, *11*, 31-42.

Goldstein, A. P., Sprafkin, R. P., & Gershaw, N. J. Structured learning therapy: Training for community living. *Psychotherapy: Theory, Research and Practice*, 1976, *13*, 374-377.

Goldstein, M. S. Academic careers and vocational choices of elite and non-elite students at an elite college. *Sociology of Education*, 1974, *47*, 491-510.

Gordon, R. E., Patterson, R. L., Bates, H. D., Weathers, L. R., Bedell, J. R., & Hatcher, M. *Modular training & treatment* (Monograph Series No. 2, Report No. 2). Tampa, Fla.: Human Resources Institute, University of South Florida, February 1979.

Gordon, W. J. J. *The metaphorical way of learning and knowing.* Cambridge, Mass.: Porpoise Books, 1971.

Gordon, W. J. J., & Poze, T. *Strange and familiar.* Cambridge, Mass.: Porpoise Books, 1972.

Gutride, M. W., Goldstein, A. P., & Hunter, G. F. The use of modeling and role-playing to increase social interaction among asocial psychiatric patients. *Journal of Consulting and Clinical Psychology*, 1973, *40*, 408-415.

Hamilton, G. V. A study in perseverance reactions in primates and rodents. *Behavior Monographs*, 1916, *3*, 11-14.

Hamilton, J. A., & Krechevsky, I. Studies in the effect of shock upon behavior plasticity in the rat. *Journal of Comparative Psychology*, 1933, *16*, 237-253.

Heppner, P. P. A review of the problem-solving literature and its relationship to the counseling process. *Journal of Counseling Psychology,* 1978, *25,* 366-375.

Hersen, M., Edelstein, B. A., Turner, S. M., & Pinkston, S. G. Effects of phenothiazines and social skills training in a withdrawn schizophrenic *Journal of Clinical Psychology,* 1975, *31,* 588-593.

Holland, J. L., Gottfredson, G. D., & Nafziger, D. H. Testing the validity of some theoretical signs of vocational decision-making ability. *Journal of Counseling Psychology,* 1975, *22,* 411-422.

Intagliata, J. C. Increasing the interpersonal problem-solving skills of an alcoholic population. *Journal of Consulting and Clinical Psychology,* 1978, *46,* 480-498.

Jacobus, K. A., & Johnson, N. F. An experimental set to adopt a set. *Psychological Reports,* 1964, *15,* 737.

Janke, P. K. The effects of a "compound set" on learning efficiency. *American Journal of Psychology,* 1962, *75,* 445-451.

Jepson, D. A. Vocational decision-making patterns among non-college-aspiring adolescents. *Journal of Vocational Behavior,* 1974, *4,* 283-297.

Kempler, H. C., & MacKenna, P. Drug abusing adolescents and their families: A structural view and treatment approach. *American Journal of Orthopsychiatry,* 1975, *45,* 223-224.

Klein, G. A., Blockovich, R. N., Buchalter, P. S., & Huyghe, L. Relationships between reflection-impulsivity and problem solving. *Perceptual Motor Skills,* 1976, *42,* 67-73.

Kleinmuntz, G. (Ed.). *Formal representation of human judgment.* New York: Wiley, 1968.

Krumboltz, J. D. Behavioral counseling: Rationale and research. *Personnel and Guidance Journal,* 1965, *44,* 383-387.

Kuchta, J. C. The differential operation and modification of certain cognitive styles in hyperactive and non-hyperactive boys (Doctoral dissertation, Catholic University of America, 1975). *Dissertation Abstracts International,* 1975, *35,* 6098. (University Microfilms No. 75-13036)

Lantz, B. Some dynamic aspects of success and failure. *Psychological Monographs,* 1945, *59* (1, Whole No. 271).

Lefcourt, H. M. Internal versus external control of reinforcement: A review. *Psychological Bulletin,* 1966, *65,* 206-220.

Levine, J., & Zigler, E. The essential-reactive distinction in alcoholism: A developmental approach. *Journal of Abnormal Psychology,* 1973, *81,* 242-249.

Levinson, M., & Neuringer, C. Problem-solving behavior in suicidal adolescents. *Journal of Consulting and Clinical Psychology,* 1971, *37,* 433-436.

Libet, J. M., & Lewinsohn, P. M. The concept of social skill with special reference to the behavior of depressed persons. *Journal of Consulting and Clinical Psychology,* 1973, *40,* 304-312.

Mahoney, M. J. *Cognition and behavior modification.* Cambridge, Mass.: Ballinger, 1974.

Maier, N. R. F. Reasoning in humans: I. On direction. *Journal of Comparative Psychology,* 1930, *10,* 115-143.

Margules, A., & Anthony, W. A. Skills training programs for psychiatric patients. *Journal of Applied Rehabilitation Counseling,* 1976, *7,* 40-49.

Marks, M. R. Problem solving as a function of the situation. *Journal of Experimental Psychology,* 1951, *41,* 74-80.

Marlatt, G. A., Kosturn, C. F., & Lang, A. R. Provocation to anger and opportunity for retaliation as determinants of alcohol consumption in social drinkers. *Journal of Abnormal Psychology,* 1975, *84,* 652-659.

Mathewson, R. H., & Orton, J. W. Vocational imagery and vocational maturity of high school students. *Journal of Counseling Psychology,* 1963, *10,* 384-388.

McDonald, J. T. Brainstorming as a training technique. *Health and Social Services Journal,* 1975, *85,* 300-301.

Mendonca, J. D., & Siess, T. F. Counseling for indecisiveness: Problem solving and anxiety management training. *Journal of Counseling Psychology,* 1976, *23,* 330-347.

Menninger, K. *Man against himself.* New York: Harcourt and Brace, 1938.

Miller, P. M., & Eisler, R. M. Assertive behavior of alcoholics: A descriptive analysis. *Behavior Therapy,* 1977, *8,* 146-149.

Nash, W. R. The effects of warm-up activities on small group divergent problem solving with young children. *Journal of Psychology,* 1975, *89,* 237-241.

Neuringer, C. Rigid thinking in suicidal individuals. *Journal of Consulting Psychology,* 1964, *28,* 54-58.

Noller, R. G. Some applications of general semantics in teaching creativity. *Journal of Creative Behavior,* 1971, *5,* 50-56.

Osborn, A. F. *Applied imagination: Principles and procedures of creative problem solving.* New York: Scribner's, 1963.

Papanek, V. J. *Design for the real world.* New York: Pantheon, 1972.

Papanek, V. J. Tree of life: Bionics. In S. J. Parnes, R. B. Noller, & A. M. Biondi (Eds.), *Guide to creative action.* New York: Scribner's, 1977.

Parnes, S. J. *Creativity: Unlocking human potential.* Buffalo: D.O.K. Publications, 1972.

Parnes, S. J. A discussion of creative problem solving institutes: A program for balanced growth. *Journal of Creative Behavior,* 1975, *9,* 23-29.

Parnes, S. J. Idea stimulation techniques. In S. J. Parnes, R. B. Noller, & A. M. Biondi (Eds.), *Guide to creative action.* New York: Scribner's, 1977.

Parnes, S. J., & Meadow, A. Effects of "brainstorming" instructions on creative problem solving by trained and untrained subjects. *Journal of Educational Psychology,* 1959, *50,* 81-83.

Parnes, S. J., & Meadow, A. Evaluation of persistence of effects produced by a creative problem solving course. *Psychological Reports,* 1960, *7,* 357-361.

Parnes, S. J., Noller, R. G., & Biondi, A. M. (Eds.). *Guide to creative action.* New York: Scribner's, 1977.

Patrick, J. R. Studies in rational behavior and emotional excitement: I. Rational behavior in human subjects. *Journal of Comparative Psychology,* 1934, *18,* 1-22. (a)

Patrick, J. R. Studies in rational behavior and emotional excitement: II. The effect of emotional excitement on rational behavior in human subjects. *Journal of Comparative Psychology,* 1934, *18,* 153-195. (b)

Patterson, G. R., Hops, H., & Weiss, R. L. A social learning approach to reducing rates of marital conflict. In R. Stuart, R. Liberman, & S. Wilder (Eds.), *Advances in behavior therapy.* New York: Academic Press, 1973.

Patterson, G. R., Hops, H., & Weiss, R. L. Interpersonal skills training for couples in early stages of conflict. *Journal of Marriage and the Family,* 1975, *37,* 295-303.

Patterson, G. R., Weiss, R. L., & Hops, H. Training of marital skills: Some problems and concepts. In H. Leitenberg (Ed.), *Handbook of behavior modification and behavior therapy.* Englewood Cliffs, N.J.: Prentice-Hall, 1976.

Phillips, L., Broverman, I. K., & Zigler, E. Social competence and psychiatric diagnosis. *Journal of Abnormal Psychology,* 1966, *71,* 209-214.

Phillips, L., & Zigler, E. Role orientation, the action-thought dimension and outcome in psychiatric disorder. *Journal of Abnormal and Social Psychology,* 1964, *68,* 381-389.

Platt, J. Addiction proneness and personality in heroin addicts. *Journal of Abnormal Psychology,* 1975, *84,* 303-306.

Platt, J. J., Scura, W. C., & Hannon, J. R. Problem solving thinking of youthful incarcerated heroin addicts. *Journal of Community Psychology,* 1973, *1,* 278-281.

Platt, J., Siegel, M., & Spivack, G. Do psychiatric patients and normals see the same solutions as effective in solving interpersonal problems? *Journal of Consulting and Clinical Psychology,* 1975, *43,* 279.

Platt, J. J., & Spivack, G. Problem solving thinking of psychiatric patients. *Journal of Consulting and Clinical Psychology,* 1972, *39,* 148-151. (a)

Platt, J. J., & Spivack, G. Social competence and effective problem solving thinking in psychiatric patients. *Journal of Clinical Psychology,* 1972, *28,* 3-5. (b)

Platt, J. J., & Spivack, G. Performance in important areas of life as a source of positive self-regard. *Proceedings of the 81st Annual Convention of the American Psychological Association,* 1973, *8,* 235-236. (Summary)

Platt, J. J., & Spivack, G. Means of solving real-life problems: I. Psychiatric patients versus controls, and cross-cultural comparisons of normal females. *Journal of Community Psychology,* 1974, *2,* 45-48.

Platt, J. J., Spivack, G., Altman, N., Altman, D., & Peizer, S. B. Adolescent problem solving thinking. *Journal of Consulting and Clinical Psychology,* 1974, *42,* 787-793.

Prince, G. M. The operational mechanism of synectics. *Journal of Creative Behavior,* 1968, *2,* 31-40.

Reed, S. K., & Abramson, A. Effect of the problem space on subgoal facilitation. *Journal of Educational Psychology,* 1976, *68,* 243-246.

Reid, J. W. An experimental study of "analysis of the goal" in problem solving. *The Journal of General Psychology,* 1951, *44,* 51-69.

Rosenberg, P. H., & Latimer, R. Suicide attempts by children. *Mental Hygiene,* 1966, *50,* 354-359.

Rotter, J. B. Generalized expectancies for internal versus external control of reinforcement. *Psychological Monographs,* 1966, *80* (1, Whole No. 609).

Safren, M. A. Associations, sets and the solution of word problems. *Journal of Experimental Psychology,* 1962, *74,* 40-45.

Sarason, I. G. Verbal learning modeling and juvenile delinquency. *American Psychologist,* 1968, *23,* 254-266.

Sarason, I. G., & Ganzer, V. J. Developing appropriate social behaviors of juvenile delinquents. In J. D. Krumboltz & C. E. Thoreson (Eds.), *Behavioral counseling: Cases and techniques.* New York: Holt, Rinehart & Winston, 1969. (a)

Sarason, I. G., & Ganzer, V. J. Social influence techniques in clinical and community psychology. In C. D. Spielberger and I. Iscoe (Eds.), *Current topics in clinical and community psychology.* New York: Academic Press, 1969. (b)

Sarason, I. G., & Ganzer, V. J. Modeling and group discussion in the rehabilitation of juvenile delinquents. *Journal of Counseling Psychology,* 1973, *20,* 442-449.

Scheerer, M. Problem solving. *Scientific American,* 1963, *208* (4), 118-128.

Schulz, R. W. Problem solving behavior and transfer. *Harvard Educational Review,* 1960, *30,* 61-77.

Shaw, C. R., & Schelkun, R. F. Suicidal behavior in children. *Psychiatry,* 1965, *28,* 157-169.

Shneidman, E. S. The logic of suicide. In E. S. Shneidman & N. L. Farberow (Eds.), *Clues to suicide.* New York: McGraw-Hill, 1957.

Shneidman, E. S. Psycho-logic: A personality approach to patterns of thinking. In J. Kagan & G. Lesser (Eds.), *Contemporary issues in thematic apperceptive methods.* Springfield, Ill.: Charles C. Thomas, 1961.

Shneidman, E. S. Logical analyses. In G. Gerbner, O. K. Holsti, K. Krippendorff, W. J. Paisley, & P. J. Stone (Eds.), *The analysis of communication content.* New York: Wiley, 1969.

Shure, M. B., & Spivack, G. Means-ends thinking, adjustment and social class among elementary school children. *Journal of Consulting and Clinical Psychology,* 1972, *38,* 348-353.

Shure, M. B., & Spivack, G. *A mental health program for preschool (and kindergarten) children and a mental health program for mothers of young children* (Final summary report MH 20372). Washington, D.C.: National Institute of Mental Health, 1975.

Shure, M. B., Spivack, G., & Jaeger, M. J. Problem solving thinking and adjustment among disadvantaged preschool children. *Child Development,* 1971, *43,* 1791-1803.

Siegel, J. M., & Spivack, G. Problem solving therapy: The description of a new program for chronic psychiatric patients. *Psychotherapy: Theory, Research and Practice,* 1976, *13,* 52-54.

Simon, H. A., & Barenfield, M. Information processing analysis of perceptual processes in problem solving. *Psychological Review,* 1969, *76,* 473-483.

Solley, C. M., & Stagner, R. Effects of magnitude of temporal barriers, type of goal and perception of self. *Journal of Experimental Psychology,* 1956, *51,* 62-70.

Spivack, G., & Levine, M. *Self regulation in acting out and normal adolescents* (Report M-4531). Washington, D.C.: National Institutes of Health, 1963.

Spivack, G., & Shure, M. B. *Social adjustment of young children: A cognitive approach to solving real-life problems.* San Francisco: Jossey-Bass, 1974.

Taylor, C. W. Panel consensus technique: A new approach to decision making. In S. J. Parnes, R. B. Noller, & A. M. Biondi (Eds.), *Guide to creative action.* New York: Scribner's, 1977.

Thorndike, E. S. Animal intelligence: An experimental study of the associative processes in animals. *Psychological Review Monographs,* 1898, *2* (4, Whole No. 8).

Thurstone, L. L. *The nature of intelligence.* New York: Harcourt-Brace, 1924.

Toolan, J. M. Suicide & suicidal attempts in children & adolescents. *American Journal of Psychiatry,* 1962, *118,* 717-724.

Torrance, E. P. Sociodrama as a creative problem solving approach to studying the future. *Journal of Creative Behavior,* 1975, *9,* 50-63.

Torrance, E. P., & Myers, R. E. *Creative learning and teaching.* New York: Dodd, Mead, 1970.

Trower, P., Bryant, B., & Argyle, M. *Social skills and mental health.* Pittsburgh: University of Pittsburgh Press, 1978.

Tuckman, J., & Cannon, H. E. Attempted suicide in adolescents. *American Journal of Psychiatry,* 1962, *119,* 228-232.

Turner, W. M., & Rains, J. D. Differential effects of "brainstorming instruction" upon high and low creative subjects. *Psychological Reports,* 1965, *17,* 753-754.

Ullmann, L. P. *Institution and outcome: A comparative study of psychiatric hospitals.* Oxford: Pergamon Press, 1967.

Upton, A., & Samson, R. *Creative analysis.* New York: Dutton, 1963.

Vincent, J. P., Weiss, R. L., & Birchler, G. R. A behavioral analysis of problem solving in distressed and non-distressed married and stranger dyads. *Behavior Therapy,* 1975, *6,* 475-487.

Walkup, L. E. Creativity in science through visualization. *Journal of Creative Behavior,* 1967, *3,* 79-82.

Weathers, L., Bedell, J., Marlowe, H., Gordon, R., Adams, J., & Reed, V. Psychotherapeutic games. *The Journal of the Florida Medical Association,* 1978, *65,* 891-896.

Weiss, R. L., Hops, H., & Patterson, G. R. A framework for conceptualizing marital conflict: A technology for altering it, some data for evaluating it. In L. A. Hamerlynck, L. C. Handy, & E. J. Mash (Eds.), *Behavior change: Methodology, concepts and practice.* Champaign, Ill.: Research Press, 1973.

Zigler, E., & Levine, J. Premorbid adjustment and paranoid-non-paranoid status in schizophrenia: A further investigation. *Journal of Abnormal Psychology,* 1973, *82,* 189-199.

Zigler, E., & Phillips, L. Social effectiveness and symptomatic behaviors. *Journal of Abnormal and Social Psychology,* 1960, *61,* 231-238.

Zigler, E., & Phillips, L. Psychiatric diagnosis and symptomatology. *Journal of Abnormal and Social Psychology,* 1961, *65,* 215-222. (a)

Zigler, E., & Phillips, L. Social competence and outcome in psychiatric disorder. *Journal of Abnormal and Social Psychology,* 1961, *63,* 264-271. (b)

Zigler, E., & Phillips, L. Social competence and the process-reactive distinction in psychopathology. *Journal of Abnormal and Social Psychology,* 1962, *65,* 215-222.

Zwicky, F. *Morphological astronomy.* Berlin: Springer-Verlag, 1957.

Chapter 2

Heterosocial Skills Training: A Behavioral-Cognitive Approach

Brian L. West
Katherine E. Goethe
William M. Kallman

Abstract

While significant attention has been afforded the complaints of shyness and social anxiety, few studies have attempted comprehensive treatment programs designed to affect the underlying cognitive and behavioral mechanisms in heterosexual-social interactions. The present report introduces a cognitive-behavioral treatment program for the training of heterosocial skills through the use of the heterosexual group process. Methods for client screening and treatment evaluation are discussed, and supportive preliminary results are provided. The Heterosocial Skills Training (HST) package is a 12-week procedure utilizing didactic, behavioral, and cognitive approaches, with in vivo application of learned skills. Through modeling, role playing, and rehearsal, clients are taught the basic nonverbal and verbal aspects of conversation (initiation, open-ended questions, parallel statements, minimal encouragements, change of topic), requests for meeting, and acceptance or rejection of requests. Data on three separate groups indicate reduced scores on Fear of Negative Evaluation and Social Avoidance and Distress scales, as well as increased scores on the Assertiveness Inventory. Coded videotaped conversations revealed increases in eye contact frequency and total client talk-time. Follow-up data indicate a maintained reduction in social anxiety. Implications for areas of future research, particularly in the cognitive realm, are discussed.

The role of interpersonal dysfunction in psychological disorders has been demonstrated in a number of studies (Argyle, Trower, & Bryant, 1974; Hersen & Bellack, 1976b; Levine & Zigler, 1973; Percell, Berwick, & Beigel, 1974; Zigler & Phillips, 1960, 1961). In these investigations particular research emphasis has been placed on the individual's social competency (i.e., social skills) and the correlation between social competency and various psychopathologies. Results have suggested that the client's prognosis for successful adjustment to daily functioning is directly related to the level of social skills development (Levine & Zigler, 1973; Zigler & Phillips, 1960, 1961, 1962). Recent psychological treatment approaches have concentrated on remediating skill deficits in training the client to produce adaptive interpersonal responses. Training in nonverbal and verbal social skills has been used successfully in a wide variety of populations including psychiatric patients (Goldsmith & McFall, 1975; Hersen & Bellack, 1976a; Zigler & Phillips, 1961), alcoholics (Levine & Zigler, 1973), and the mentally retarded (Doljanec, Schrader, & Christian, Note 1). According to Frank (1974), the use of social skills training has been reported as one of the most beneficial approaches to short-term psychotherapy.

In a recent review of outcome studies Marzillier and Winter (1978) pointed out that although empirical evidence supports the relationship of skills acquisition to positive behavioral change, the absence of adequate follow-up studies makes it difficult to evaluate whether in vivo behavioral styles have been significantly altered following social skills training. Marzillier and Winter concluded that the impetus of future research should be (a) to construct means of maintaining and generalizing skills to real-life situations and (b) to investigate the relevant contribution of various components of training, including cognitive processes.

This chapter will present a therapy program involving the systematic use of cognitive procedures and behavioral skills rehearsal for the maintenance and generalization of a specialized class of social skills—heterosocial behaviors. *Heterosocial skills* are defined here as behaviors that enable the individual to obtain social reinforcement from interpersonal contact with a member of the opposite sex. Based on social learning principles, the present model draws from the socialization process of group interactions to construct a more natural setting for acquisition of targeted social behaviors.

The rationale of the present program is threefold: (a) that skills training (through modeling, role playing, and behavior rehearsal) is necessary to increase the proficiency of interpersonal behaviors, (b) that the client learns to maintain his own social behavior through examination of self-evaluations and the use of self-reinforcement, and (c) that a group setting provides a means for a more natural socialization process to occur.

Previous social skills research has focused on particular deficits in heterosexual-social behaviors. Treatment programs aimed at remediating these deficits have appeared on many of that nation's campuses and clinics, in psychiatric residential centers, and in do-it-yourself pop psychology publications. Despite the burgeoning interest in the area, few researchers have attempted to delineate and standardize treatment procedures. Treatment approaches have traditionally attempted to reduce social discomfort by one of two training strategies: (a) the acquisition and rehearsal of heterosocial skills (Caldwell, Calhoun, Humphreys, & Cheney, 1978; Curran & Gilbert, 1975; McGovern, Arkowitz, & Gilmore, 1975; Minkin, Braukmann, Minkin, Timbers, Timbers, Fixsen, Phillips, & Wolf, 1977; Twentyman & McFall, 1975) or (b) the use of cognitive coping strategies (Rehm & Marston, 1968). Noticeably absent is the integration of cognitive and behavioral training.

In a study by Minkin et al. (1977) specifically defined interpersonal skills such as asking questions, providing positive feedback, and proportion of time spent talking were validated by observer ratings as necessary components for competent interpersonal interactions. Results showed that behavioral rehearsal and feedback procedures were effective in increasing the frequency of such skills in a population of adolescent females.

Caldwell et al. (1978) also used behavioral rehearsal for targeted specific conversational skills. These skills were defined as open-ended questions and various types of statements communicating interest and information exchange in a conversation. Two females reporting social anxiety and displaying low frequency of the targeted verbal skills were instructed to practice the skills with a male confederate. The skills were operationally defined, with examples given to each subject. The therapist provided immediate positive feedback to the subject on her performance in an analog situation. An increase in the targeted verbal skills was noted in the posttreatment assessment phase, during which the subjects interacted with male confederates who were blind as to the purpose of the study. While these results were promising in that the skills generalized to new males, the small number of subjects involved and the lack of any real-life observations makes it difficult to adequately evaluate the generalizability of the training.

McGovern et al. (1975) evaluated a social skills training program for college men with dating inhibitions. The three treatment approaches consisted of (a) using a dating manual, (b) placing clients in a group discussion of behavioral strategies, or (c) placing clients in a skills training group which met in different settings. Each group met for five sessions and was conducted by a trained female therapist. In the first treatment approach (dating manual) subjects attended groups in triads and discussed information about effective dating behaviors, concerns about dating, and strategies for managing heterosocial situations. The second and third treatment groups, one in a therapy room and the other in the natural environment, rehearsed predetermined behavioral situations with a female group leader, from whom each subject received feedback on his performance. The authors reported that all three treatment procedures were effective in reducing self-reported discomfort in heterosexual interactions. They concluded that fulfilling informational needs, such as through the use of the manual, may have been the most efficacious mode of treatment. These authors further concluded that the contribution of behavior rehearsal to the effectiveness of skills training is uncertain, and they called for an experimental analysis of treatment components in heterosocial skills training.

Twentyman & McFall (1975) utilized a behavioral training approach that included a cognitive rehearsal technique in the treatment of shy college males. The training involved modeling, role playing, and rehearsal of responses, both covert and overt, in a series of progressively more difficult and presumably more anxiety-producing situations. Subjects role played standardized situations. A decrease in self-reported anxiety and an increase in the frequency of desired behaviors were reported following treatment, although the targeted outcome (an increase in dating) did not occur. The authors concluded that while behavioral training holds promise as an effective treatment approach, additional research is needed to determine the relative contribution of covert and overt procedures in the experimental program. The failure of the treatment to generalize to the dating situation questions the clinical utility of these procedures.

Using only covert positive self-reinforcement training procedures, Rehm and Marston (1968) instructed their subjects to become involved in a standardized hierarchy of progressively more difficult social situations and to evaluate and reinforce themselves for each situation. The authors reported a greater reduction in anxiety for this treatment group than for nondirective and self-help control groups. Concurrently, the subjects improved verbal output. The implication is that self-evaluations which are reinforcing actually help to strengthen and maintain skills and to reduce anxiety. The importance of the role of negative self-evaluations in inhibiting performance is supported here, since performance improved even though skills training was not part of the treatment.

The influence of cognitive variables such as belief systems and self-statements on behavior is well established (Ellis & Harper, 1961; Mahoney, 1974; Meichenbaum, 1976). Bandura (1969, 1977) has underscored the importance of cognitive variables in the acquisition and maintenance of new behaviors. Although environmental contingencies are usually correlated with increased performance, the relative gains above some base rate may be minimal or easily extinguished because of inappropriate self-administered contingencies. These inappropriate cognitions may be termed negative self-evaluation statements and may stem from a person's family or peer interactions. For example, the explicit or implicit message

"You are inferior and cannot take care of yourself" may be established and conveyed into adulthood. This message, having been internalized by the receiving individual, then becomes "I am inferior or disliked," along with a myriad of negative assumptions. The etiological mechanism behind social inhibition may be, and indeed probably is, of a dual nature. That is, social inhibition may be a result of both a lack of important skills and an inhibition of behavior through maladaptive cognitive patterns. The research already reviewed indicates that behavioral training alone may not be sufficient in remediating heterosocial dysfunction and that cognitive approaches may need to be included in therapy. We suggest that self-reinforcing cognitions learned in therapy may provide a means for maintenance of the newly acquired skills after treatment termination. Kazdin (1975) has suggested that self-evaluation and self-reinforcement skills be taught to clients for the purpose of improving in vivo performance. According to Rathjen, Rathjen, & Hiniker (1978), an integration of cognitive and behavioral treatment procedures would be the most efficacious approach in social skills training.

In addition to problems of skills maintenance, the lack of generalizability of skills from therapy situations to complex real-life situations has been underscored by several authors (Eisler, 1976; Hersen & Bellack, 1977; Hersen & Eisler, 1976). Eisler (1976), in a review of the assessment procedures of social skills treatment, has pointed to the artificiality of the analog situations commonly employed in skills training. The significance of this argument is far reaching with regard to response generalization. The contrived situation used in the therapy encounter may prove to be unusual or atypical for the client, thus restricting the learning of any meaningful discrimination for performing in real-life situations. Trower, Bryant, and Argyle (1978) have noted a difference in American and English skills training. They feel that American skills trainers have tended to be "too rigid in their methods, with a subsequent loss in appropriateness and relevance of situations to daily life" (p. 125). The consequences of applying such inflexible programs would restrict the generalizability of skills utilization outside of therapy, since real-life situations are not under the strict control of programmed responses by confederates. The use of the group therapy setting allows clients to practice newly acquired skills with one another; giving them the opportunity to respond to new and different situations as they occur in the group is assumed to promote generalization to other situations (Martinson & Zerface, 1970).

Treatment Rationale

The present analog program utilizes the heterosocial event of a mixed-sex therapy group for promoting training and approximating the real-life social setting. In the heterosexual group setting opportunities for direct interaction with the opposite sex are increased, and the potential for avoidance is decreased. This provides an in vivo situation for desensitizing the socially anxious client to interpersonal contact. The progressively more difficult nature of the verbal skills, homework assignments, and integration of self-evaluation and skill performance resembles hierarchical desensitization procedures. Through positive feedback, a reduction of both anxiety and fear of evaluation becomes associated with social intercourse.

Availability of opposite sex group members satisfies the need for an opposite sex confederate in role-playing situations. The opportunity for direct interactions with the opposite sex negates the use of a standardized response for rehearsal activities which had been used in past studies. In addition, the group provides each individual with a greater number of opportunities for experiencing a variety of new situations and behaviors than does individual therapy. Clients can vicariously learn appropriate responses from peer interactions (McGovern et al., 1975). This increase in exposure to a number of relevant situations aids the client in her attempts to add useful skills to her repertoire. The successful outcome from this predicts a greater flexibility of responses in real-life situations.

The use of the group setting may present problems in the uniform presentation of any training program because of distinctive differences in client skill development at the outset of therapy (Trower et al., 1978). Such person variables may be handled in one of two ways. Either the assessment procedure for client selection can be designed to produce a highly homogeneous population, or the treatment can be flexible enough to accommodate client differences. This problem of treatment rigidity may be resolved by using staging procedures based on client skill capabilities and development. According to Cashdan (1973), the therapeutic relationship can be defined in two ways: (a) in terms of achieving goals of therapy (i.e., increased social contact, reduction of interpersonal discomfort) and (b) in terms of permitting assessment of individual client development. Predetermined behavioral goals (i.e., client behavior criteria) for each stage assist the therapist in determining when to change treatment emphasis. This is particularly significant when basic skills, either cognitive or behavioral, must be developed before additional ones can be learned and a proper integration of skills can be obtained. In the present program the delivery of skills training is segregated into stages according to the difficulties of the skills required, the complexity of skills integration, and the development of interpersonal contact within the group. Definition of the staging process was derived from reported group therapy process studies elaborated upon in the methodology section of this chapter.

While the review provided here does not begin to survey all the research available, the intent has been to demonstrate the need for an integrated cognitive-behavioral treatment program in a heterosexual group setting. The studies reviewed here provide evidence of the importance of cognitions in maintaining social inhibition. We know of no programs, other than that described in the present study, for heterosocial problems which utilize cognitive contingencies within a standardized heterosexual group behavioral training structure.

METHOD

Client Selection and Assessment

Client selection was based on a homogeneity of presenting problems and test results, along with significant information obtained through a structured interview. The following paragraphs describe three scales used to assess clients both for screening and posttesting. Each description includes score criteria for determining a client's problem area. Negative self-statements and anxiety are the problems which are identified through these inventories.

The Assertion Inventory (AI). A 40-item measure, the Assertion Inventory (Gambrill & Richey, 1975) assesses the client's (a) degree of discomfort or anxiety on a 5-point scale in specific situations (1 = none, 5 = very much) and (b) probability of displaying the behavior (1 = always do it, 5 = never do it) if actually presented with a situation. The 40 items fall into the following 8 categories: turning down requests, expressing personal limitations (such as admitting ignorance in some area), initiating social contact, expressing positive feelings, handling criticism, differing with others, being assertive in service situations, and giving negative feedback.

Social Avoidance and Distress Scale (SAD). A 28-item true-false questionnaire, the Social Avoidance and Distress Scale (Watson & Friend, 1969) assesses the degree to which a person avoids social situations due to social anxiety. Social avoidance is defined as "avoiding being with, talking to, or escaping from others for any reason. Both actual avoidance and the desire for avoidance are included" (p. 449). Social distress is defined as "the reported experience of a negative emotion, such as being upset, distressed, tense or anxious, in social interactions or the reported lack of negative emotion, such as being relaxed, calm, at ease or comfortable" (p. 449). A score greater than 9 on this scale would suggest a person who is generally uncomfortable in social situations and prefers to be alone.

Fear of Negative Evaluation (FNE). Fear of Negative Evaluation (Watson & Friend, 1969) is a 30-item inventory that assesses the avoidance of social situations caused by the fear of negative evaluation from other sources and the fear of loss of social approval. A score of 15 or above on this scale would suggest an attempt to gain social approval by avoiding social disapproval in various situations.

Interview session. Following the completion of the inventories, a 30- to 45-minute interview session is held. Basic demographic information such as age, current living environment, and employment status is obtained for the client and immediate family members. In addition to the client's present dating and social outlets, the therapist needs information on the client's history of social contact and relationships. A history of humiliating experiences or a lack of or abundance of friendships and intimate relationships is extremely helpful in assessing a client's problem areas. A structured interview format is an efficient method for obtaining the most information. A list of suggested questions applicable to the client's adolescent, college, and current situation follows:

1. Number of dates (per week, month, year).
2. Number of close friends (males and females) with whom the client reports feeling comfortable.
3. Number of acquaintances with whom the client engages in social activities (number of males and females).
4. Types and frequency of social activities participated in.
   a. How does the client feel, in general, during the activity (anxious, comfortable, etc.)?
   b. How does the client perceive her or his performance in the activity?

Cognitions, perceptions of performance, and emotional reactions in social situations are quite important. For instance, although a client may engage in some social activities, anxiety or discomfort during these activities may lead to diminished participation.

In addition to data derived from the client's history and present situation, the therapist needs to be observant of nonverbal behavior during the interaction. Indeed, the interview is a form of interpersonal interaction for the client, and any behaviors exhibited during that time may also be elicited during more casual interpersonal exchanges. Duration of eye contact, loudness of voice, and overt signs of anxiety or discomfort (e.g., shakiness, quivering voice, rigid posture, shallow breathing, excessive giggling) are the simplest and most obvious behaviors that the therapist may observe while maintaining the interview continuity. A 5-point rating scale of intensity, or a frequency rating of the behavior(s), is useful in assessing these overt signs of distress. Such a record of observation would yield data that supplement the self-report inventories and personal history.

Therapists

Male and female cotherapists, in this case clinical psychology graduate students, are used as training leaders. The use of male and female cotherapists is based on the finding that opposite sex coleaders facilitate gender role performance (Novotny, 1977). Role performance is modeled by the same-sex therapist when introducing each new skill. Social interaction between sexes is also modeled, with an emphasis on conversational style and personal feedback.

Procedure

Training components. The treatment program presented here is currently ongoing at the Clinical Services Center of Virginia Commonwealth University in Richmond. The heterosocial skills training is a 12- to 15-session group therapy procedure utilizing didactic, behavioral, and cognitive approaches with in vivo application of learned skills. Didactic procedures consist of therapists' instructions in basic skills. Along with these mini-lectures, handouts which summarize the material are given to each client.

Through the behavioral techniques of modeling, role playing, and rehearsal, clients are taught the basic skills of conversation. The components of a conversation are divided into nonverbal and verbal categories. Nonverbal behaviors and verbal behaviors, shown in Table 2.1, are specifically defined for the clients. At least one session is devoted to each behavior, and role playing never extends beyond the level of skills being taught. For instance, the role playing of initiating a conversation would be stopped by the therapists once one client made a remark to and received a nonverbal or verbal response from her partner. Relevant social situations that have been offered by members are used for the role playing. Situations brought up either by the therapist or group members are acted out to a successful conclusion as determined by the therapists and group members. Situations used in modeling and role playing are carried over from session to session. In this way clients learn to build a conversation through successive approximations, and the practice effects enhance skill acquisition.

Cognitive skills are learned through the client's increasingly accurate self-perceptions of performance and the production of self-reinforcing statements. Cognitive restructuring is initially implemented by the co-therapists and subsequently by the group members. The initial sessions focus on popular misperceptions of heterosocial contact, and as members begin offering their own views, specific attention is given to personal misconceptions, overgeneralizations, and self-defeating attitudes.

The learning process is enhanced by homework assignments which require group members to practice behaviors in vivo. In addition, these in vivo experiences require only the level of skill that is currently being trained. In other words, clients are expected to do only those things which they are able to do and have practiced in the training sessions.

Therapy stages. Based on the finding that imposed structure during initial sessions results in greater group interaction (Crews & Melnick, 1976), the first three stages emphasize skills training, rehearsal, and feedback. Subsequent sessions are progressively less structured to promote

Table 2.1
Conversational Skills Used in Training with Definitions for Coding

NONVERBAL BEHAVIORS

1. Eye Contact: Used to show interest and attention.
2. Gestures: Can accent or emphasize a point the speaker is making.
3. Head Nods: A sign of comprehension, empathy, interest, and attention.
4. Facial Expression: Used to clarify and emphasize the speaker's point.
5. Voice (tone and volume): Used to bring attention to what the speaker is saying.
6. Interpersonal Distance: Helps direct attention to what the speaker is saying. Also a sign of comfort (a person sits or stands closer to those people with whom she is more comfortable).
7. Posture: A sign of comfort. Also used to bring attention to what the speaker is saying.
8. Body Orientation: Increases attention and shows interest in what the other person is saying.

VERBAL BEHAVIORS

1. Initiation of Conversation: Any client's statement or question which is used to begin a conversation or offer a new topic that is relevant but distinctly different from the previous explicit topic.
2. Minimal Encouragements: Interjection of a word or short phrase which indicates the listener's interest and encourages the speaker to continue. The listener's remarks may be in the form of questions, statements of interest, or exclamations of affect which are consistent with the affect expressed by the speaker.
3. Open-Ended Questions: Questions intended to elicit information from the listener, typically requiring some elaboration on the part of the listener rather than requiring only a yes or no or single-word reply. Should be directly related to the listener's previous statements.
4. Parallel Experiences: Listener's statements about personal experiences, the content of which is within the boundaries of the explicit topic of the partner's previous statement. Points out either similarities or dissimilarities between the listener's own experiences and those related by the dyad partner.
5. Reflective Statements: Listener's statements which mirror or attempt to be an interpretation of what the dyad partner has just said. Encourages the speaker to continue talking.

client responsibility for interactions. Liberman's (1970) finding that client behavior is, in part, shaped by the comments of peers in the group necessitates initial structure of client-to-client feedback with an emphasis on positive comments. In the final stages, with the reduction of structured practice, the development of pertinent skills, and the increase of personal comments, the group forms a greater cohesive unit from which the members can derive social reinforcement for their successful responses.

Stage 1 is an introductory phase in which the therapists underscore the common goals of the group members by defining the purpose of the group. During *Session 1* the therapists and clients introduce themselves and provide some personal information, such as employment, age, and so forth. The therapists then take turns introducing the common functions of the group therapy, including the reason for the homogeneity of the group, the basic learning principles behind skills training, and the responsibility of each member to participate in the group. Identifying existing appropriate skills and providing positive feedback to others is emphasized to clients as a major therapy procedure.

The description of client responsibilities includes several important points: (a) group reinforcement for the appropriate skills can be influential in changing behaviors; (b) each client's input is important for structuring the specific situations for training in the group; (c) the group's goals are to increase the comfort of the members in social situations, so changes clients wish to make should be proposed in the group and appropriate alternative behaviors offered; and (d) expectancies of the outcomes of situations and self-statements have effects on behaviors. Particular attention is given to the fact that each person, though able to perform a social conversational skill, may expect negative feedback from others or themselves and therefore may avoid social situations. The basic techniques of modeling, role playing, behavior rehearsal, and reinforcement are briefly described. Emphasis is placed on the element of practice and reinforcement for successive approximations of targeted behaviors. Following a period set aside for questions, the clients are asked to engage in a 5- to 7-minute conversation with a confederate of the opposite sex. These conversations are videotaped in private therapy rooms to minimize any fear of peer evaluation. The confederates (clinical psychology graduate students, in this study) are as nondirective as possible and initiate the conversation with "Tell me a little about yourself." This analog situation provides a baseline measure for several behaviors (see Table 2.2). After the conversations, clients are asked to discuss, as a group, their performance, expectancies, and self-evaluations during the

Table 2.2
Coded Behaviors of Clients during Pre- and Posttreatment Confederate Conversations

1. Face-to-Face Time: Amount of time client and confederate were looking at one another.
2. Head Nods: Frequency of head movements that were demonstrating attention and agreement.
3. Gestures: Frequency of hand movements that coincided with client speech.
4. Talk-Time Ratio: Percentage of client talk-time to confederate talk-time.

analog situations. The discussion serves two purposes: (a) to provide a baseline for group conversation (with minimal therapist prompts) and (b) to assess specific types of clients' self-statements and expectancies. (It should be noted that the clients are fully aware that they are being videotaped during these conversations and have consented to the measure.)

Homework assignments are given during the session in an attempt to strengthen the connection between thoughts and behaviors discussed and performed in therapy and outside of therapy. For the first session each client is asked to write out, in detail, three interpersonal interactions in which the client has, or expects to have, problems performing sufficiently well. Clients are asked to provide information concerning what they do, where the interaction occurs, what they are thinking about, what the outcome of their actions is, and whether the results are satisfactory.

The function of the treatment procedure, as it is explained to the group, is to provide: (a) proper discriminative/interpretive abilities (i.e., nonverbal cues, etc.), (b) a repertoire of behaviors that can be used in most settings, and (c) a capability to accurately evaluate one's own performance.

The second half of the homework assignment is to keep a behavioral diary. Each client receives a booklet of seven sheets representing the days of the week. Clients are requested to record the approximate duration of each conversation with a member of the opposite sex which lasts for at least 2 minutes, the number of participants, a summary of what was talked about, and how the client felt about the event. Again, this type of assessment attempts to evaluate transfer of skills development (both cognitive and behavioral) to everyday situations. This initial assessment provides a baseline of the client's social contact with the opposite sex. Ongoing evaluation involves having the client keep a weekly diary every fourth week throughout therapy.

Stage 2 aids in the development of basic social skills and exchange of reactions by members about their being in the group. Criterion behaviors for terminating this stage are (a) successful identification and use of targeted verbal and nonverbal behaviors and (b) demonstration of member-to-member feedback. The second stage consists of four sessions (Sessions 2 to 5). In *Session 2* the communicational significance of nonverbal behaviors is introduced. Clients are asked to provide information about what they noticed other people doing in addition to the verbal content of their conversation. Clients' responses are written on a blackboard, with the therapists including any significant nonverbal cues that were overlooked by the group (see Table 2.1). All responses are verbally reinforced by the therapists. Possible reluctance by the group to offer responses is dealt with by asking individuals to respond and/or to point out the nonverbal behaviors of therapists and members of the group.

The videotapes of client/confederate conversations are played back without sound, with the permission of each client. This exercise further elaborates the impact of nonverbal behavior in a conversational setting. The group's instructions are to note the nonverbal behaviors of both client and confederate. The therapists model positive feedback for the group by commenting on each client's appropriate nonverbal behaviors. The other members in the group are then asked to make positive remarks on the performance. The clients who are receiving feedback are then asked to tell the group what they expected to happen in the conversation, the positive parts about their performance, and whether the appraisal of their performance was congruous with feedback from the group.

At the end of the session the therapists discuss areas of improvement needed by the group as a whole (e.g., "Perhaps we will practice more eye contact next session"). The homework assignment again relates the therapy practice to nontherapy settings by having the clients monitor their own and others' nonverbal behavior during conversations. The group is asked to discuss this during the next session.

During *Session 3,* client/confederate videotapes are played, and the members of the group are asked to give feedback on the performance. Appropriate task-oriented comments are reinforced by the therapists. Inappropriate comments are either ignored or reviewed as not being productive in the group. Incongruity between self-appraisals and group comments are followed up by the therapists. Clients are asked to give evidence to support their negative self-evaluations. Rational reappraisals of behavioral performance are reinforced, and suggestions for use of self-reinforcement are offered. These consist of positive self-statements which follow appropriate social behavior. For instance, when a client said "Hello" to a co-worker, if the co-worker smiled and responded with a greeting, the client was to say to himself, "I did that very well, because I looked at her and greeted her and she answered back."

The homework assignment is for clients to monitor nonverbal behaviors, particularly their own, and to evaluate their conversation partner's reactions (specifically nonverbal).

Session 4 marks the beginning of conversational skill training. The therapists model initiation of conversations in several predetermined situations (see Table 2.3 for examples). A client is then asked how he might begin a conversation in the same situation. Any response that is introductory or points out some commonality of topics between the participants is reinforced by "That's good" or "I would respond to that," etc. Proper nonverbal behaviors such as eye contact, gestures, and full body orientation are also positively reinforced.

Using his own initial response, or possibly another one, the client is then asked to role play the situation with another client. The group

Table 2.3
Examples of Situations Used in Role Play of the Various Aspects of Conversations

1. Initiation: "You are on a break at your job. You see a guy who is about your age. He works in another part of the store, and consequently you don't know him very well. You would like to start talking to him. What would you do?"
2. Maintenance: (a) "At a party you go over to a girl and ask her to dance. She says, 'I'm not really much of a dancer.' You would say . . . ." (b) "You are talking to a guy at work, and he is explaining how he missed the bus, thus getting to work late. He pauses. You want to continue the conversation. What would you say?"
3. Request for Change of Topic: "You are at a party talking to a guy you've just met. He is explaining how he installed his carburetor earlier in the day. Although you would like to continue talking to him, you know nothing about cars and would like to change the topic of conversation. What would you say?"
4. Request for a Meeting: "You've been talking to a fellow employee at a coffee break and would like to meet with her later (e.g., at lunchtime). What would you say?"

members are encouraged to make positive comments about the performance. The client is then asked by the therapists (eventually other group members begin to fill this therapeutic role) to give his reactions to role playing the situation. Particular attention is given to the congruity of the self-appraisal and the actual performance. Thus, if the client states that he "can't start a conversation with a girl" or that "she wouldn't talk to me anyway," other members or the therapists point out that in fact, he had just been successful in the role playing. Many times this negative self-evaluation or self-defeating belief is tested in a dialogue like the following one.

First Client:	Well, she had to talk to me. We're in the group.
Therapist:	Have you ever tried to start a conversation like this before?
First Client:	Sure, but girls don't want to talk to me.
Therapist:	Can you give me an example of a situation where that happened?
First Client:	Uh . . . well . . . Yeah, this morning I tried to talk to this girl, but she didn't want to talk to me.
Second Client:	What did she say?
First Client:	Hello. I'm okay, or something like that.
Therapist:	Then what did you do?
First Client:	I walked away. I knew she didn't want to talk to me.

Second Client: How do you know that? You walked away from her.
Therapist: Were you expecting her to say something else?
First Client: (Laughs) Yeah, get away . . . or something.
Therapist: Does it sound to you like you expected something worse to happen?
First Client: Yeah, I guess so. I'm not really sure what to expect, you know.
Therapist: Well, what happened with Kay here in the group? Kay, what feedback can you give him?
Role Play Partner (Kay): I think you did a great job. I felt like talking because you looked right at me and you were polite.
First Client: Would you really have talked to me?
Role Play Partner: Sure.

In this way the situationally specific recall or reenactment of an event permits a more accurate appraisal of behaviors and self-statements. This deliberate intervention allows observation of the direct connection between behaviors and cognitions and the resulting outcome. In addition, it also provides a means to promote self-assessment with peer group reinforcement (such as Kay's statement of confidence in the First Client as the end of the foregoing excerpt). The session is ended with a request that behavioral diaries be kept for the next 7 days.

Session 5 focuses on listening skills, termed elsewhere as "minimal encouragements" (Caldwell et al., 1978). The therapist begins the training session by asking the group "How do you know someone.is listening to you or is interested in what you are saying? How do you let others know you're listening to them?" Group responses (including verbal and nonverbal behaviors) are written on the blackboard. A sample of these includes "uh-huh," "tell me more," "that's interesting," "I understand," nodding, smiling, and looking concerned (by concentrating on what's being said). Conversations are then role played with clients beginning and maintaining a conversation for about 1 minute. Feedback and self-appraisal procedures are done as before, with each member participating and videotape playbacks of the role play being reviewed. In the playback, feedback (mainly positive) is given again. Constructive criticism by other members, in the form of indicating alternative behaviors or absent skills, is permitted if accompanied by positive remarks.

Stage 3 (Sessions 6 and 7) is defined as an intermediate skill development phase. The group members are given didactic training in interviewing and conversation skills, with homework assignments requiring the client to attempt in vivo use of the skills. Time is set aside during each session to discuss social situations which have occurred outside of therapy.

The first intermediate skill which is introduced in *Session 6* is asking open-ended questions. Therapists define *open-ended questions* as those inquiries which require the other person to reply with more than a simple yes or no. Therapists explain to the clients that these questions elicit more information from other persons and keep them talking.

At this point, the situations used in the initiation of conversation session are brought up again. Clients are asked what open-ended questions would be appropriate to use in those situations, and these questions are written on a blackboard. Clients are then asked to role play each situation to the point that at least one open-ended question has been asked and answered. Group and therapist feedback is given, and the situation is rehearsed until each client reports that he or she feels comfortable with the responses.

For homework clients are requested to begin a conversation with two people they know very little about or don't know at all. Therapists emphasize that full conversations are not necessary and that just beginning an interaction is enough. This is emphasized to relieve performance anxiety. Often clients will ask, "What do I do if they keep talking to me?" The therapists might answer, "Then you succeeded in starting a conversation. That's all you're supposed to do right now."

In *Session 7* the clients learn to maintain a conversation through the use of reflective statements and parallel experiences. Therapists define *reflective statements* as those which mirror what someone has just said. The purpose of reflective statements is checking with the other person to make sure you understood what she said. For example:

Speaker: I just had my car fixed for $150, and it still doesn't run right!

Listener: It's really upsetting not to have your car running the way you want it.

Next the therapists define *parallel experiences* as those experiences which the listener has in common with the speaker. For example:

Speaker: I just had my car fixed for $150, and it still doesn't run right!

Listener: I know how you feel. I had my car fixed last month, and I had to take it back three times.

Situations discussed in previous sessions are used to elicit from group members appropriate responses in the form of reflective statements or parallel experiences. Group members make suggestions for appropriate responses, with feedback from other clients and the therapists. Situations are then created by therapists and clients, with role playing and feedback.

The next skill discussed is changing the topic. The therapists present the group with a predetermined situation such as: "You know nothing about and are not interested in mechanical subjects. Someone has been talking to you for the last 5 minutes about his car's tune-up procedures. You'd like to change the topic. How?" Clients' suggestions are written on the blackboard. If the clients don't know what to do, the therapists may prompt by using the following suggestions:

1. Wait for some information that can be applied to you and can be directed into a new topic.
2. Say something like "I really don't know much about that, so it's really difficult for me to understand what you mean."
3. Redirect the topic with questions.

Using previously discussed and new situations, clients role play full conversations with at least one change of topic. Feedback, with suggestions for improvement, is given and then the situations are rehearsed.

Stage 4 (Sessions 8-10) includes more advanced skills of interaction, such as requesting meetings with people of the opposite sex. Before proceeding with these skills, however, the therapists must make an effort to confront group members on negative self-evaluations and behavioral skills problems still remaining. In guided discussions group members are encouraged to use a more constructive (positive and negative) manner to give each other feedback on their expectations of outcomes on skills of conversation and interpretation of social interaction cues (verbal and nonverbal).

Session 8 is used primarily to confront group members on negative cognitions still remaining and on special problems which affect competency in conversing with others. The therapists ask the clients to describe any difficulties they are having with homework assignments, giving or receiving feedback, and any behavioral skill practiced. Role playing is utilized for problem situations and for practice of conversational skills, after which group members provide positive and negative feedback to each other. The therapists ask specific questions about how the clients feel and what they are thinking during and after the role play and feedback.

Full conversations are practiced during *Session 9*. Feedback is withheld until clients have changed topics at least once during the interactions. In addition, three-way and four-way conversations add a new dimension for clients so that they can learn to engage in an ongoing conversation. Feedback is provided, with suggestions on appropriate ways to include other people in dyadic interactions. The homework assignment is to list at least 10 places or ways to meet people. Therapists may provide a few suggestions, such as a singles club in an apartment complex, YWCA activities, or through friends.

The group discusses this homework assignment at *Session 10.* Clients are asked to read their lists and to give feedback to each other on list items. Some questions the therapists may ask are: "Is this an appropriate place to meet and get to know someone? What kind of topics might you talk about? How would you go about meeting people at this place?"

The therapists then direct the discussion into the next skill topic, making requests. The therapists say, "Now you have the basic conversation skills. You have some ideas on how and where to meet new people, so what do you do if you want to see them again?" Clients are asked to role play a situation in which they have been conversing with someone for quite some time and want to make a request to get together again. Group members give suggestions on what might be said, and then each group member role plays both requesting and accepting a request for a meeting.

The next topic is rejection of requests. Group members discuss their feelings and perceptions of rejection; the group also confronts each member on negative self-statements occurring as a result of rejection. Previous situations for requesting meetings are role played again, but group members turn down the requests and also role play having their requests turned down. Group members are asked to give feedback and to give their perceptions of the rejection. The therapists ask the clients to give suggestions about how to receive a rejection and then ask for a meeting at another time. Clients' self-statements after being rejected and during their preparation to ask for another meeting are dealt with by examining the possible reasons behind the rejection and replacing negative self-statements with plausible explanations and/or "being turned down happens to everyone."

During *Stage 5* (Sessions 11 and 12) the therapists are able to observe group interaction more than at any other time. Group members more readily initiate feedback and discussions. Therapists usually monitor clients' statements and behaviors for incongruities but point out these incongruities only when other group members do not.

Session 11 begins with personal feedback. Group members and therapists give feedback to each other on the following questions: (a) How does the member or therapist affect you interpersonally now? (b) Has that changed in any way from the first group meetings? (c) What are the individual's strengths? (d) What role does the client or therapist fit into as a group member (e.g., active, passive, comic, comforter)?

During this session members frequently comment that they feel at ease with the group members but are still uncertain around strangers. The therapists constantly point out that the group members were strangers too until through conversations and self-disclosure they formed relatively close relationships with each other. Therapists also confront members on their perceptions of other people outside the group and ask how they would get to know more about these people.

The homework assignment is discussed during *Session 12*. Each client is asked to describe the situation, how she felt and what types of thoughts occurred during the interaction. Any problems the client reports are corrected by role play and group feedback. At the end of the session the behavioral diary is assigned as homework.

The final session of the therapy is used as a posttesting and closure session. The FNE, SAD, and AI are re-administered. Each client engages in another 5- to 7-minute videotaped conversation with a confederate of the opposite sex. The confederate is unknown to the group members, so generalization of the group learning experience is tested.

As for group closure, the therapists ask the clients to discuss what changes they have perceived in themselves since therapy began. Group members' perceptions of therapy termination and expectations for their social lives are discussed. In addition, the therapists pose the question "Although the therapy group is terminated, do the relationships between the group members terminate also?"

RESULTS

The intent of this chapter has been to introduce a therapy approach that integrates cognitive and behavioral social skills training in an appropriate environmental setting. By utilizing the current knowledge about various treatment modalities for heterosocial problems, we have attempted to construct a program that offers a standardized cohesive regime. The following sections report the preliminary results of an ongoing investigation and some discussion of these findings.

Three therapy groups have been conducted so far. Two groups consisted of four individuals, and the third group consisted of five. All groups had male and female members. An attempt was always made to keep the male-female ratio relatively equal. The members ranged from 20 to 32 years in age (mean age was 26 years). Two clients were full-time undergraduate students, one was in a professional school, and the remaining ten held full-time jobs. Client referrals were either from the university psychology center, other mental health agencies, or self-referrals. Client selection and treatment procedures were carried out as indicated in the method section. Four of the clients terminated therapy prematurely (primarily because of scheduling problems), and two clients did not return for posttreatment assessment.

Data for seven subjects who completed the entire training were analyzed by using a repeated measures analysis of variance. The results are divided into two categories of dependent variables: (a) self-report inventories and (b) behavioral codings.

Self-Report Inventories

The three self-report questionnaires were used to assess the client's perception of her self-statements, probability of performing, and overall distress. As can be seen in Figure 2.1, all self-report inventory scores decreased after treatment. Treatment effects were statistically significant for the Fear of Negative Evaluation scale (FNE) ($F(1,6) = 21.57, p < .01$) and Social Avoidance and Distress scale (SAD) ($F(1,6) = 9.982, p < .025$). The mean scores changed from 27.14 to 21.14 on the FNE, and 22.43 to 15.86 on the SAD. This change in the direction of the norm, although neither posttreatment mean score approximates the normal population mean. This may be due to the extremely elevated pretreatment scores. In comparison, McGovern et al. (1975) reported lower scores in their pretreatment phase for the SAD and FNE, 16.12 and 20.87, respectively.

Figure 2.1. Pre-Post Self-Report Inventory Scores for the FNE and SAD Scales

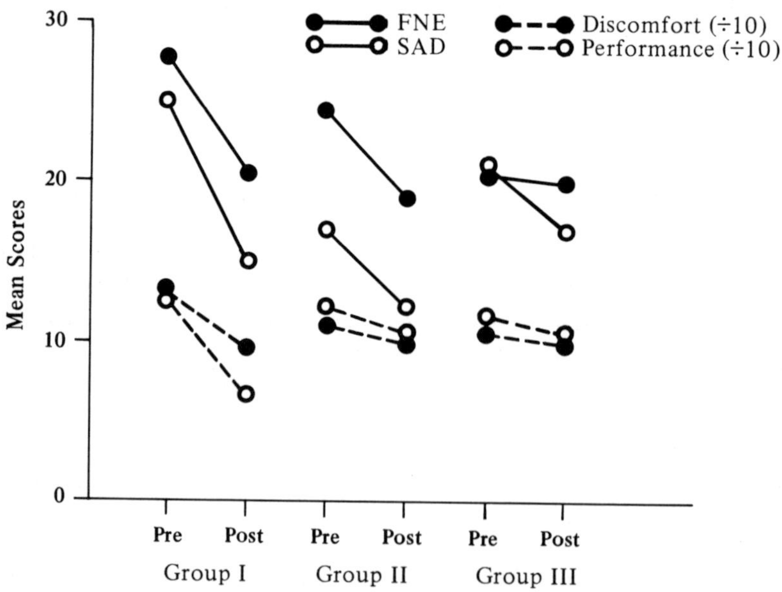

Note: Lower numbers represent lower anxiety (discomfort) and greater probability of making the appropriate response (performances).

On the AI, treatment effects were significant for the Discomfort scale ($F\ (1,6) = 12.693$, $p < .025$) and the Probability scale ($F\ (1,6) = 7.016$, $p < .05$). The mean pretreatment scores were 128 and 121.14 for the Discomfort and Probability indices, respectively. Posttreatment mean scores of 101.29 and 102.86 (Discomfort and Probability) reflect a significant change that approaches the normal range.

Verbal and Nonverbal Codings

Videotaped 5- to 7-minute conversations with confederates were coded for 4 minutes. The first minute of the exchanges was not coded since it tended to vary from simple instructions to humorous remarks about the nature of the situation. Tapes were coded for talk-time ratios (client/confederate), frequency of hand and arm gestures, face-to-face orientation time, and head nods.

For the nonverbal codings, percentages were computed, based on the ratio of confederate or client talk-time to total talk-time. Both of these percentages were significantly affected by treatments (confederate: $F\ (1,6) = 12.45$, $p < .025$; client: $F\ (1,6) = 9.72$, $p < .025$). That is, confederates talked less of the time since clients talked more of the time after treatment. Concurrently, of course, confederate talk-time decreased significantly between assessment sessions ($F\ (1,6) = 11.64$, $p < .025$). Posttreatment increased talk-time and percentage of speaking time have been reported to be inversely related to anxiety (Borkovec, Stone, O'Brien, & Kaloupek, 1974).

The number of client gestures and duration of face-to-face orientation to confederates increased after treatment, although those changes were not statistically significant. Frequency of head nods, increasing in some clients and decreasing in others, did not change overall. These results are not surprising, since no feedback or emphasis was given specifically for head nods; it was given for gestures and eye contact.

Correlated Data

Change scores were computed for the self-report inventories, percent client talk-time, gestures, and face-to-face orientation by subtracting post- from pretreatment scores. Pearson product-moment correlations were computed on these scores and are shown on Table 2.4.

Changes in gestures and face-to-face orientations did not significantly correlate with changes in any of the self-report measures or with percent client talk-time. This suggests that a change in a person's discomfort or behavioral avoidance of social situations is not necessarily related to a change in nonverbal behaviors. Nonetheless, trends were noted for an increase in gestures with a decrease in reported social anxiety (SAD) and an increase in face-to-face orientation. Changes in the desired direction

Table 2.4
Correlation of Test Scores and Frequency Coding of Behaviors

	SAD	*Disc.*	*Prob.*	*% Talk*	*Face*	*Gestures*
FNE	.605	.446	−.810**	−.455	.066	−.221
SAD		.776**	.642	−.838**	.320	−.326
Discomfort			−.811**	−.993**	.407	———
Probability				.744*	.172	———
% Client Talk-Time					−.300	.428
Face-to-Face						.227

*p < .10
**p < .05

of the SAD and the Discomfort and Probability scales of the AI did correlate or approached significant correlation with an increase in percentage of client talk-time.

As anticipated, self-report inventories negatively correlated with the Probability of Performing scores, indicating a reduction in anxiety and a fear of negative evaluation with an increase in assertive and social behaviors. Not surprisingly, the self-reported assertiveness discomfort and social anxiety were positively correlated.

Preliminary follow-up results of the self-report inventories were collected on four subjects at either 4 or 9 months posttreatment. Changes in SAD scores from the pretreatment assessment to the follow-up measures were significant (F (1,4) = 14.65, $p < .01$), suggesting a continued reduction of social anxiety. Scores on the FNE and the AI remained below the pre- and posttreatment measures.

Two subjects who did not complete the treatment program were also tested during the posttreatment phase. All scores were equivalent with the pretreatment assessment. It is difficult to ascertain the validity of these findings because of the small sample. Nevertheless, the results are promising. More extensive follow-ups, using both group participants and subjects who dropped out (as controls), are currently being done.

Discussion

Preliminary results suggest that the heterosocial skills training program significantly reduced self-reported anxiety and negative evaluation fears as demonstrated by the scores on the SAD and FNE. It was anticipated that treatment emphasis on cognitive restructuring of self-evaluations and expectations would lower the FNE scores. The hypothesis that a hierarchical approach to acquiring and practicing conversational skills, along with changes of negative self-statements and self-reinforcement, would reduce social anxiety was supported by reduction in the SAD scores.

Likewise, specific overt behaviors (i.e., gestures of hands and arms to accent verbal comments, eye contact) that were modeled and rehearsed during the group increased in the confederate conversation sessions. The fact that these measures approached significance while head nods did not increase supports the contention that skill changes were effected by the training, since head nods were not part of feedback and specific role play. The low correlation between nonverbal behavioral skills development (i.e., posture orientation, eye contact, smiling) and a decrease of anxiety (SAD) has been reported elsewhere (McGovern et al., 1975). The inability to specify nonverbal behaviors necessary for appropriate conversations guided us to emphasize training of the verbal conversational skills. Frequency coding of these specific skills was not performed in the preliminary investigation. The significant increase in client talk-time, though, suggests a greater ability to perform social conversation. This is further supported by the negative correlation of talk-time to social anxiety. The impact of conversational skills training for remediating social anxiety does not seem surprising since verbal skills provide a means for receiving attention and other types of social reinforcement, which subsequently relieves anxiety. In any event, further elaborative research is needed to determine the importance of specific conversational components.

The maintenance and proper use of skills was hypothesized to be dependent on the individual's ability to evaluate her performance accurately and to reinforce herself. Posttreatment and follow-up data indicate a reduction in fear of negative evaluation that is concomitant with a lessening of social anxiety, an increase in the display of skills, and an increase in the probability of performance. This evidence supports the hypothesis that cognitive restructuring can reduce anxiety, thus permitting acquisition and performance of new skills. Additional research is needed to isolate this phenomenon.

The need for in vivo measures to assess the generalizability of skills training to everyday interpersonal functioning was discussed previously. A behavioral diary to evaluate this relationship has been used (McGovern et al., 1975), albeit with minimal success. The recording of social meetings by the subjects used in this study was found to be inconsistent. In many cases diaries were not filled out, some had only one or two days completed, and some were never turned in to the therapists. The measuring device (diary) may be found to be a reliable measure. We recommend that monetary or other value contingencies be used as an incentive for the careful completion of these instruments. Another problem with the behavioral diaries was inadequate specification of the types of interactions the clients had. It was difficult to determine whether the interactions recorded were of a business, work-related, or purely social nature. Further refinement of the use of diaries is necessary for accurate and adequate assessment information to be obtained.

The significance of using a heterosexual group setting was not directly tested since the first goal of this project has been to establish the efficacy of the cognitive-behavioral approach. Obviously, comparison studies utilizing matched populations (i.e., populations matched on measures of social skills deficits and level of anxiety) should be carried out to evaluate the additional treatment advantages, if any, of this particular type of group composition. In the present study, group members interacted outside of therapy and even upon termination of therapy. Some members called one another to discuss personal problems, several reported an increase of dating, and one group had several parties after therapy terminated. The implication is that the therapy provides an important social outlet in addition to a teaching environment.

Reference Notes

1. Doljanec, R. F., Schrader, S. J., & Christian, J. G. Development of verbal interaction skills in the mentally retarded. Paper presented at the 11th Annual Meeting of the Association for Advancement of Behavior Therapy, Atlanta, December 1977.

References

Argyle, M., Trower, P., & Bryant, B. Explorations in the treatment of personality disorders and neuroses by social skills training. *British Journal of Medical Psychology,* 1974, *47,* 63-72.

Bandura, A. *Principles of behavior modification.* New York: Holt, Rinehart & Winston, 1969.

Bandura, A. *Social learning theory.* Englewood Cliffs, N.J.: Prentice-Hall, 1977.

Borkovec, T. D., Stone, N. M., O'Brien, G. T., & Kaloupek, D. G. Evaluation of a clinically relevant target behavior for analog outcome research. *Behavior Therapy,* 1974, *5,* 503-513.

Caldwell, A. T., Calhoun, K. S., Humphreys, L., & Cheney, T. H. Treatment of socially anxious women by a skills training program. *Journal of Behavior Therapy and Experimental Psychiatry,* 1978, *9,* 315-320.

Cashdan, S. *Interactional psychotherapy.* New York: Grune & Stratton, 1973.

Crews, C., & Melnick, J. The use of initial and delayed structure in facilitating group development. *Journal of Counseling Psychology,* 1976, *23,* 92-98.

Curran, J. P., & Gilbert, F. S. A test of the relative effectiveness of a systematic desensitization program and an interpersonal skills training program with date anxious subjects. *Behavior Therapy,* 1975, *6,* 510-521.

Eisler, R. M. The behavioral assessment of social skills. In M. Hersen & A. Bellack (Eds.), *Behavioral assessment: A practical handbook.* New York: Pergamon Press, 1976.

Ellis, A., & Harper, R. A. *A guide to rational living.* Hollywood, Calif.: Wilshire Book Co., 1961.

Frank, J. D. Therapeutic components of psychotherapy: A 25-year progress report of research. *Journal of Nervous and Mental Disease,* 1974, *159,* 325-342.

Gambrill, E. D., & Richey, C. A. An assertion inventory for use in assessment and research. *Behavior Therapy,* 1975, *6,* 550-561.

Goldsmith, J. B., & McFall, R. M. Development and evaluation of an interpersonal skills-training program for psychiatric inpatients. *Journal of Abnormal Psychology,* 1975, *84,* 51-58.

Hersen, M., & Bellack, A. S. A multiple-baseline analysis of social skills training in chronic schizophrenics. *Journal of Applied Behavior Analysis,* 1976, *9,* 239-245. (a)

Hersen, M., & Bellack, A. S. Social skills training for chronic psychiatric patients: Rationale, research findings, and future directions. *Comprehensive Psychiatry,* 1976, *17,* 559-580. (b)

Hersen, M., & Bellack, A. S. Assessment of social skills. In A. R. Ciminero, K. S. Calhoun, & H. E. Adams (Eds.), *Handbook of behavioral assessment.* New York: Wiley, 1977.

Hersen, M., & Eisler, R. M. Social skills training. In W. E. Craighead, A. E. Kazdin, & M. J. Mahoney (Eds.), *Behavior modification: Principles, issues, and applications.* Boston: Houghton Mifflin, 1976.

Kazdin, A. E. Covert modeling, imagery assessment, and assertive behavior. *Journal of Consulting and Clinical Psychology,* 1975, *43,* 716-724.

Levine, J., & Zigler, E. The essential-reactive distinction in alcoholism: A developmental approach. *Journal of Abnormal Psychology,* 1973, *81,* 242-249.

Liberman, R. P. A behavioral approach to group dynamics: I. Reinforcement and prompting of cohesiveness in group therapy. *Behavior Therapy,* 1970, *1,* 141-175.

Mahoney, M. J. *Cognition and behavior modification.* Cambridge, Mass.: Ballinger, 1974.

Martinson, W. D., & Zerface, J. P. Comparison of individual counseling and a social program with nondaters. *Journal of Counseling Psychology,* 1970, *17,* 36-40.

Marzillier, J. S., & Winter, K. Success and failure in social skills training: Individual differences. *Behaviour Research and Therapy,* 1978, *16,* 67-84.

McGovern, K. B., Arkowitz, H., & Gilmore, S. K. Evaluation of social skill training programs for college dating inhibitions. *Journal of Counseling Psychology,* 1975, *22,* 505-515.

Meichenbaum, D. Toward a cognitive theory of self-control. In G. E. Schwartz & D. Shapiro (Eds.), *Consciousness and self-regulation: Advances in research and theory* (Vol. 1). New York: Plenum Press, 1976.

Minkin, N., Braukmann, C. J., Minkin, B., Timbers, G. D., Timbers, B. J., Fixsen, D. L., Phillips, E., & Wolf, M. M. The social validation and training of conversational skills. In C. M. Franks and G. T. Wilson (Eds.), *Annual review of behavior therapy theory and practice.* New York: Brunner/Mazel, 1977.

Novotny, H. R. Acquisition of social competency skills. In G. G. Harris (Ed.), *The group treatment of human problems: A social learning approach.* New York: Grune & Stratton, 1977.

Percell, L. P., Berwick, P. T., & Beigel, A. The effects of assertive training on self-concept and anxiety. *Archives of General Psychiatry,* 1974, *31,* 502-504.

Rathjen, D. P., Rathjen, E. D., & Hiniker, A. A cognitive analysis of social performance. In J. P. Foreyt & D. P. Rathjen (Eds.), *Cognitive behavior therapy.* New York: Plenum Press, 1978.

Rehm, L. P., & Marston, A. R. Reduction of social anxiety through modification of self-reinforcement: An instigation therapy technique. *Journal of Consulting and Clinical Psychology,* 1968, *32,* 565-574.

Trower, P., Bryant, B., & Argyle, M. *Social skills and mental health.* Pittsburgh: University of Pittsburgh Press, 1978.

Twentyman, C. T., & McFall, R. M. Behavioral training of social skills in shy males. *Journal of Consulting and Clinical Psychology,* 1975, *43,* 384-395.

Watson, D., & Friend, R. Measurement of social-evaluative anxiety. *Journal of Consulting and Clinical Psychology,* 1969, *33,* 448-457.

Zigler, E., & Phillips, L. Social effectiveness and symptomatic behaviors. *Journal of Abnormal and Social Psychology,* 1960, *61,* 231-238.

Zigler, E., & Phillips, L. Social competence and outcome in psychiatric disorder. *Journal of Abnormal and Social Psychology,* 1961, *63,* 264-271.

Zigler, E., & Phillips, L. Social competence and the process-reactive distinction in psychopathology. *Journal of Abnormal and Social Psychology,* 1962, *65,* 215-222.

Chapter 3

Changes in Assertiveness and Marital Satisfaction after Participation in an Assertiveness Training Group

Judith S. Blau

Abstract

This study measured selected effects of participation in assertiveness training groups for married women. Changes in assertiveness for female participants and their husbands were assessed by using the Adult Self-Expression Scale, the Assertion Inventory, and four behavioral measures. Changes in their marital satisfaction were also assessed by using the Dyadic Adjustment Scale. The effects of participation in differentially focused (i.e., general or marital assertiveness) situations were compared. Subjects were 40 married couples. The wives were assigned to one of four treatment conditions: a general situations assertion training group; a marital situations assertion training group; a waiting-list control group; or a no-treatment control group. The groups met for five 2-hour sessions over a 5-week period. Techniques used in the groups included modeling, role playing, and behavior rehearsal. Participants were also given homework assignments. Data analyzed from pre- and posttreatment assessments—and a 6-week follow-up assessment—tend to support the hypotheses that women do become more assertive after participation in an assertiveness training group, that their spouses do also, and that a positive relationship exists between changes in assertiveness and marital satisfaction. The findings, the influence of the methodological limitations of the study on the results, and the implications for clinical work in the area of assertiveness training and marital therapy are discussed.

Overview of the Problem

A growing social and psychological phenomenon within the American middle class is that many women are striving to become increasingly independent and assertive individuals. They are joining assertiveness training groups in the hope of achieving these goals. The focus of this study is the effect on married couples of participation in such a group. Three consequences that were thought to result from participation in an assertiveness training group for married women have been described and measured. These consequences are changes in assertiveness for the woman, changes in assertiveness for her husband, and subsequent changes in their marital satisfaction.

Although approximately 44,000,000 women are in the United States labor force (U. S. Department of Labor, Bureau of Labor Statistics, Nov. 1979), it is apparent that sex discrimination persists. One study, for example, found that working women were paid an average of $3,458 less than men who were equally qualified in terms of experience and education (Levitan, Quinn, & Staines, 1971). Discrimination is also partially responsible for the fact that the advancement of women usually stops short of top level professional and managerial positions. A *Harvard Business Review* survey found that in 20 major corporations less than 1% of officials, managers, and professionals were women (Osborn & Harris, 1975).

Another hindrance to the advancement of women in the job market is the traditional role that women have fulfilled. They have been trained to be passive and submissive, and to have marriage and motherhood as their major goals. A well-known study by Broverman, Broverman, Clarkson, Rosenkrantz, and Vogel (1970) indicated that male and female clinicians have different concepts of desirable traits for men and women and that the male sex-role stereotype was consonant with attributes considered to constitute good health. Male-valued items included traits such as being aggressive, independent, dominant, logical, direct, adventurous, self-confident, ambitious, easily able to separate feelings from ideas, and not at all dependent. Female-valued traits included being gentle, quiet, interested in appearance, easily able to express tender feelings, and having a very strong need for security.

Increasingly, middle-class women are coming to view their roles as unsatisfactory. Many women choose to live with a sexual partner without being married. Many women turn to careers for personal fulfillment, as well as for financial reasons. Some find, however, that once they become involved in a job, they experience difficulty in handling many interpersonal situations satisfactorily. Many find themselves fearful of appearing too strong or competent, lest they seem "unfeminine." The research of Horner (1970) has shown that the achievement motives of college women are challenged by their own conflicting motives to avoid success. Many women are caught in a double bind in which they are convinced that men are threatened by successful, intelligent females. Therefore, they will change their behavior in such a way as to avoid success. An added problem is that many working women come home to partners who continue to make the same demands on them as if they were not working. The mate may not care to increase his help in doing household chores, so the woman may come to feel overburdened and angry. Phelps and Austin (1975) noted:

> Inevitably there is an imbalance as women begin to choose new alternatives and attempt to integrate them into the security of established conventions. Husbands may object to a wife choosing to attend a class rather than staying home to cook dinner; [and] prospective employers may hesitate to hire a qualified woman because of her "family obligations." (p. 1)

Many women choose not to work, yet find themselves with too much spare time, too few interesting activities, and too little companionship from their husbands. These alternatives are increasingly considered to be unsatisfactory by many women. Osborn and Harris (1975) pointed out that like their employed counterparts, these women

> aspire to be strong and effective. They see themselves as unassertive, passive, and placating . . . Often they constantly placate others because they are afraid to offend them or because they feel they have no right to put their own interests above the interests of others. (p. 37)

There is some disagreement in the literature as to a specific definition of the term *assertive*. However, in all discussions of assertiveness training it is assumed that the goal is to create an individual who is aware of his needs, wants, and feelings and is able to express them directly without injury to other individuals. The first written formulation of the concept of assertiveness was by Salter (1949), who noted that the spontaneous expression of feelings of tenderness, anger, and aggression leads to better interpersonal relations and to subjective feelings of well-being. Definitions include: having the ability to express oneself and to feel good (not guilty) about doing so, as long as others are not hurt in the process (Alberti & Emmons, 1974); "the proper expression of any emotion other than anxiety towards another person" (Wolpe, 1973, p. 81); and "an open and direct, honest, and appropriate expression of what a person feels and thinks" (Lazarus, 1972, p. 34).

Many individuals experience difficulty in being assertive within the marital relationship. Typical patterns in which partners lack assertiveness include: one partner is usually demanding and angry, while the other is submissive; members of a dyadic relationship share little of their needs, wants, feelings, and thoughts with each other; and mates tend to argue with little resolution.

A woman involved in assertiveness training may wish to maintain a satisfactory relationship with her husband but may also desire to be more assertive than before. The husband may not react favorably to the changes in his wife. If there is no resolution of the conflict, emotional disturbance and an increasingly unsatisfying marriage may follow (Bell, 1972). However, a positive outcome can be predicted if as the woman becomes more assertive (and maintains this change), her mate learns to respond similarly. The social learning perspective (Bandura, 1969) supports this prediction in the sense that the wife can serve as a model of assertive behavior for her husband. A necessary condition is that the wife begin to act assertively with the husband, rather than being assertive only in other types of interpersonal situations. "The basic assumption is that if both . . . [partners] were trained to become more assertive in their interactions . . . , their relationship would become closer to what is implied as a 'healthy relationship' " (Lehman-Olson, 1976, p. 93), and it would be experienced more positively. One of the purposes of this study was to assess changes in marital satisfaction, in either a positive or negative direction, after a wife has been in an assertiveness training group. This information could be used to more accurately inform potential clients as to the possible risks and benefits of participation in such training.

A crucially relevant issue is that of generalizability of training. Kirschner (1976) has noted: "There are two contrasting views regarding the nature of assertiveness, each leading to a different expectancy concerning the generalizability of training effects" (p. 117). Salter (1949) considered assertiveness to be a pervasive trait and felt that increased assertiveness in one area is highly generalizable to other situations. Other authors (Lazarus, 1971, 1972; Wolpe, 1973; Wolpe & Lazarus, 1966) have contended that assertive behavior is highly situation-specific and have expected little transfer of training effect. Wolpe and Lazarus also emphasized the necessity of teaching a client to be assertive in situations that are difficult for that particular individual.

A review of the literature shows that few empirical investigations of the generalizability of assertiveness training effects to in vivo situations exist. Available reports (Kazdin, 1974; Kirschner, 1976; McFall & Lillesand, 1971; McFall & Marston, 1970; McFall & Twentyman, 1973) have indicated that the effects of behaviorally oriented assertiveness training procedures do generalize to situations of the same type as those used in training. However, there is no reliable evidence demonstrating generalization of gains in assertiveness to areas not explicitly dealt with in training.

Thus, one could expect that a relatively efficacious way to teach women in assertiveness training groups to be assertive with their husbands would be to concentrate on those issues which are particularly relevant to each woman's marriage. The literature supports the view that the transfer of training effect would thus be greater than if the women were trained to handle issues outside the relationship.

PREVIOUS RESEARCH LINKING ASSERTIVENESS TRAINING WITH THE MARRIED COUPLE

Few studies have been reported concerning assertiveness training for one partner of a marital relationship and the resultant effects on the spouse. Moreover, existing studies have not used control groups and have rarely reported follow-up data.

Eisler and Hersen (1973) reported one of the few studies of assertiveness training and its effects on a marriage relationship. These authors investigated short-term family-oriented crisis intervention in which several behavioral techniques that are considered to be components of assertiveness training were used. They presented three case studies in which family members were treated together. Subsequent to the assertiveness training, the families more willingly expressed complaints, were more successful in problem solving skills, and were otherwise more likely to communicate. Follow-ups of two of the three cases indicated continued improvement in problem areas for those families. For the third case, Eisler and Hersen reported that the marital couple responded more openly and positively to one another at the end of 8 weeks of treatment.

Research more germane to the present investigation was reported by Eisler, Miller, Hersen, and Alford (1974). Evaluations were made on the effects of four 45-minute assertiveness training sessions on passive male patients. Three couples were videotaped while discussing their marital conflicts, before and after the training. Various aspects of the husbands' assertiveness deficits were identified on the basis of observations of the couples' initial interactions. Each patient was instructed to rehearse specific component behaviors, such as response latency, duration of speech, and loudness of voice. Role playing involving interpersonal situations with a female research assistant was also utilized.

In the first case reported, the husband received relatively unstructured scenes related to his pretreatment marital interaction. A behavioral analysis of the pretreatment interaction revealed that the husband did not express either positive or negative feelings about any of his wife's behaviors. In the few instances that he expressed an opinion, he failed to elicit her viewpoint about the subject. Posttreatment results showed that the husband's duration of speech and visual contact with his spouse increased greatly. The behavior of his wife remained constant on these measures. Both spouses showed increases in positive expression and decreases in negative expression and increases in smiling; they also reported increases in marital satisfaction.

In the second case reported by Eisler et al. (1974), the husband was given assertion training based on relatively structured interpersonal situations that were not related to his marital interaction. He was asked to role play scenes such as someone cutting in front of him in a long line. Based on the assessment of his pretreatment unassertive marital responses, the treatment focused on duration of speech, loudness of voice, duration of visual contact, and requests for his interpersonal partner to change her behavior. The results indicated that the duration of the husband's looking at his wife increased substantially but that she did not reciprocate. Requests by the husband for changes in his spouse's behavior increased, and such requests initially increased but then declined for the wife. Loudness of speech evidenced almost no change for either partner. Contrary to the authors' expectations, the husband's duration of speech decreased slightly, while the wife's duration of speech showed almost no change.

The third case differed in that the husband was trained through the use of simulated marital encounters typical of his interaction with his wife. The deficient behaviors selected from the pretreatment videotape included: duration of looking at his wife, duration of speech, requests for his wife to change her behavior, and latency of response. As in the previous cases, he received instructions, behavioral rehearsal, and performance feedback. Results indicated a change in response in the assertive direction for all four measures when the patient was asked to role play the 10 simulated marital scenes without his wife being present. The first three measures showed an increase, while the fourth (response latency) showed a decrease in score. The assessment of the generalization of assertiveness to the marital situation, as observed on the posttreatment videotape, indicated that the husband's duration of speech and visual contact increased considerably, while the wife's decreased considerably on these measures. In addition, the husband had developed a pattern, before treatment, of drinking after experiencing marital conflict. Posttreatment results showed that statements related to the husband's drinking decreased sharply for both spouses. For the husband, requests for changes in his spouse's behavior increased from a very low to a moderate rate, while the same behavior as practiced by the wife went from a high to a moderate frequency. Eisler et al. (1974) do not explain why the changes in the wife were positive, but it seems likely that she was overly aggressive before the treatment.

Based on their results, the authors suggested that:

> teaching patients to be more assertive in role-played encounters generalizes to interactive behaviors with their real life marital

> partners. Also, under some conditions, changes in the husbands' behavior elicits different patterns of response from their wives (Cases 1 and 3). Further, it appears that there is greater generalization to the spouse when husbands are trained on simulated interactions highly relevant to their present marital problems (Cases 1 and 3) than when trained on situations requiring assertiveness unrelated to their present marital difficulties (Case 2). (p. 647)

However, the validity of drawing conclusions based on this research and of generalizing to other client populations is limited, given that only three case studies and no follow-up data were reported.

Shoemaker and Paulson (cited in Lehman-Olson, 1976) reported a study evaluating the effect on both the marital and the parent-child relationship when a mother receives assertiveness training. Sixteen women, each experiencing trouble with child rearing, underwent 5 weeks of assertiveness training in a group setting. The wives significantly increased their assertive responding from pre- to posttreatment, as did their husbands who had not been in treatment. No long-term follow-up data are reported for this study, however.

The Present Investigation

The studies done to measure the effects of assertiveness training have often been methodologically poor, lacking control groups, adequate sample sizes, and follow-ups. There are few studies measuring whether assertiveness does increase as a result of participation in an assertiveness training group for women. Also, there are no data purporting to measure familial effects of such participation at a time when separation and divorce are frequent. An assessment of the latter would seem to be important, given that some people are fearful that women's groups may be contributing to the dissolution of marriages (Osborn & Harris, 1975).

The current study was designed to make a contribution in the following ways: (a) by providing preliminary information as to whether assertiveness training groups help women to be more assertive; (b) by assessing what, if any, effect this has on the satisfaction couples experience in their relationship; (c) by assessing the relative efficacy of treating women in groups that focus on general as opposed to marital situations; and (d) by evaluating the above questions in a controlled fashion with a 6-week follow-up on the subjects.

Subjects

The subjects of the study were 40 married couples. Twenty of the female partners participated in assertiveness training groups for married women offered by the Psychiatry Department of Temple University.

The participants were recruited through a newspaper advertisement and were randomly assigned to the two treatment conditions. Ten of these women constituted a *general situations* experimental group (W_G); their husbands were designated H_G. The remaining 10 women were in a *marital situations* experimental group. They and their husbands were designated W_M and H_M, respectively. Ten other women, who had also responded to the newspaper advertisement, constituted a waiting list control group (W_{WL}); their husbands were designated as group H_{WL}. There was also a second no-treatment control group (W_{NC}) for females, consisting of 10 women who stated that they were not interested in joining an assertiveness training group and who had not received any such training prior to the study. These subjects were obtained through neighbors and friends of friends of the experimenter. This group was included in order to assess on measures of assertiveness any differences between the population of women who request assertiveness training and those who do not. Subjects in group W_{NC} were matched with groups W_G, W_M, and W_{WL} on demographic variables that were obtained from an intake form filled out by all subjects. The mean ages of women and men were 34 and 35, respectively. The couples averaged 11 years of marriage and two children. Educational levels for both sexes ranged from eighth grade to postgraduate training. Incomes ranged from \$5,000 to over \$20,000 per family. Analyses of the data indicated no significant differences among groups concerning these characteristics.

The assertiveness training groups met once per week for 5 weeks. Each session lasted 2 hours. Two groups of each experimental treatment were conducted in order to obtain the requisite number of subjects. The major treatment techniques included modeling, role playing, and behavior rehearsal. Subjects were required to keep a daily record of situations in which they had been assertive and asked to share these experiences with the group. They were also asked to record situations in which they had not been appropriately assertive and to work on these in the group.

Training of the Group Leaders

Two female advanced graduate students in the doctoral program in clinical psychology at Temple University were trained as group leaders. Each leader ran one general and one marital situations group. Neither woman had prior experience in leading an assertiveness training group. The training was conducted by the experimenter, who was experienced with group and individual assertiveness training and who was supervised by a senior staff member of Temple University's Psychiatry Department.

The two leaders met with the experimenter three times, for approximately 6 hours of training. They were shown parts of two videotapes that explained the rationale and various techniques of assertiveness training. They were asked to read *Your Perfect Right* (Alberti & Emmons,

1974), a handbook on assertiveness training. Also, role playing and modeling of assertive behaviors were practiced. The training was considered adequate when both leaders felt comfortable role playing the group situation with the experimenter and the other leader and when they felt a strong familiarity with the group format prescribed in this study.

Every session of the four groups was tape-recorded and listened to by the experimenter to verify that the format was being followed satisfactorily. In addition, the investigator conferred with the leaders after each group meeting to discuss any problems or suggestions.

Assessment Instruments

Two paper and pencil measures of assertiveness were utilized: the Adult Self-Expression Scale (ASES) (Gay, Hollandsworth, & Galassi, 1975) and the Assertion Inventory (AI) (Gambrill & Richey, 1975). The AI has two subscales that ask respondents to rate their degree of discomfort and probability of engaging in particular behaviors. A third subscale asks subjects which of the items they would like to handle more assertively.

A behavioral test was also included to directly assess changes in verbal communication. Each couple was given one of three joint problem solving questions, one at each assessment. The questions were randomly distributed among couples to control for order effects. Subjects were allotted 5 minutes to discuss each of the following:

1. If the two of you could plan anything at all to do together, what would that be?
2. If you could spend any amount of money on a dream vacation, where would you go and what would you do?
3. If I were to give you $50, how would you spend it?

The discussions were tape-recorded, with the subjects' consent, and they were scored by independent raters on four measures: explicit agreement, explicit disagreement, the number of statements using the word *I,* and the latency of response from the end of the partner's verbal response to the beginning of the subject's next statement.

One self-report paper and pencil measure of marital satisfaction, the Dyadic Adjustment Scale (DAS) (Spanier, 1976), was used.

Each of the 5-minute behavioral tape-recorded segments was scored on the four behavioral measures by one of four raters. The raters were graduate students in the clinical psychology doctoral program at Temple University. They were blind as to the hypotheses and design of the study and to the treatment group from which each recording was taken.

Descriptions of the behavioral measures were studied by each of the raters, who then met with the researcher to help clarify any ambiguities. The group listened to and rated four pilot segments. The ratings were discussed and repeated, with fair to moderate agreement among raters as to scores. Each rater was then given two sample segments to score at home, still yielding only moderate agreement. The raters met again with the researcher to more explicitly define the measures and to rate several pilot segments. These ratings were closer in score. The raters were then given the same five pilot segments to score individually, which yielded an acceptable interrater reliability. For the four measures, Kendall's coefficient of concordance was: explicit agreement, $W = .69$; explicit disagreement, $W = .70$; *I* statements, $W = .58$; and response latency, $W = .89$.

Experimental Treatment I (W_G)

The session outline for the general situations treatment group follows.

Session 1:

1. Orientation to the group and overall purpose were covered.
   a. The therapist introduced herself and introduced the group members to one another.
   b. The leader explained what is meant by the concept of assertiveness (modified from Alberti & Emmons, 1974).
   c. The leader explained the differences between nonassertive, assertive, and aggressive behavior (modified from Alberti & Emmons, 1974) and gave examples.
2. The leader presented a situation that called for assertive behavior by saying:

   > You are in a very nice restaurant. You order what promises to be a nice, juicy piece of steak done rare. The waitress brings your order and you cut into it, only to find that the meat is well done and tough. What would you do?

   Members of the group were invited to respond with what they would answer. Each woman was given an opportunity. Then the leader asked which response seemed most assertive, and why. The therapist modeled an appropriate response:

   > Excuse me, waitress. I ordered my steak rare, and this one is well done. May I have another piece of meat, please?

   The group members then practiced making similar responses. They were invited to critique one another's suggestions. By the end of the session each member was able to make an assertive response to that situation.
3. A homework assignment was given. Subjects were asked to remember one situation in which they had acted assertively during the week so that they would be able to describe it to the group the next week. They were also asked to remember one event in which they did not act assertively. The therapist introduced this assignment by saying:

> It is very important that you begin to practice on a daily basis what you learn in this group. You are all asked to remember one situation in which you acted assertively during the week and to be ready to tell us about it next week. Also, remember one situation in which you did not act assertively and be ready to tell us about it next week so that we can work on it together. I find that it's very helpful for you to write down these situations after they happen so that you can more clearly remember what happened and how you felt.

Session 2:

1. Each subject was asked to report on one situation in which she had acted assertively. If she was pleased with how she had handled the situation and the therapist felt that she had handled it well, the leader said something like, "You handled that very well." If the subject did not feel that she had handled the situation as well as she could have, the group was invited to suggest assertive responses. Other group members were asked to critique these responses.
2. Each woman was asked to report on a situation in which she had not acted assertively. If mention was made of a situation involving the husband, the leader helped the participant with that situation. To discourage the presentation of marital situations again, the therapist said, "How about an example from outside the home now?" The members were encouraged to role play various situations and to comment on whether behaviors seemed assertive, and why. The group member presenting the scene was asked to incorporate the suggestions of the group and to practice an appropriate response until she felt comfortable doing so.
3. The same homework assignment as in Session 1 was given.

Sessions 3, 4, and 5:

These sessions followed the same format as Session 2. However, 15 minutes at the end of Session 5 were devoted to a recapitulation of general assertiveness principles and a discussion of how group members would continue their efforts at being assertive.

Experimental Treatment II (W_M)

The session outline for the marital situations group follows.

Session 1:

1. The session began the same way as Session 1 for Group W_G (see item 1 of Session 1 for Group W_G).
2. The leader introduced the focus of the group by saying:

> Some of the most emotional experiences tend to be those connected with our husbands. We tend to get locked into behavior that is relatively fixed and hard to change, although some of it is unsatisfying. For example, you and your husband may con-

stantly argue about the temperature of the house or about the restaurant in which you want to have dinner. If these kinds of disagreements happen often enough, they can lead to an unhappy relationship. One of you may constantly get his or her way to the dissatisfaction of the other, and then neither partner wins. Therefore, we will be focusing the attention of our group meetings on these types of situations.

3. A group member was then asked to volunteer a situation and suggest ways to be more assertive. If she did not provide an appropriate reply, the leader modeled an assertive response. The subject was asked to role play the scene with another group member until she was comfortable being assertive in that situation. Members of the group were asked to provide feedback on whether responses seemed appropriate, and why.
4. Each member was asked to provide a situation, which was worked with in the same manner already outlined.
5. The same homework as for Group W_G was assigned. However, subjects were asked to select only those incidents that involved their husbands.

Sessions 2, 3, 4, and 5:
These sessions followed the same format as described for Group W_G, except that the large majority of situations discussed involved each subject's husband.

Waiting List Control Group (W_{WL})

Subjects in this group were informed at the pretreatment assessment that the assertion training group would not begin for approximately 5 weeks. At the end of that time they were telephoned and asked to come with their husbands for another assessment. They were told that it was necessary to obtain another set of responses just prior to the beginning of the group. After this assessment, they were informed that their participation in the study was complete, and they were paid $25. The women were offered assertiveness training at the Behavior Therapy Unit at a later date. If a woman did not wish to wait, she was referred to another agency that had such groups available. Five of the waiting list women decided to wait for a Behavior Therapy Unit group, one woman was no longer interested, and four asked to be referred elsewhere.

No-Treatment Control Group (W_{NC})

The 10 female subjects in this group had no desire for participation in an assertiveness training group and had never experienced such training prior to the study.

Results

A major concern of this study was whether women who participated in an assertiveness training group (W_G or W_M) would be significantly more assertive on the posttreatment assessment than women in no-treatment control groups (W_{WL} and W_{NC}). One-way analyses of variance (Winer, 1971) of the pretreatment assertiveness data were performed. No significant group differences were observed in the AI response probability, the "number better" scales, and the four behavioral measures. Significant differences among group means were found on the ASES (F (3, 36) = 9.36, $p < .01$) and the discomfort scale of the AI (F (3, 36) = 8.64, $p < .01$). Post hoc analyses using the Newman-Keuls procedure revealed that on both of these measures group W_{NC} was significantly more assertive than the other three groups. Similar directional differences were observed on the other assertiveness measures.

Group differences on posttreatment scores were therefore assessed by using analyses of covariance (Winer, 1971), with the pretreatment scores on each measure as the covariates. Scores on the response probability and number better scales and the four behavioral measures did not change significantly. A significant group main effect was found for the ASES (F (3, 35) = 5.11, $p < .01$). The adjusted posttest scores of the ASES showed that the waiting list control group (W_{WL}) mean was lower (less assertive) than the means of the other three groups, which were comparable. A Newman-Keuls analysis of the adjusted group means revealed that the W_{WL} mean was significantly lower than the means of the other three groups. The group main effect of the discomfort scale approached significance (F (3, 35) = 2.64, $p < .06$). The adjusted posttest means of Groups W_G, W_M, and W_{NC} were lower (more assertive) than the mean of the waiting list control group (W_{WL}).

A second major concern of this study was whether women who participated in the assertion training groups would show significantly greater changes on the measure of marital satisfaction on the posttreatment assessment than no-treatment control group women. A one-way analysis of variance of pretreatment data collected on the DAS showed a significant difference among the means of the four groups (F (3, 36) = 3.24, $p < .05$). A Newman-Keuls analysis of the DAS data was performed. Groups W_G and W_M had comparable pretreatment means, as did Groups W_{WL} and W_{NC}, but the differences among these two subsets of groups were not large enough to be detected by the post hoc test. As a control for these pretreatment differences, a one-way analysis of covariance was performed on the pre- and posttreatment DAS data, with the pretreatment scores as the covariates. No significant differences among the adjusted posttest means were found.

Four 3 (pre, post, follow-up) × 2 (W_G and W_M) repeated measures analyses of variance (Winer, 1971) were performed to assess whether participation in the general assertiveness training group would result in significantly greater changes in general assertiveness—as measured by scores on the ASES and the three subscales of the AI—than participation in the marital assertiveness training group. No significant group main effects or group by time interaction effects were found on any of the four measures. On each variable, however, the time main effect was significant. The *F* ratios for the ASES, the discomfort scale, the response probability scale, and the number better scale were respectively: $F(2, 36) = 10.02, p < .01$; $F(2, 36) = 23.97, p < .01$; $F(2, 36) = 4.75$, $p < .05$; and $F(2, 36) = 18.17$, $p < .01$. Newman-Keuls analyses were performed on each of the four measures to assess at what points in time the changes occurred. Scores on each measure changed significantly from pre- to posttest and remained stable from posttest to follow-up.

The fourth major question was whether participation in the marital assertiveness training group would result in significantly greater changes in marital assertiveness—as assessed by scores on the measures of explicit agreement, explicit disagreement, the number of *I* statements, and mean response latency—than participation in the general assertiveness training group. A 3 × 2 repeated measures analysis of variance of the pre, post, and follow-up assessment data on the four measures was performed to compare the extent of change of groups W_G and W_M. None of the scores on the four measures changed significantly, nor was any directional trend evident.

A 3 × 2 repeated measures analysis of variance of pre, post, and follow-up treatment data collected on the DAS was also performed to compare the extent of change of W_G and W_M. The group main effect and the group by time interaction effect were not significant. The time main effect did show significant differences over time ($F(2, 36) = 3.57$, $p < .05$). A Newman-Keuls analysis revealed that the changes in scores from pre- to posttest were comparable to the changes from posttest to follow-up assessment.

One-way analyses of the pretreatment data collected from husbands in all four groups were performed. No significant differences among group means were found on any of the measures, nor were any consistent directional differences evident for any of the groups. For the assessment of posttest differences, one-way analyses of covariance were performed at pre- and posttest, with pretest scores covaried. No significant differences among the adjusted group means at posttest were found.

The extent of change in assertiveness of Group H_G compared to Group H_M was assessed. A 3 (pre, post, follow-up) × 2 (H_G and H_M) repeated measures analysis of variance was performed on data collected on each of the dependent assertiveness measures. For the ASES, the discomfort and number better scales of the AI, the number of explicit agreements, disagreements, and *I* statements, and the mean response latencies, no significant changes in score were found. The response probability scale showed a significant time main effect ($F\ (2, 36) = 6.32, p < .01$), indicating that the scores on this measure for treatment group husbands as a whole changed significantly. A Newman-Keuls analysis revealed that the scores changed significantly in the assertive direction from pre- to posttreatment and remained stable through the follow-up assessment.

A 3 X 2 repeated measures analysis of variance of the pre, post, and follow-up treatment scores on the DAS showed a significant time main effect ($F\ (2, 36) = 6.76, p < .01$), again indicating that scores for treatment group husbands as a whole changed significantly. A Newman-Keuls analysis revealed that significant changes in scores did not occur from pre- to posttreatment assessment. Rather, they were evident at the 6-week follow-up test.

For testing the assumption that a significant relationship would be found between change scores on the assertive and marital satisfaction measures, changes (posttest minus pretest scores) on the assertive measures were correlated with similarly computed DAS change scores. The scores of wives and husbands were analyzed separately. For wives a significant correlation between change scores on the DAS and the ASES ($r = +.39$, $p < .05$), the discomfort scale of the AI ($r = -.44$, $p < .01$), and the number of *I* statements ($r = +.50$, $p < .01$) was found. For husbands a significant correlation between changes in score on the DAS and the ASES was found ($r = +.44, p < .05$).

Discussion

The major purpose of this study was to measure selected effects of participation in assertiveness training groups for married women. Changes in assertiveness and marital satisfaction for women participants and their husbands were measured and compared to demographically matched no-treatment control groups.

To a limited extent, women who participated in an assertiveness training group were significantly more assertive at posttest than were nonparticipating women. At preassessment, treatment group women had scores comparable to waiting list control group women. After treatment, participants were markedly more assertive on the ASES and the discomfort subscale of the AI. Thus, treatment group women reported decreased discomfort and increased likelihood of responding assertively in a large variety of situations after treatment. The reported changes in behavior included increases in: giving and receiving compliments; openly expressing annoyance or anger to a spouse or boss; refusing unreasonable requests from a boss, parent, friend, or spouse; openly expressing disagreement to a boss, close friend, or spouse; asking for help or favors from friends or relatives; and speaking up in conversations. In addition, subjects reported increased ability to perform the following actions: to refuse pressure from salespeople; to confront strangers who had pushed ahead of them in a line; to send improperly prepared food back to the kitchen in a restaurant; and to insist that a negligent landlord make promised repairs. These gains were maintained at the 6-week follow-up assessment.

No-treatment control group women were significantly more assertive than women in the other three groups at the study's beginning. The adjusted posttest mean scores of treatment group women approached those of Group W_{NC} on the ASES and the discomfort scale of the AI, but they still indicated less assertiveness on the other measures. The unadjusted ASES and AI scores of treatment group women at posttest were comparable to those of the female and male samples on which these measures were normalized. None of the scores on the assertiveness measures changed significantly for the waiting list control group.

The results did not support the hypothesis that assertiveness training group participants would show significantly greater changes in score on the measure of marital satisfaction at posttest than control group women. The scores on the DAS for women in all four groups were comparable at postassessment. However, DAS scores of women in the two treatment groups, when considered together, did change significantly by the follow-up assessment, indicating some increased marital satisfaction. Thus, there is some evidence that the gains in assertiveness discussed above were accompanied by changes in the marital satisfaction experienced by treatment group women. Marital satisfaction improved due to perceived changes in the following areas: there was less disagreement among spouses on such issues as handling family finances, demonstrations of affection, and ways of dealing with parents or in-laws; spouses quarreled and "got on each other's nerves" less often; and spouses calmly discussed their concerns more frequently.

Greater changes in overall assertiveness were predicted for general situations (W_G) than for marital situations (W_M) assertiveness training participants. Greater changes in marital assertiveness for marital situations (W_M) participants than for those in the general situations (W_G) treatment were also expected. Neither of these hypotheses was supported by the data. Scores on the ASES and AI changed similarly for women in the two treatment groups. No consistent pattern of change in score on the four behavioral measures was evident for either treatment group.

Scores on the dependent assertiveness measures were expected to change significantly more from pre- to posttest for husbands of women in the two treatment groups (H_G and H_M) than for husbands in the no-treatment control groups (H_{WL} and H_{NC}). This hypothesis received some support from the data. Response probability scale scores on the AI changed significantly for men in H_G and H_M. Thus, these husbands, who were aware of their wives' participation in assertiveness training, reported being more likely to make assertive responses at posttest than at pretest. The data also indicated that these changes were maintained at the 6-week follow-up assessment.

DAS scores for men in the four groups showed no significant changes from pre- to posttest. However, scores on this measure did change significantly from posttest to follow-up test for husbands of treatment group women. Thus, their marital satisfaction showed an increase after their wives' participation in the assertion training.

Several significant positive relationships were found between change scores on the assertiveness and marital satisfaction measures. For wives significant correlations between changes in score were found between the ASES and the DAS, the discomfort scale of the AI and the DAS, and the number of *I* statements and the DAS. For husbands the correlation between changes in score on the ASES and DAS was significant. These data provide some support for the notion that changes in assertiveness are positively correlated with changes in marital satisfaction.

From the results obtained, the following conclusions about group assertiveness training for married women seem tenable. First, participation in an assertiveness training group is likely to increase one's ability to behave assertively in a variety of situations and with decreased discomfort. Second, there is little differential effect between participation in a general or a marital situations treatment. Assertive behavior learned in a marital situations group seems to increase one's ability to be assertive in a variety of circumstances, but not necessarily in the marital relationship. Third, husbands of participants become more likely to behave assertively in many different situations, but not necessarily within the marriage. Last, if a wife and husband practice assertive behavior over a period of several months, their marital satisfaction seems to increase. These conclusions, however, may have been influenced by the methodology of this study in several ways.

A major limitation of the study concerns the behavioral assessment instruments. For example, for the purpose of reaching a workable operational definition and an acceptable interrater reliability, the measures of explicit agreement and disagreement were made extremely parsimonious; their expression had to be very direct. The word *I* had to be used in order to be scored. Examples were: "Yes, I think that's a good idea" and "No, I don't agree with you." Raters did not count statements that might more realistically be expected in a spontaneous discussion, such as: "That's all right," "Okay," or "That's a good idea." Therefore, measures of agreement and disagreement may be useless as measures of assertiveness until more workable operational definitions are found.

The measure of response latency may also be an ineffective and relatively meaningless measure of assertiveness for most people. This measure was selected from the work of Eisler et al. (1974). The third case they described involved a man who was trained through the use of simulated encounters typical of his marital interaction. Several deficient behaviors were selected from his pretreatment videotape, including latency of response. After treatment, the subject showed a marked decrease in response latency, along with increases on the other measures. Although decreased response latency indicated increased assertiveness for that subject, it did not prove to be a useful indicator in the present study. The important major difference was that response latency was an assessed deficit for that patient but not for the present subjects. The patient described by Eisler et al. (1974) had a pretest mean score on 10 trials of approximately 1.95 seconds. His posttreatment mean score was down to approximately .85 seconds. The mean score at pretest of all 40 women in this sample was 1.07 seconds. The mean score for their husbands was also 1.07 seconds. Eisler et al. did not report what criterion they used to determine that their subject was deficient in speed of response. It is possible that their patient did indeed have a longer response latency than the norm. Also, a ceiling effect may have operated in this study so that subjects did not have much room for change in the downward direction. The wives' posttest mean score was 1.21 seconds, and for husbands it was 1.04 seconds. The fact that the Eisler et al. patient had a posttreatment score of .85 seconds may have been an artifact of his being trained on that particular measure; his latency of response score may have regressed toward a mean that is more like the sample in this study. No follow-up assessment of the subject that would provide information about this issue was reported.

Thus, the finding that Group W_M women did not show significant changes in marital assertiveness may have been an artifact of the measures used to assess such change. The fact that the scores of Group W_M changed in a manner comparable to Group W_G on the ASES and AI tends to support Salter's (1949) contention that assertiveness in one area is generalizable to other situations. The fact that no consistent pattern of change in score on the behavioral measures was evident for either treatment group tends to support the specificity of training concept (Kirschner, 1976; Lazarus, 1971, 1972; Wolpe, 1973; Wolpe & Lazarus, 1966), which holds that changes in assertiveness can be expected only for the behaviors specifically attended to in training.

This study did not focus on teaching the specific skills that would have been assessed by the behavioral measures. Indeed, the importance of performing careful pretreatment behavioral analyses of clients, even when they are to be members of a group, seems to be underscored. When doing research it seems desirable to use measures that assess the most salient assertiveness deficits of each participant. The problem seems to lie in developing a sophisticated understanding of the interrelationship between specific target behaviors, the means of their assessment, and their applicability to different populations. In addition, a careful matching of clients who have trouble being assertive in a set of overlapping types of situations (i.e., with spouse, with children, at work) may not prove as useful ultimately as matching clients according to specific skill deficits. A possible reconciliation of the generalizability and specificity of training concepts indicated by the present results is one which suggests that if people learn to feel comfortable with certain specific assertive skills taught in a group, they will be able to use these skills in an increasingly wide variety of situations.

A no-treatment control group of women was included in the design to assess differences in assertiveness between women who request assertiveness training and those who do not. It was assumed that this group would be representative of the demographically matched population of women. However, scores on the ASES and AI indicated markedly more assertiveness for no-treatment control group women than for women in the other three groups. Scores for W_{NC} were more than one standard deviation away from the norms established for females and males on these measures in the direction of greater assertiveness. It is questionable whether the scores of this group are truly representative of the assertiveness level of demographically similar women. Possibly, their scores were artificially inflated for two reasons related to the methodology. First, by definition, the no-treatment control group consisted of women who had stated that they were not interested in obtaining assertiveness training. These women may have tried to indicate on the measures that they were already assertive, without need of such training. Second, the women were neighbors or friends of friends of the researcher, who they knew would be reviewing their questionnaires. The scores of W_{NC} may have been less extreme if the following methodological differences had occurred: (a) the women had been total strangers to the experimenter; (b) the women had been selected totally at random (i.e., from a telephone book); or (c) if they had not been asked whether they would actually participate in an assertiveness training group.

The correlations between scores on the ASES and the discomfort and probability scales of the AI were significant. These results support the idea that the two measures are assessing similar constructs of assertiveness. Depending upon the assessment goals of researchers in the future, it may be useful to include only one of these measures in the design of a study. A saving of time and energy could thus be effected for subjects and researchers.

The results of this study have other important implications for work in assertion training. Although there were gains in the assertive direction for treatment group women, the changes did not continue after treatment. Informal talks with many of the women at the study's end revealed that for the most part they felt the training was too short. Many of them were only beginning to feel comfortable as active participants in the group and with being assertive outside of the group when the meetings ended. They reported a desire for additional modeling and encouragement from the group in order to continue their efforts at assertiveness. The number of sessions in the present study (five) was based on an informal telephone survey of agencies in the Philadelphia vicinity, which indicated that the majority of assertiveness training groups for women meet between four and six sessions. The results of this study suggest that groups of this length do not maximize the potential for change in their clients. Furthermore, if a husband is not supportive of his wife's participation in assertion training, unlike the population in this study, a woman may particularly need the encouragement and training additional sessions would supply.

The present findings suggest that training in assertiveness for one marital partner will lead to some changes in assertiveness in the spouse. Furthermore, if the assertive behaviors are maintained, increased marital satisfaction may follow. However, one must be cautious about generalizing these findings to a clinical setting in which people seek help for more severe problems than lack of assertiveness. The subjects in this study are not necessarily representative of the population that comes for individual or marital therapy. Nevertheless, increased assertiveness may enhance the marital relationships of adults who function adequately. In addition, assertiveness training may prove to be a useful component of marital therapy. It is hoped that future studies will address these issues.

REFERENCES

Alberti, R., & Emmons, M. *Your perfect right: A guide to assertive behavior* (2nd ed.). San Luis Obispo, Calif.: Impact Press, 1974.

Bandura, A. *Principles of behavior modification.* New York: Holt, Rinehart & Winston, 1969.

Bell, J. E. A theoretical position for family group therapy. In G. D. Erikson & T. P. Hogan (Eds.), *Family therapy: An introduction to theory and technique.* Monterey, Calif.: Brooks/Cole, 1972.

Broverman, I., Broverman, D., Clarkson, F., Rosenkrantz, D., & Vogel, S. Sex role stereotypes and clinical judgments of mental health. *Journal of Consulting and Clinical Psychiatry,* 1970, *34,* 1-7.

Eisler, R. M., & Hersen, M. Behavioral tasks in family oriented crisis interventions. *Archives of General Psychiatry,* 1973, *28,* 111-116.

Eisler, R. M., Miller, P., Hersen, M., & Alford, H. Effects of assertive training on marital interactions. *Archives of General Psychiatry,* 1974, *30,* 643-649.

Gambrill, E. D., & Richey, C. A. An assertion inventory for use in assessment and research. *Behavior Therapy,* 1975, *6,* 550-561.

Gay, M., Hollandsworth, J., & Galassi, J. P. An assertiveness inventory for adults. *Journal of Counseling Psychology,* 1975, *22,* 340-344.
Horner, M. Femininity and successful achievement: A basic inconsistency. In J. M. Bardwick (Ed.), *Feminine personality and conflict.* Monterey, Calif.: Brooks/Cole, 1970.
Kazdin, A. E. Effects of covert modeling and modeling reinforcement on assertive behavior. *Journal of Abnormal Psychology,* 1974, *83,* 240-252.
Kirschner, N. Generalization of behaviorally oriented assertive training. *Psychological Record,* 1976, *26,* 117-125.
Lazarus, A. A. *Behavior therapy and beyond.* New York: McGraw-Hill, 1971.
Lazarus, A. A. *Clinical behavior therapy.* New York: Brunner/Mazel, 1972.
Lehman-Olson, D. Assertive training: Theoretical and clinical implications. In D. Olson (Ed.), *Treating relationships.* Lake Mills, Iowa: Graphic Publishers, 1976.
Levitan, T., Quinn, R., & Staines, G. Sex discrimination against the American working woman. *American Behavioral Scientist,* 1971, *15,* 252-253.
McFall, R. M., & Lillesand, D. B. Behavior rehearsal with modeling and coaching in assertion training. *Journal of Abnormal Psychology,* 1971, *77,* 313-323.
McFall, R. M., & Marston, A. An experimental investigation of behavior rehearsal in assertion training. *Journal of Abnormal Psychology,* 1970, *76,* 295-303.
McFall, R. M., & Twentyman, C. T. Four experiments on the relative contributions of rehearsal, modeling, and coaching to assertion training. *Journal of Abnormal Psychology,* 1973, *81,* 199-218.
Osborn, S., & Harris, G. *Assertive training for women.* Springfield, Ill.: Charles C. Thomas, 1975.
Phelps, S., & Austin, N. *The assertive woman.* San Luis Obispo, Calif.: Impact Press, 1975.
Salter, A. *Conditioned reflex therapy.* New York: Capricorn Press, 1949.
Spanier, G. Measuring dyadic adjustment: New scales for assessing quality of marriage and similar dyads. *Journal of Marriage and the Family,* 1976, *38,* 15-28.
U.S. Department of Labor, Bureau of Labor Statistics. *Employments and earnings,* 1979, *26,* 2. Washington, D.C.: Government Printing Office, 1979.
Winer, B. J. *Statistical principles and experimental design* (2nd ed.). New York: McGraw-Hill, 1971.
Wolpe, J. *The practice of behavior therapy.* New York: Pergamon Press, 1973.
Wolpe, J., & Lazarus, A. A. *Behavior therapy techniques.* New York: Pergamon Press, 1966.

Chapter 4

Group Training of Problem Solving with Psychiatric Patients

Barry A. Edelstein
Eugene Couture
Mary Cray
Pamela Dickens
Nancy Lusebrink

Abstract

*Chronic institutionalized adult psychiatric patients participated in a 5-week problem solving program which addressed the following components of problem solving: problem definition, alternative solution generation, evaluation of solution adequacy, selection of best solution, and implementation of solution. Problem content areas included banking, budgeting, medication, health, telephone usage, meal planning, transportation, and community resources. Performance was assessed in a multiple baseline design across problem solving components. Generalization from classroom to analogue situations was assessed. Statistically significant improvement in performance was obtained for all problem solving skills taught. Considerable generalization to an analogue situation was observed. The results are discussed with respect to the use of relatively untrained staff for enhancing problem solving and coping skills of patients prior to discharge.**

*The authors appreciate the cooperation and support of J. R. Daniell, Superintendent of Spencer Hospital. Appreciation is also expressed to the COPE Unit staff, especially Dennis Lowe and Sheila Emerson.

Chronic mental patients and their associated problems have been described as a national crisis (Huey, 1978), and these persons have been termed as major victims of institutionalization and the lack of adequate training, treatment, and aftercare (The Task Force Reports to the President's Commission on Mental Health, 1978). During the current trend toward deinstitutionalization in mental health, pressures from legal, political, and civic groups have forced many hospitals to discharge thousands of patients, many of whom are unprepared for community living. Recidivism rates have been extremely high in spite of the rapidly growing community mental health center system. In spite of the fact that variables predictive of return to mental hospitals have been identified (Alevizos & Callahan, 1977; Miller & Willer, 1976), most large mental hospital programs have addressed institutional behaviors while ignoring behaviors and skills which require attention for adaptation in the community.

The magnitude of these problems associated with the chronic mental patient is of little surprise in light of the general characteristics of these individuals. Test and Stein (1978) have described these individuals as exhibiting high vulnerability to stress, deficiencies in coping skills, extreme dependency, difficulty with working in competitive markets, and difficulty with interpersonal relationships. What is surprising is the shortage (if not total absence) of programs designed to provide patients with the practical skills needed to function in an ordinary community (Place & Winer, Note 1). In his review of outcome studies Erickson (1975) has noted that "virtually unexplored is the value of a curriculum teaching specific content to the patient on how to handle specific problems in a more constructive way" (p. 535). The Task Force Reports to the President's Commission on Mental Health (1978) have also asserted that "we still do not know very much about how people of diverse sociocultural backgrounds solve problems and cope with stress and crisis" (p. 160). Moreover, we are not even sure how people of the same sociocultural backgrounds solve problems and cope with crisis.

Although problem solving has been a topic of research inquiry for many years, there is no consensus on how a person problem solves or how anyone should go about studying problem solving. Approaches have ranged from learning (Gagne, 1964; Kantor & Smith, 1975; Kendler & Kendler, 1962; Skinner, 1974), to more cognitive orientations (Kohler, 1925; Maier, 1970; Mayer, 1977), to mathematical models and computer simulation of problem solving (Newell, Shaw, & Simon, 1958; Newell & Simon, 1972; Scandura, 1977; Simon, 1969). Dirkes (1978) has noted that the process of problem solving has been examined as an instance of creative thinking (Guilford, 1968; Travers, 1972), cognitive learning (Torrance & Myers, 1970), discovery and invention (Getżels & Jackson,

1962), fluency, flexibility, originality, and elaboration (Guilford, 1967), and hypothesis generation (Gagne, 1970).

From the perspective of a teacher or therapist, accurate knowledge on precisely how one problem solves is not essential. A suitable analogue, model, algorithm, and/or heuristic may suffice for teaching one how to solve a problem more expediently or effectively. Mayer (1977) has noted that "any definition of 'problem' should consist of the three ideas that: (a) the problem is presently in some state, but (b) it is desired that it be in another state, and (c) there is no direct, obvious way to accomplish that change" (p. 5). An *algorithm* is a set of rules that automatically sets forth a way or method for accomplishing the desired change. An *heuristic* is a rule of thumb or "general plan of action" (Mayer, 1977). Heuristics are used to choose among these algorithms. One heuristic is "means-end analysis," which has been employed by Spivack and Shure (1974) in teaching children problem solving skills. This analysis has two parts: (a) comparing the present state of affairs with another state of affairs following the application of an heuristic or set of rules and (b) operating on the situation accordingly.

Problem solving models consisting of heuristics and algorithms must be specific, elaborating steps which will ultimately lead to problem solutions. The steps typically employed by researchers have included the following: (a) problem identification, definition, or formulation, (b) generation of alternative solutions or responses, (c) selection of the best alternative, (d) performance of the responses constituting the solution, and (e) evaluation of the outcome.

What little empirical evidence exists with clinical populations suggests that psychiatric patients are deficient with respect to these problem solving steps or procedures in addressing and solving problems that reflect situations of daily living (e.g., Platt & Spivack, 1972a, 1972b, 1973, 1974). These data are consistent with more basic research, cited by Heppner (1978), which has indicated that emotionally charged stimuli reduce flexibility of problem solving patterns in animals, as well as human subjects (Everall, 1935; Hamilton, 1916; Hamilton & Krechevsky, 1933; Patrick, 1934a, 1934b).

In light of the foregoing considerations of the problems of chronic psychiatric patients and the utility of problem solving training with other populations and problems (e.g., Shaftel & Shaftel, 1967; Spivack & Shure, 1974; Kifer, Lewis, Green, & Phillips, Note 2), the present study was conducted to examine the ability of chronic psychiatric patients to solve problems of everyday living and to examine the effects of teaching problem solving skills. Procedures similar to those proposed by D'Zurilla and Goldfried (1971) and Featherstone, Willis, Cole, Blousefield, and

Smith (Note 3) were used. Tests of the generalization of problem solving skills across problem areas and from classroom to analogue situations were also included.

METHOD

Subjects

The subjects were six male and six female psychiatric patients who were selected for a new program involving problem solving skills training, as well as other community survival skills. Ages ranged from 28 to 63 years, with a mean of 48 years. The mean total length of hospitalization was 6 years and 6 months, while the mean length of the present hospitalization was 1 year and 3 months. The mean number of admissions was 7, with a range of 1 to 24 admissions. All patients were voluntary admissions. Nine of the patients carried a diagnosis of schizophrenia, one was diagnosed as manic-depressive, and two carried a diagnosis of organic brain syndrome. The mean educational level was 10 years and 4 months.

Teachers and Materials

For four day and evening shifts one MA-level psychologist and one MA-level social worker served as teachers. Classes were conducted in 1-hour segments. Evening shift aides worked individually with residents who were having problems in understanding the daily classroom material and those who were unable to meet the minimal performance criteria during day classes.

Teaching materials were totally self-instructional for staff members and so sufficiently detailed that virtually any staff member could teach the classes with little practice. The materials consisted of five training modules, four of which were devoted to basic problem solving skills and the last of which was a summary of previous modules.

Procedure

Before classes began, patients were assessed for their ability to demonstrate problem solving skills. Written or oral pretests were conducted by staff in individual patients' rooms or in classrooms. Each patient was read a description of a problematic situation and asked to do the following: (a) identify the main problem, (b) generate possible solutions, (c) state the consequences of each solution, (d) select the best solution, (e) justify the choice of the solution, and (f) describe how one would implement the solution.

Test instructions. The general test administration instructions for the teachers read as follows.

You will find an attached sheet describing four problem situations, situation 1, 2, 3, and 4. You are to read each situation to the patient and, after each situation is read, you will ask the patient the questions on the scoring sheets (see Figure 4.1 at the end of test instructions).

Step 1. The patient should identify the main problem.

a. Ask the question on the scoring sheet and record the patient's response.

b. If the response was incorrect, tell the patient the correct answer.

Step 2. The patient should try to generate as many possible solutions to the problem as he can.

a. Ask the question on the scoring sheet and record the patient's response.

b. If the patient provides less than two different solutions, ask him once: "What else can you do?" Record any additional responses. Do not repeat the question.

Step 3. For each solution, the patient should state the consequences of carrying out the solution. For example, some effects of punching out an individual might be: (1) you would feel bad, (2) the other person would be angry, or (3) the other person would hit you back.

a. Just ask the question on the scoring sheet and record anything the patient says.

Step 4. The patient should select the best solution to the problem.

a. Ask the question on the scoring sheet and record the patient's response.

Step 5. The patient should tell you why he or she thinks his or her solution is the best.

a. Ask the question on the scoring sheet and record the patient's response.

Step 6. The patient should describe how he or she would actually go about doing what he or she says he or she should do.

a. Ask the question on the scoring sheet and record the patient's response.

Procedure for Module 1. The procedure for each module is similar to the following outline for Module 1.

Teacher's Overview

In this module the patient will be taught the first step in problem solving, identifying the main problem. At first, the problematic situation will be presented along with several possible problems; the patients will pick out the main problem. Problematic situations will then be presented where only a couple of possible problems are stated, and patients must supply other possible problems. Finally, the main problem will have to be selected with no help from the material.

Figure 4.1. Scoring Sheet for Problem Solving

Name of Patient: ________________ Date: ______________________________

Recorder's Name: ________________ Problem Situation No.: ____________

Step 1. "What is the main problem?"

Step 2. "What are some possible things you could do to solve the problem?" (Remember, if the patient gives you only one solution or alternative, ask, "What else can you do?")

Step 3. Read each of the solutions that the patient gave in Step 2 back to him. For each of these solutions, ask, "What would happen if you actually did this?"

Step 4. "What do you think is the best solution?"

Step 5. "Why do you think this is the best solution?"

Step 6. "How would you actually go about doing what you think you should do?"

Note. For more information on scoring, an expanded procedural guide is available upon request from the first author.

Method of Presentation

Whenever you present the material, ask one person for the answer. Explain at the beginning of the group that it is all right for the persons to ask other individuals for help. This is an option available to them outside the hospital and should be encouraged. Be sure that other group members are quiet until asked to help. Remind the group that they are permitted to ask other members for help.

Be sure to be positive in your feedback to each member of the group. Try pairing a positive comment with negative feedback whenever negative feedback must be given. The patient must be told of the adequacy of his choice in a pleasant way and in a way that permits him to then make the correct response.

Patient Introduction and Overview

In your words, tell the patients that they will all be learning how to solve problems when they come up in everyday life. They will learn how to look at a situation when they are in a problematic situation, to identify the problem, to think of many ways of solving the problem, and to act on the problem. In this module, they will learn the first step in problem solving: identifying the main problem.

In order to solve a problem, they must first know what that problem is. They must ask: (1) Is this my problem or someone else's? (2) Can I do anything to deal with it? One important thing to learn is that it is okay to ask another person for his opinion or help if one is stuck. It is also important that the person try to act by himself. In this group, you can ask other people for help after you have at least tried to figure out the answer. Patients who were not asked a question should be quiet until the patient asks for help.

Identifying the Problem

1. Preparation:
   Before class, print the following situation and problem selection on a poster or on the blackboard.

   (Situation 3)

   Your landlord is always crabby when he comes to collect the rent. One of the windows in your bedroom was cracked when you moved in and during a cold night, it cracked all the way and a section is now out. When he comes to collect the rent, you mention it to him. He screams at you and accuses you of breaking the window, saying it was never cracked. Whenever he was crabby before you simply let it go, but this time he says you must pay to have the window replaced.

   (Possible Problems):
   a. The problem is that your landlord is a real grouch.
   b. The problem is that his screaming has made you feel real down and hurt.
   c. The problem is that your window is cracked and should be repaired by the landlord. *(correct)
   d. The problem is that you don't have enough money to pay for a new window.
2. Presentation – Structured:
   a. Tell the class (group) that you are about to begin showing them how to problem solve. Tell them:
      1. That first you will read to them a situation.
      2. They they are to identify the main problem.
      3. That to help them to identify the main problem you will read to them several possible problems and ask which one they think is the main one.
   b. Slowly read the situation and possible problems to the group. Read a second time if necessary.
   c. Ask one person for his opinion on which is the main problem. Remind that person that he can ask for help from another member of the group. Thank him for answering.
   d. Ask each person in turn for his opinion and note it on the blackboard or on a piece of paper. Be sure to say something positive after each person answers.
   e. Read problem "a" [under Possible Problems] to the group

and ask anyone who chose "a" for her reasons for this choice. Ask her the following questions:
1. Whose problem is it?
2. Can you reasonably expect to change it?
3. Will changing this change the situation?

f. Read problems "b, c, d" and review them in the same manner as with problem "a."
g. Point out that "c" is the main problem. Point out that the selection of main problem involves judgment. There is no absolute right answer, only answers that one thinks are best in his estimation.
h. Ask one group member to discuss why "c" is probably the main problem. Remind him that he can ask other members for help.

3. Semi-structured Presentation:
   a. Preparation:
   Write the first two possible problems on the board, leaving space to write other possible problems.
   b. Tell the class now that they have pointed out the main problem in one situation, you would like to practice with other situations. This time you will supply only a couple of possible problems. They will be asked to supply other possible problems and then to choose the one best main problem.
   c. Read the following situation to the group and read the two associated possible problems and ask them to supply several more. Be sure to ask for problems until the main one is revealed. It is permissible to re-read the situation and to prompt responses by asking "Is this a problem?" Try to use as few prompts as possible. It is not necessary to ask everyone for a possible solution.

   (Situation 28)

   You have recently been to the doctor's office, where he wrote you a new prescription for your medication. He also told you that you had to lose some weight. You stopped at the pharmacy on the way home and had your prescription filled. You took the pills just like the instructions said to, beginning as soon as you got home that day. You soon felt ill and vomited. You have felt ill each day since, but you are still eating well and have gained 2 pounds. What are the possible problems in this situation?
   1. The doctor is going to be really angry with you for gaining weight.
   2. You are being punished with the illness.
   3. Your medication is making you ill. (Do not write this one down. It is the main problem.)

   d. When you have enough possible problems (including the main problem, which has been supplied by a group member), go through the possible problems one by one as you did the first time, asking:

1. Whose problem is it?
2. Can you reasonably expect to change it?
3. Will changing this problem change the situation?

Again, try to arrive at a choice where the main problem:

1. Is identified as the problem of the person himself.
2. Is changeable.
3. Will change the situation for the better.

e. Point out the main problem and ask someone to discuss why it is considered the main problem. Remind the person that it is all right to ask other group members for help.

(Situation 29)

You wrote a check to cover the bill from the phone company. As it turns out, the check bounced. You did not have enough money to cover the check. The phone company sends you a notice saying that your service will be cut off if you don't pay the bill within 5 days. Your next payday is 7 days away. You are very angry that you are in this fix. What is the main problem here?

(Possible Problems):

1. Your check has bounced and now the bank will fine you $5 for an overdraft.
2. Your phone is going to be shut off.

4. Unstructured Presentation:

a. Tell the group that now that they have pointed out several possible main problems in a couple of situations, you would like them to point out just the one main problem in a situation. You would like them to state why they chose that as the main problem. Remind them that they can ask each other for help if they need it.

b. Read the following problem situations to the class, one situation at a time. After you read one problem situation, ask one person to pick out the main problem and to justify her or his answer. Then ask if another person agrees or disagrees, and why. Continue asking until one person gets the correct main problem, and then ask one final person for agreement or disagreement. You can help the person if he or she asks by asking questions about the problem (for example, "Well, whose problem is that?").

c. Begin with Situation 3, which was the situation that was used in the very first section of this module. This time do not show it to the class; just read it to them. Do not read the possible problems.

(Situation 4)

As usual, you are boarding the bus to go downtown on a Saturday. You are in a hurry because you are late and you trip going up the steps. The bus driver makes a rude remark about you. What is more a problem, he says it very loudly so that everyone on the bus can hear him. You feel like everyone is looking at you. (Main problem: The bus driver has made a rude remark and embarrassed you.)

(Situation 30)

It is Tuesday, and you are preparing to make your third meal of the week. You decide to make spaghetti. After you have eaten, you find that you still have enough spaghetti to last for two more meals. You already have food thawed in the refrigerator for the next 2 days. If you do not cook the thawed food it will spoil. (Main problem: You have too much food thawed, including spaghetti.)

d. Be sure to praise everyone's performance before leaving.

Posttest

1. The posttest should be done within 1 day of completing the module. It may be done in class or at other times.
2. Use the same procedure and data sheets used in the pretest.
3. Test each patient alone.
4. Use situation numbers: 3, 4, 5, and 6.

Two problem situations were duplicated between pre- and posttests to assess both learning and generalization.

As noted in the foregoing instructions, group training began with very structured presentations and progressed through semi-structured and unstructured presentations. The amount of teacher participation in identifying problems is gradually faded out. Even though this fading from structured to unstructured presentation is stated at three levels of difficulty, fading can occur within each style of presentation to increase the probability of errorless learning.

Module 1, as already described, was followed by three more modules for teaching the remaining steps of problem solving and an additional module reviewing all steps of problem solving. Each module was organized and presented in a similar fashion.

Generalization was assessed in an analogue situation prior and subsequent to the training with all modules. The pretest assessment involved bringing individual residents into a simulated grocery store in a large classroom. The walls of the room were lined with shelves filled with a large variety of canned food and staple goods of different brands, with the price marked on each item. A cashier and cash register were positioned near the door. Upon entering the store, each patient was handed a shopping basket and asked to purchase groceries which would provide nutritious meals for approximately 3 days. Each patient was then permitted to select food items. Once the patient selected all the necessary groceries, she brought them to the cashier for check out. In the process of registering the prices of items the cashier overcharged for several items. This overcharge represented the problem situation. If the patient didn't recognize the problem, she was prompted by being asked whether she had enough money or whether she had noticed the prices of groceries. Correct identification was scored if the patient stated the problem without a prompt. Once the problem was established, each patient was asked to state all the solutions to the problem which he considered. The number and adequacy of these responses were later scored. Finally, the adequacy of the solution implementation was scored by the cashier.

Several procedures were employed during training to capitalize on the group format. To occasion the generation and statement of alternative solutions, we divided patients into equal-sized teams within each group. Teams competed with each other in attempting to generate the most alternative solutions to each problem. This was particularly helpful with relatively withdrawn chronic patients. Implementation was scored as either adequate or inadequate. Considerable disagreement was initially demonstrated among the raters because the data were stated so generally. Following discussions of ratings, agreement of .81 was obtained.

RESULTS

The following analysis applied to the 12 subjects who completed the problem solving training. The data were compiled and analyzed in a multiple baseline design across skill areas (see Figure 4.2). Repeated measures analyses of data obtained at each pretest and following each skill training module were conducted to evaluate performance changes across skill training modules. Each analysis of variance (ANOVA) obtaining significant differences across modules was followed by simple two-tailed t tests.

Skill 1: Main Problem Identification

The pretest mean percentage correct identification of the main problem was 62.5%. Problem identification was trained in Module 1. The mean percentage correct identification following intervention in Module 1 was 85.4%. The ANOVA yielded a significant value of F for performance across training modules, $F(5, 55) = 13.53$, $p < .001$. The difference between pretest and intervention in Module 1 was statistically significant, $t(11) = 3.07$, $p < .02$. No significant difference was found between the mean of the intervention test score for Module 1 and the mean of tests administered following training on the remaining skills. A significant difference was found between pretest means (62.5%) and final posttest means (100%), $t(11) = 2.99, p < .02$.

Figure 4.2. Mean Performance of Subjects across Pretest, Training Modules, and Posttest

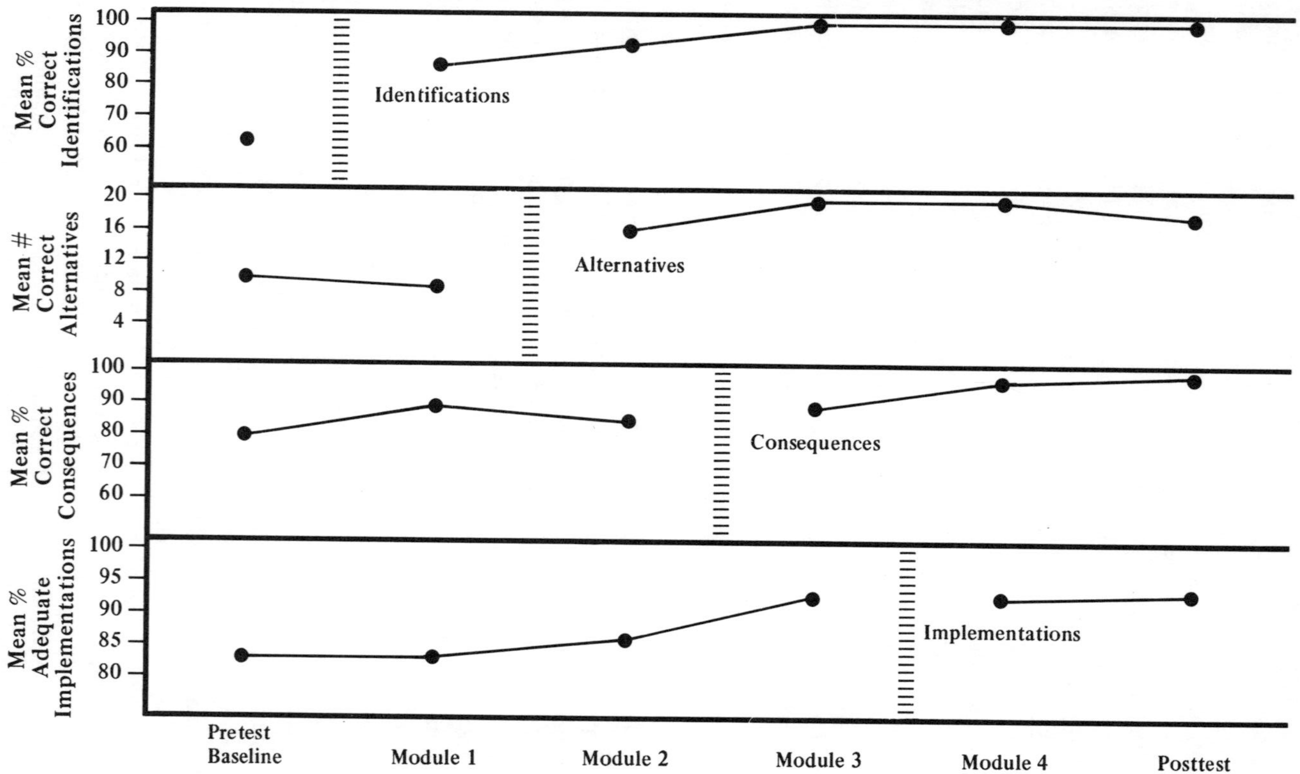

Skill 2: Generating Alternative Solutions

The mean total number of alternative solutions generated across the four problem situations in the pretest was 10 (2.5 per situation). The generation of alternative solutions was trained in Module 2. The mean number of alternative solutions for the test following training of this skill in Module 2 was 16 for four problem situations (4 per problem). The ANOVA yielded a significant F value for performance across training modules, $F(5, 55) = 33.47, p < .001$. There was a significant increase in number of alternatives from pretest to training of this skill, $t(11) = 4.16$, $p < .0001$. The mean number of alternative solutions increased significantly from Module 2 to Module 3, an increase from 16 to 20, $t(11) = 2.4$, $p < .05$. There was also a significant increase between the pretest mean of 10 (2.5 per situation) and the posttest mean of 18 (4.5 per situation), $t(11) = 7.61, p < .001$.

Skill 3: Adequacy and Realism of Consequences

Each consequence stated for each solution generated was rated as either realistic or unrealistic. This skill was trained in Module 3. A proportion of realistic responses was calculated for each subject by dividing the number of realistic consequences by the total number of consequences stated. This was done for all four problem situations presented during each assessment. The ANOVA yielded a significant value of F for performance across training modules, $F(5, 55) = 6.42, p < .001$. Proportion of realistic consequences increased significantly from a pretest value of 79.5% to 88.31% following training in this skill, $t(11) = 2.76, p < .02$. There was also a significant increase from the pretest mean of 79.5% to the posttest mean of 99.3%, $t(11) = 3.14, p < .01$.

Skill 4: Implementation

Each stated implementation was scored as correct or incorrect. Implementation was trained in Module 4. Proportion of correct implementations was calculated for each subject by dividing the number of correct implementations by the total number of implementations. The ANOVA yielded a significant effect across training modules, $F(5, 55) = 1.73, p < .05$. Performance significantly increased from a pretest mean of 83% to a mean of 93% following training in Module 4, $t(11) = 2.29, p < .05$. Performance also increased significantly from a pretest mean of 83% to a posttest mean of 93.75%, $t(11) = 2.87, p < .02$.

In summary, the results indicated a significant increase in skills in pretest to posttest comparisons. This increase occurred following intervention in three of the four target skills, with no prior significant increase over pretest level (i.e., identification of the main problem, generation of alternative solutions, and implementation). On two of these skills (identification of main problem and implementation), there was no significant change in the mean of the tests administered in succeeding modules. Significant improvement was observed in generating alternative solutions and stating realistic alternatives both during and following the training of each of these skills.

Analogue Assessment

Analogue assessment was conducted in the grocery store where patients were exposed to problems characteristic of those they would encounter outside the hospital. These assessments were conducted prior to problem solving training and following the last training module. As shown in Table 4.1, the mean number of alternative solutions generated increased from 1.5 during the pretesting to 1.8 during posttesting. The mean percentage of correct choice of alternatives increased from 27% to 73%, and the mean percentage of correct implementations rose from 18% to 73%, both clearly significant changes.

DISCUSSION

The results clearly support Platt and Spivack's (1972a, 1972b, Note 4) finding that chronic psychiatric patients are frequently deficient in day-to-day problem solving skills. When comparing beginning with terminal performance, the present data suggest that chronic psychiatric patients are most deficient in the skills of identifying the main problem in a problematic situation and selecting the best alternative solution. These deficits are closely followed by the skills of generating alternative solutions, stating realistic consequences, and stating the steps required for implementation of the chosen solution. The results of the assessment of implementation skills probably do not as accurately represent what the patients could actually do in a problem situation as the data might suggest. The criteria for determining the adequacy of one's implementation skills were rather general and probably yielded inflated pretest and posttest scores for this skill.

Table 4.1
Analogue Assessment

	Pretest	*Posttest*
Percentage Correct Identification of Problem	100%	100%
Mean Number of Alternatives Generated	1.5	1.8
Percentage of Adequate Alternatives	27%	73%
Percentage of Adequate Implementations	18%	73%

The data suggest that improvement in the generation of alternative solutions to problems also occasions improvement in the patients' selection of the best alternative solution. Improvement in the identification of the main problem apparently improved the generation of response alternatives to each problem situation. In spite of the improvement of two skills when intervening with one, the large magnitude of changes following intervention with several of the skills preserved the control features of the multiple baseline design. The improvement from pre- to posttest for each of the skills dramatically demonstrates the potential for improvement in skill areas which typically would be ignored in traditional large-scale treatment programs.

The classroom data also suggest that the patients learned a generalized skill which was not specific to the problem situations on which the skills were trained. This is evidenced by the maintenance of performance even when new problem situations were introduced in each session.

The simulation assessment probably offered a more externally valid measure of problem solving skills than measures used by previous researchers. Considerable improvement was evidenced on three of the five skills assessed in the in vivo simulation. The means of assessing correct identification of problems and number of alternatives generated were apparently insensitive to the pretraining skill levels. This problem was, to some extent, a function of the unstructured nature of the assessment task, as well as rather general scoring criteria and rather apparent problem situations. Subsequent training sessions have incorporated more subtle problem situations and more stringent scoring criteria. In spite of the shortcomings, the in vivo simulation assessment provided a reasonable test of generalization from a structured classroom training situation to one more closely approximating a real problem situation in the community.

The generalization across situations (imagery and analogue) is perhaps the most significant finding of the present study. It was hoped that the problem solving strategies would be incorporated as high-probability responses which are emitted in all problem situations.

Several difficulties occurred in the process of conducting the present study. Subjects were initially very reluctant to perform before the group. This reluctance was alleviated by the competition generated in discussing problem situations. A second difficulty occurred when subjects were asked to identify the main problem. Many subjects were unaccustomed to breaking down the problem solving process into steps. Considerable time was spent shaping the subjects into articulating the problem rather than merely jumping into the solution. The group format helped considerably by affording peer-generated models of appropriate problem identification behaviors. If a subject immediately responded to a problem with a solution, she was asked why one would want to do that. This tended to prompt the person to define the problem. Some subjects had difficulty understanding the instructions and problem situations when they were read orally by the teacher. This confusion was reduced somewhat by providing written problems for those subjects who could read. The trade-off on time came because some group members were very slow readers.

Limitations of the present study include a limited sample of analogue assessments, an insufficiently detailed scoring procedure for evaluating the adequacy of alternative solutions generated, and an insufficiently detailed analysis of implementation strategies proposed by patients. All these deficiencies have been corrected in subsequent training programs, with a concomitant increase in the sensitivity of assessment measures. In all cases the results have paralleled those in the present study.

The present study demonstrated that problem solving skills can be taught to a group of chronic psychiatric patients by staff with very little, if any, training in problem solving or classroom instruction. The procedures described can be employed by any large psychiatric hospital to provide badly needed skills required for adaptation in the community (Erickson, 1975; Place & Winer, Note 1). This training should enhance the ability of patients, who are vulnerable to stress, to solve everyday problems involving emotionally charged stimuli.

Reference Notes

1. Place, D. M., & Winer, S. *Re-entering the community: A pilot study of mentally ill patients discharged from Napa State Hospital.* Menlo Park, Calif.: Stanford Research Institute, June 1974.
2. Kifer, R., Lewis, M., Green, D., & Phillips, E. *The S.O.C.S. model: Training pre-delinquent youths and their parents in negotiation responses to conflict situations.* Paper presented at the 81st Annual Convention of the American Psychological Association, Montreal, August 31, 1973.
3. Featherstone, P., Willis, J., Cole, R., Blousefield, M., & Smith, L. *A self instructional unit on problem solving for disadvantaged adults.* Unpublished manuscript, University of British Columbia, 1977.
4. Platt, J. J., & Spivack, G. *Dimensions of interpersonal problem solving thinking in adult normals and psychiatric patients.* Unpublished manuscript, Department of Mental Health Services, Hahnemann Medical College and Hospital, 1974.

References

Alevizos, P. N., & Callahan, E. J. Assessment of psychotic behavior. In A. R. Ciminero, K. S. Calhoun, & H. E. Adams (Eds.), *Handbook of behavioral assessment.* New York: Wiley, 1977.

Dirkes, M. A. The role of divergent production in the learning process. *American Psychologist,* 1978, *33,* 815-820.

D'Zurilla, T., & Goldfried, M. Problem solving and behavior modification. *Journal of Abnormal Psychology*, 1971, *78*, 107-126.

Erickson, R. C. Outcome studies in mental hospitals: A review. *Psychological Bulletin*, 1975, *82*, 519-540.

Everall, E. E. Perseveration in the rat. *Journal of Comparative Psychology*, 1935, *19*, 343-368.

Gagne, R. M. Problem solving. In A. W. Melton (Ed.), *Categories of human learning.* New York: Academic Press, 1964.

Gagne, R. M. *The conditions of learning.* New York: Holt, Rinehart & Winston, 1970.

Getzels, J., & Jackson, P. *Creativity and intelligence.* New York: Wiley, 1962.

Guilford, J. P. *The nature of human intelligence.* New York: McGraw-Hill, 1967.

Guilford, J. P. *Intelligence, creativity and their educational implications.* San Diego: Knapp, 1968.

Hamilton, G. V. A study in perseverance reactions in primates and rodents. *Behavior Monographs*, 1916, *3*, 11-14.

Hamilton, J. A., & Krechevsky, I. Studies in the effect of shock upon behavior plasticity in the rat. *Journal of Comparative Psychology*, 1933, *16*, 237-253.

Heppner, P. P. A review of the problem-solving literature and its relationship to the counseling process. *Journal of Counseling Psychology*, 1978, *25*, 366-375.

Huey, K. The APA conference on the chronic mental patient. *Hospital and Community Psychiatry*, 1978, *29*, 257-260.

Kantor, J. R., & Smith, N. W. *The science of psychology.* Chicago: Principia Press, 1975.

Kendler, H. H., & Kendler, T. S. Vertical and horizontal processes in problem solving. *Psychological Review*, 1962, *69*, 1-16.

Kohler, W. *The mentality of apes.* New York: Harcourt, Brace, 1925.

Maier, N. R. F. *Problem solving and creativity in individuals.* Belmont, Calif.: Brooks/Cole, 1970.

Mayer, R. E. *Thinking and problem solving.* Glenview, Ill.: Scott, Foresman, 1977.

Miller, G. H., & Willer, B. Predictors of return to a psychiatric hospital. *Journal of Consulting and Clinical Psychology*, 1976, *44*, 898-900.

Newell, A., Shaw, J. C., & Simon, H. A. Elements of a theory of human problem solving. *Psychological Review*, 1958, *65*, 151-166.

Newell, A., & Simon, H. A. *Human problem solving.* Englewood Cliffs, N.J.: Prentice-Hall, 1972.

Patrick, J. R. Studies in rational behavior and emotional excitement: I. Rational behavior in human subjects. *Journal of Comparative Psychology*, 1934, *18*, 1-22. (a)

Patrick, J. R. Studies in rational behavior and emotional excitement: II. The effect of emotional excitement on rational behavior in human subjects. *Journal of Comparative Psychology*, 1934, *18*, 153-195. (b)

Platt, J. J., & Spivack, G. Problem solving thinking of psychiatric patients. *Journal of Consulting and Clinical Psychology,* 1972, *39,* 148-151. (a)

Platt, J. J., & Spivack, G. Social competence and effective problem solving thinking in psychiatric patients. *Journal of Clinical Psychology,* 1972, *28,* 3-5. (b)

Platt, J. J., & Spivack, G. Studies in problem solving thinking of psychiatric patients: Patient-control differences and factorial structure of problem solving thinking. *Proceedings of the 81st Annual Convention of the American Psychological Association,* 1973, *8,* 463-464. (Abstract)

Platt, J. J., & Spivack, G. Means of solving real-life problems: I. Psychiatric patients versus controls, and cross-cultural comparisons of normal females. *Journal of Community Psychology,* 1974, *2,* 45-48.

Scandura, J. M. *Problem solving: A structural process approach with instructional implications.* New York: Academic Press, 1977.

Shaftel, F. R., & Shaftel, G. *Role playing for social values: Decision making in the social studies.* Englewood Cliffs, N.J.: Prentice-Hall, 1967.

Simon, H. A. *The sciences of the artificial.* Cambridge, Mass.: Massachusetts Institute of Technology Press, 1969.

Skinner, B. F. *About behaviorism.* New York: Knopf, 1974.

Spivack, G., & Shure, M. B. *Social adjustment of young children: A cognitive approach to solving real-life problems.* San Francisco, Calif.: Jossey-Bass, 1974.

The task force reports to the President's Commission on Mental Health (Vol. 2, Appendix No. 040-000-00391-6). Washington, D.C.: U.S. Government Printing Office, 1978.

Test, M. A., & Stein, L. I. Community treatment of the chronic patient: Research overview. *Schizophrenia Bulletin,* 1978, *4,* 350-364.

Torrance, E. P., & Myers, R. E. *Creative learning and teaching.* New York: Dodd, Mead, 1970.

Travers, R. M. *Essentials of learning.* New York: Macmillan, 1972.

Part Two

Descriptions of Institutional Programs

The three chapters which are included in Part Two are significant because they represent several of the earliest attempts to change from a medical model of treatment within an institutional setting to a learning or psychological model. These programs are unique in that the expectations which clients are given for treatment focus on a learning context rather than a clinical context, which makes coming for help less stigmatized. Thus, nonclinical names such as Life Skills Program, Couples Communication Center, and Adult Development Program are used, while informal consumer-provider and teacher-student relationships are fostered. In Chapter 6, for example, Hubert E. Armstrong, Jr., tells his clients (students) that the Life Skills Program is a "learning place, not a get well place."

Besides describing institutional programs based on a curriculum approach to skills training, the three chapters are related more intimately in the sense that two of the present authors, Cornelius Bakker and Hubert E. Armstrong, Jr., first set up the Adult Development Program described in Chapter 7 in 1970. Johan Verhulst observed these procedures while on sabbatical from the University of Leuven in 1971 and then designed a similarly structured program for distressed couples which began operation in 1972. Finally, Hubert E. Armstrong, Jr., implemented the Life Skills Program in 1977 although it was already well on its way to becoming a curriculum for skills training as a partial hospitalization program modeled after the Adult Development Program. These three chapters, then, represent the systematic growth and expansion of a basic concept which was developed within one setting and expanded to two other settings. As will be seen later, Chapter 9 represents a further extension to a different population.

Chapter 5 by Johan Verhulst and Cornelius Bakker describes the Communication Center which provides help for distressed couples. This program is a highly structured, intensive time-out from children, job, and other concerns for the couples to focus on their relationships for 3 weeks. The couples represent a wide variety of socio-cultural-economic backgrounds, problem severities, and problem types. Interestingly, the authors report that there were virtually no dropouts among the first 150 couples treated.

In Chapter 6 Hubert E. Armstrong, Jr., presents the life skills curriculum for seriously impaired outpatients of an inner-city community mental health center. Of particular interest to the clinician will be the seven-level hierarchy of interpersonal skills or tasks within which students and teachers work as an individual progresses through the program. The teaching methods—the use of classes, individual study, and GATE groups serving as homerooms within the learning day—will also be helpful to those clinicians interested in this approach.

The last section, Chapter 7 by Cornelius Bakker, Marianne Bakker-Rabdau, and Johan Verhulst, describes the original effort to establish a curriculum for skills training in this country, a "school for life" approach. During its 10-year evolution the program has developed two major emphases, general coping skills and skills designed to ameliorate a specific problem such as depression. The authors' background interest in ethology contributes to additional unique elements of the general skills curriculum. For example, a course on aggressiveness training is offered in addition to the more usual assertiveness training offered by most programs. When one reads the definitions of assertion and aggression, however, the two courses appear to be entirely consistent with behavioral goals and methods. The reader will find a rich source of ideas for clinical care, program administration, and research in this chapter.

Chapter 5

Marital Therapy in an Educational Group Setting: The Communication Center in Leuven, Belgium

Johan Verhulst
Cornelius B. Bakker

Abstract

In recent years there has been a growing interest in psychoeducational programs complementary to, or as an alternative to, traditional psychotherapy. In 1972 an experimental educational program for couples was established at the University of Leuven in Belgium. This program features separate sessions for each individual couple, coached by a dual-sex staff team, as well as group sessions in a classroom setting. Both types of interventions are based upon an educational model. This chapter describes the characteristic qualities of the setting, the interventions used, and the relationships developed, which account for the educational nature of the program. Results of evaluation research are cited. Further developments, based upon this research and upon clinical experience, are discussed.

Teaching and learning have traditionally been acknowledged to be major ingredients of all psychotherapy. Recently this educational aspect has especially been emphasized and developed in the field of marital therapy. Experimental programs consisting of cognitive learning and group training of communication skills have been successfully established (Ely, Guerney, & Stover, 1973; Wells, Figurel, & McNamee, 1975). Other programs which focus more upon the direct teaching of problem solving and contracting techniques (Jacobson, 1977; Rappaport & Harrell, 1972) have also been organized.

In 1972 a comprehensive psychoeducational program for couples, the Communication Center, was established at the University of Leuven in Belgium. This experimental program has the following characteristics which distinguish it from the aforementioned developments: (a) it addresses itself to clients in serious distress who are seeking professional help; (b) the staff consists entirely of qualified mental health professionals (the core staff consists of four psychologists, two physicians, and one sociologist); (c) although the program is based upon a nonmedical, educational model, the Center belongs to a university hospital system and has functional ties with various clinical departments which provide consultation services when needed, e.g., psychiatric crisis intervention and drug treatment, gynecological and andrological workups; (d) the focus is on the intimate partner relationship, with individual problems, as well as marital discord, being dealt with in the context of the dyadic partner relationship; (e) the program favors an intensive, short-term approach (four to five couples are enrolled each month in daily activities, Monday through Friday, for 3 weeks); and (f) group activities for all couples alternate with more individualized marital therapy sessions conducted by male-female teams of cotherapists.[1] The approach is cognitive-behavioral.

In describing this experimental program the following topics are addressed: its goals and basic assumptions, the field setting in which it takes place, the techniques and interventions used, and the characteristics of the educational relationship which is established. Finally, there is a short review of the built-in evaluation research and of the clinical developments which have occurred during the first 7 years of the program's existence.

Goals and Basic Assumptions

It has become more and more customary in the last decade to view marital problems as properties of the relationship system, rather than as linear processes "caused" by the individual characteristics of both partners (Berman & Lief, 1975; Olson, 1975). It is our assumption that

[1] The marital therapy sessions follow an educational approach, transforming complaints and conflicts into goals for change and providing cognitive information and training skills which are relevant to the educational goals. However, for practical and linguistic purposes, it was decided to continue speaking of *therapy* and *therapists* rather than introducing somewhat awkward and illogical sounding terms such as *the dual-sex team of teachers* or *the individual couple class.* The fact that the therapists and clients call one another by their first names, which in Belgium is highly unusual in a patient-therapist relationship, sufficiently emphasizes the educational nature of the sessions.

most individual psychological problems, "neurotic" behaviors, or sexual dysfunctions can also best be approached in the context of one's intimate partner relationship. Indeed, few individuals can be "neurotic" on their own, and personal problems are often experienced most intensively at the level of one's intimate relationships. Furthermore, problem behaviors are usually maintained by interactional dynamics and serve a function in the relationship homeostasis. Incidentally, the notion that problems can reside *in between* people is alien to the medical model, which is based upon the concept of diagnosis and treatment of illnesses located in the individual patient.

A second basic assumption is that dysfunctional interactional patterns are the consequence of social learning, operant conditioning, and interactional processes. It has been stated that dysfunctional interaction patterns tend to perpetuate themselves and even tend to increase when both partners, in an attempt to force change upon one another, consistently resort to using more of the same manipulations based upon the same personal perceptions, cognitions, and reaction patterns (Watzlawick, Weakland, & Fisch, 1974). Hence, there is a need for the rationale for systematic cognitive restructuring and for teaching conceptual frameworks to facilitate training in alternative interaction modes.

The goals of the Communication Center can be summarized as follows: (a) to establish a concrete, effective, short-term therapeutic program, based upon an educational model, for the treatment of any psychological problem which can be affected by a conjoint marital approach; (b) to induce change in a planned, structured way; and (c) to establish a setting which facilitates change, one which has strong "change appeal" (Verhulst, 1975) for the clients.

THE FIELD SETTING

The University of Leuven's psychiatric hospital is located in several buildings, scattered around in a park which was formerly the arboretum of the Lords of Lovenjoel. The Communication Center occupies the top floor of the hospital's laboratory building, originally constructed to accommodate resident physicians. It has a large living room, which is used for informal gatherings as well as for the educational groups. The room is furnished with armchairs, bookshelves, and a blackboard. A radio/recorder gives background music. There is a fully-equipped kitchen. The smaller consultation rooms for the marital therapy sessions are located around this central area.

The atmosphere is homelike and informal. Only the therapists and the clients occupy the space. Administrative business is taken care of in another building. Therapists and clients address one another by their first names. There is a sense of being bonded by a common goal. Conflicts and problems are defined as interesting challenges, requiring creative solutions. There is a tendency to tune in to the humorous and more ironic side of life.

The program is semiresidential: Most couples lodge in an independent center in the neighborhood, namely a convent, where a number of guest rooms are reserved for the clients of the Communication Center. They are encouraged to leave their children with family and friends, in order to be involved full time with one another for 3 weeks. Each client couple has a daily session of marital therapy. To allow maximum utilization of the staff, these sessions are scheduled in the morning, allowing each therapist couple to see two client couples for 1½-2 hours each. Group activities take place in the afternoon from 2:00-6:00 p.m. Evenings and weekends are to some degree reserved for homework assignments.

At the beginning of the program the clients are instructed about the fact that classes and training groups serve an educational function and are not "group therapy" in the traditional sense. The intimate conflicts of each couple are their private concern, and one can easily participate in all the group activities without sharing core problems of the relationship with the rest of the group. However, after the first week, during which the client couples become acquainted with one another, the group usually becomes very cohesive and starts functioning as a mutual support system. During class discussions couples spontaneously apply the theoretical insights offered to their own situation. However, when such a discussion becomes emotionally intense, with partners trying to confront one another in front of the total group, the staff will structure their conflict within the cognitive framework of the particular class and refer the couple to the marital therapy sessions for further elaboration.

As a social support group, the clients often arrange for parties after hours, outings into town, or evening discussions. Sometimes the group gets very involved with each other's problems during these after-hours encounters. One group, for instance, even arranged a conference for a member couple to which the wife's extramarital lover was invited. The staff does not attempt to actively intervene in such evolving group phenomena. The clients are seen as being able to manage their mutual relationships and to apply the basic principles of rational, no-fault interactional analysis and personal assertiveness taught in the program. The assumption is that these supplementary group activities will be sufficiently controlled by set-induced restraints. Indeed, in the 5 years of the program's existence, during which the first author served as director of the program, these activities have never seemed to be detrimental to any of the couples.

There is no active selection of the clients. Any couple who wants to enroll in the program is accepted. There is no real financial barrier to enrollment, for the program is largely funded by the National Health Insurance Companies, and additional charges are based on a sliding scale of \$30-\$500 per couple, depending on their income. In Belgium this therapeutic program must be prescribed by a physician, thus assuring that the income of the patients continues (through the social security systems) while they are absent from their work during the 3 weeks. Twenty percent[2] of the clients have had a college education, as opposed to 16% with only a primary school education. Obviously, the program has a biased attraction toward the better educated. Another important characteristic of the client population is the high occurrence of serious psychiatric disorders. One-third of all the individual clients have previously been in psychiatric treatment. Nearly 10% have been hospitalized in a psychiatric ward before. Reliable psychopathological data about these patients are not available since no formal psychiatric examination is performed. Using self-report data provided by the clients at the intake interview, we can divide the kinds of problems which bring people to the Center into three groups:

1. Fifty percent of the clients come because they experience the relationship as being in trouble. The complaints range from an inability to communicate, to problems of physical violence and potential divorce.
2. Thirty percent of the couples declare that their relationship is basically sound but that they want to learn to deal with the illness of one partner. Such illness may be acute psychotic breakdowns, depressions, anxiety attacks, hypochondriacal complaints, psychosomatic illnesses, or emotional instability.
3. The remaining 20% of the couples are referred for specific sexual dysfunctions. Indeed, until October, 1976, a formal Masters and Johnson therapy was offered in the course of this 3-week training program (see later discussion).

The staff of the Communication Center has been a relatively stable group during these first 5 years. New staff members are chosen on the basis of liking, i.e., after an initial selection on the basis of their primary qualifications, candidates are invited to a luncheon with the total staff group; interviews follow. Staff members then rate the candidates on a 0-10 scale of personal liking, i.e., how much would one like to closely work with this person? The candidate who obtains a high compound score with low-score variability is offered the job. This system has proven to be very valuable in creating positive staff relationships, avoiding co-therapist incompatibility, and inducing a sense of supportive collaboration in the therapeutic setting.

[2] The data bear upon the first 100 couples enrolled in the program. For a more detailed analysis, see Verhulst (1975).

Further fostering staff cohesion, a biannual faculty retreat is organized and a monthly staff dinner is held at the end of the 3-week course. There is a strong commitment to continuously update and change the program content and to experiment with alternative intervention techniques. These are discussed with the total staff group during a monthly day-long, in-service meeting. The purpose is to create a living program and a stimulating work environment. Indeed, we assume that therapists who are committed to and enthusiastic about their therapeutic approach tend to be more effective. Also, the direct and intense involvement with clients during 3 full weeks and the repeated confrontation with often extremely destructive interpersonal behavior represent serious personal stress for the therapists. Pleasant staff relationships, stimulating discussions, and therapeutic innovations help prevent the occurrence of staff discouragement and burnout.

Another contextual aspect which may produce set-induced changes is the inherent value system of the program. This value system is present in the general philosophy and attitude toward relationships, as well as in the propagated theories and ideas. It can be summarized in a number of assumptions which are implicitly and explicitly presented to the clients throughout the program.

For example, the program strongly supports the individual's right to maximally extend and expand personal freedom and autonomy in the management and the direction of her life. What is right or wrong is each person's individual prerogative. When others, including one's partner or the therapists, would succeed in influencing this decision by power of authority or by subtle coercion and pressures, they are infringing on this prerogative, and they consequently can expect a hostile or destructive response, possibly even the breakdown of the relationship. Others can be helpful only by clarifying the consequences which each option is likely to entail (Bakker & Bakker-Rabdau, 1973).

Assertiveness is thus a basic interactional skill. There is value in having things for oneself and in defending one's personal territory. Indeed, a sense of ownership is a prerequisite for social sharing: One can give to others only what one really owns. One needs to be able to say no before one can say yes. Furthermore, conflicts can best be solved by negotiation and by quid pro quo bargaining, in order for each partner to attain a subjective sense of territorial equity.

All this can best be achieved in a relationship of equivalence, i.e, a relationship in which each person perceives that his own time, thoughts, and feelings are as important as those of the partner. Such a relationship can be brought about only by a continuous effort of planning and learning.

Marriage is a gradual project. When this project threatens to fail, people usually approach the problem from a linear-causal point of view: Who was wrong? Who is to blame for this development? This approach only adds another focus of discussion and conflict to an already difficult situation. It is, therefore, essential to learn to analyze these problems as a circular process. From an interactional point of view, the partner's behavior is elicited by one's own behavior, which is in turn a reaction to the partner's behavior.

This set of assumptions is in line with the predominant values of modern Western society. The values of individual freedom, autonomy, and ownership are cornerstones of the neocapitalistic system. The quid pro quo principle is used and propagated by most contemporary marital therapists (Sager, 1976; Weiss, Birchler, & Vincent, 1974). Even in modern sex therapy the clients are taught to "give to get" (Masters & Johnson, 1970). Our society, which is becoming more and more aware of the interdependency of different systems, is currently more accepting of a circular, no-fault analysis of interpersonal problems.

Treatment Techniques and Interventions

The daily sessions of marital therapy will be discussed before the group activities are delineated. The schedule of activities for the 3 weeks is represented in Figure 5.1.

Marital Therapy Sessions

Each couple is assigned to a male-female cotherapist team. The sessions start with separate interviews for obtaining a personal history, a history of the development of the relationship, a sexual history, and an assessment of the major problems and complaints. On the first day the male therapist interviews the husband; the female therapist, the wife. On the second day therapist-and-patient teams are switched. The third day is reserved for any indicated physical examination, for instance, in cases of sexual dysfunction. When all the information is collected, the therapists arrange a round-table discussion (Masters & Johnson, 1970) which takes place on the third day. They prepare a 3-4 page typewritten report, summarizing in plain language how the therapists perceive the clients' problems, how these problems developed in the relationship, and which strategies for change one could consider. At the beginning of the round-table discussion husband and wife each receive a copy. The report is read and the clients are asked to react immediately when they feel that the information is incorrect or that a theoretical explanation or interpretation is wrong. These statements are then immediately withdrawn by the therapists, even if they think that the clients are only refusing to accept the truth. Although this procedure was imposed in the first place to model the no-mind-reading rule and to set the tone for the educational therapeutic relationship in which the client is the ultimate

Figure 5.1. Class and Therapy Schedule

	Monday	*Tuesday*	*Wednesday*	*Thursday*	*Friday*
First Week A.M.	History Taking: male therapist with male client; female therapist with female client	History Taking: crossed interviews	Class (9-12 a.m.): Basic Communication Skills (exercises)	Round-table Session	Marital Therapy Session
P.M.	Class (2-6 p.m.): Basic Philosophy	Class (2-5 p.m.): Basic Communication Skills (theory)	Physical Exam	Class (2-5 p.m.): Territoriality and Marriage (introduction)	Class (2-6 p.m.): Territoriality and Marriage (weaponry and conflict resolution)
Second Week A.M.	Marital Therapy Session	Marital Therapy Session	Marital Therapy Session	Marital Therapy Session	Marital Therapy Session
P.M.	Class (2-5 p.m.): Myths and Marital Roles	2-6 p.m.: Fight Training and Conflict Resolution	Class (2-6 p.m.): Aggression and the Management of Emotions	2-6 p.m.: Fight Training and Conflict Resolution	Class (2-5 p.m.): Parent Effectiveness
Third Week A.M.	Marital Therapy Session	Marital Therapy Session	Marital Therapy Session	Marital Therapy Session	Marital Therapy Session
P.M.	Class (2-5 p.m.): Planned Change	2-6 p.m.: Fight Training and Conflict Resolution	Class (2-5 p.m.): Sexuality and Marriage	2-6 p.m.: Fight Training and Conflict Resolution	Course Evaluation and Summary Session

judge over her mental process, it is interesting to note that the clients nearly always prove to be right. Often the clients propose an alternative hypothesis which afterward proves to be of much greater value.

Although the preparation of this report takes a long time (4-6 hours), the effort is offset by substantial benefits: (a) it provides feedback to the clients after hours of history-taking and thus often reveals misunderstandings or even prejudiced opinions; (b) it gives the clients a cognitive grasp of their problem, which helps them to concentrate on the here-and-now change projects, rather than continue to ruminate about why and how all this happened; (c) it provides a cognitive restructuring of the problems into a language of circular causality, learning, and planned change, and it relieves feelings of guilt and resentment; (d) it helps the clients in personalizing the theoretical information given in the classes, since abstract information is only useful when the clients can learn to relate it to specific situations of their everyday reality (Mahoney, 1974); and (e) it exemplifies and consolidates the structured, goal-oriented, educational nature of the therapeutic relationship.

The content of the therapeutic sessions during the second and third weeks varies greatly with each individual case, according to the agreed-upon strategy for change. Some clients analyze conflict situations and problems and try to work out specific agreements; others are primarily interested in exploring alternative interaction patterns or in understanding critical incidents from the past, using a new and more interactional frame of reference. Still others may receive sex education and Masters and Johnson training. Often homework assignments, such as further discussion of a certain problem or writing down alternative solutions, are given and then reviewed the next day during the session.

During the last days of the program most couples are encouraged and helped in writing a relationship plan outlining their goals for change for the next year. They fill out a goal attainment scaling form for each goal and list the means and the time perspective considered. The relationship plan also contains the list of agreements made and of conflicts remaining unresolved. Couples also summarize the educational insights which they consider most applicable for their particular case. Follow-up is usually restricted to five sessions of conjoint marital therapy. The sessions take place after 1 week, 1 month, 3 months, 6 months, and 12 months. If new problems are encountered, the therapists may grant some additional follow-up interviews, or they may assist the couple in locating other psychotherapeutic resources.

Group Sessions

This section will elaborate on the group sessions which take place with the total number of clients. These sessions include classes and training groups. Typical classes are shown in the following list.

1. Basic Philosophy–A class explaining the assumptions and values favored in the program, thus allowing the clients to confront these with their personal value system.
2. Territoriality and Marriage (Bakker & Bakker-Rabdau, 1973)–A class offering a complete and comprehensive way of analyzing and solving interpersonal conflicts.
3. Myths and Marital Roles–A class pointing out the consequences of having rigid concepts about oneself, the partner, and the relationship; also teaching techniques of changing such concepts.
4. Parent Effectiveness–A class based upon concepts and techniques developed by Gordon (1970) and by Patterson (1976).
5. Aggression and the Management of Emotions–A class showing ways of recognizing and of responding efficiently to one's emotional signals.
6. Sexuality–A class dealing with sexual information, sexual attitudes, and sexual interaction patterns.
7. Planned Change–A class dealing with the techniques of formulating and implementing concrete change projects in a relationship.

The remaining afternoons are used for training in problem solving behavior. The Center developed its own adaptation of the fair fight training technique of Bach and Bernhard (1971), complemented by territorial bargaining (Bakker & Bakker-Rabdau, 1973). An outline of this approach is presented in Table 5.1.

The classes have an informal format. Clients receive handouts or make notes. They are free to interrupt the teacher. Theoretical information is alternated with experiential exercises and with group discussions, focusing on the applicability of the concepts and techniques offered. The teacher illustrates the theories with numerous examples, often referring in a humorous way to problems and patterns from her own intimate relationship. The purpose of the teacher's self-disclosure is to dedramatize interpersonal conflicts, to structure them as universal human experiences, and to show how they can be managed and changed.

A major advantage of the educational model is that it provides patients and therapists with a mutual frame of reference. The fact that both partners learn to use a common language and a common way of looking at their problems enhances their chances to understand one another and to reach satisfactory agreements. Equally important is that the therapists teach and share this same language. In doing so, they define themselves as being experts, yet not as having access to any special hidden knowledge, superior insight, or magical secret. This helps to establish a trusting relationship of equivalence and therapeutic collaboration.

The Communication Center uses existing theoretical models, such as Bakker and Bakker-Rabdau's (1973) territoriality approach, and it also

Table 5.1
Outline of the Structured Problem Solving Approach

The partners sit opposite one another. Each partner chooses two members of the group as assistants to give advice in between the rounds. The therapist structures the discussion, gives the assignments for each round, insists upon further clarification when necessary, models more direct communication, etc.

Phase I: Exploratory Phase

Goals: (a) Explore and clarify the issue, (b) understand one another's position and feelings, (c) avoid premature compromises.

Procedure:

Round 1: The partner who brought up the issue describes the problem, explores how he feels about it, and states his wishes for an ideal solution (for example, he would wish for the partner to do something spontaneously and feel good about it, even though he has learned that spontaneity and feeling are beyond anyone's direct control).

Round 2: The other partner gives her point of view, feelings, and wishes. The first partner paraphrases.

Phase II: From Wishes to Requests

Goal: Transform wishes into concrete requests for behavior change (for example, request for a regular time-space in which certain spontaneous behaviors are more likely to occur).

Procedure:

Round 3: After brainstorming with the assistants, the first partner states his requests. The second partner paraphrases.

Round 4: The second partner makes concrete requests, which are paraphrased.

Phase III: Bargaining

Goal: Reach a satisfactory compromise. The negotiation continues until both partners feel (a) that the solution is equitable and (b) that the planned change requires an equal effort from both sides.

Procedure: There are several rounds in which proposals and counterproposals are exchanged and paraphrased.

Phase IV: Consolidation

Goal: Operationalize the compromise.

Procedure: The partners write down the terms of the contract, the practicalities (who will do what, when, and how), and the duration of the trial period after which the situation will be reviewed.

has developed original ones, such as those dealing with the management of emotions. In order to be used as a common framework in an educational therapeutic program, theories should meet the following requirements:

1. Clarity–Simplicity coupled with unambiguous logic is necessary. The model should be able to draw the interest of people of various sociocultural backgrounds.
2. Relevancy–The theory should have a broad enough range of convenience to cover the problem areas the clients want to deal with.
3. Acceptability and newness–For the theory to be easily incorporated, its inherent value system should sufficiently link with the morals and values of both clients. However, since it is intended to offer a mutual and refreshingly new point of view, the model should also sufficiently differ from each client's personal way of thinking. We consider this to be of major importance. Indeed, in dealing with their problems, most clients have repeatedly tried out their own constructs and insights, with negative results. Therapy failures often seem to be based upon a fundamental incompatibility between the proposed value system and the clients' personal morals and values. When this becomes obvious, the clients are encouraged and helped to locate more compatible therapists.
4. Pragmatic applicability–The model should be conducive to better ways of problem solving. This may be either because it provides explicit or implicit guidelines for effective interaction or because it helps in delineating concrete behavioral skills, which can then be trained.

The staff regularly reviews the content of the classes from the point of view of these requirements. As a consequence, the clients have consistently rated most classes in a positive way, indicating relevancy and personal applicability of the content (Verhulst & Verstraeten, 1977).

THE EDUCATIONAL RELATIONSHIP

The educational model may be best characterized by the nature of the helping relationship. In a psychoanalytical approach the role of the therapist is usually seen as being a "neutral mirror" to the clients. In the humanistic-experiential approach of Carl Rogers the helping relationship is characterized by intimacy and empathy; the therapist models the sharing of feelings and facilitates self-disclosure and sharing experiences among the clients (Rogers, 1970). The communications and systems approach favors directive therapist interventions, such as "prescribing the symptom" to disrupt the system's homeostasis and to abolish destructive, self-generating communication patterns (Haley, 1977; Watzlawick et al., 1974). The relationship bears a quality of magical power giving the message "If you do the things I prescribe–though they may seem silly–you'll improve" In the *traditional* behavioral approach the therapist is the expert, who manipulates the system of reinforcers, teaching people to record pleasing and unpleasing behaviors and to shift the balance of interactional reinforcers toward a more satisfactory one. In the cognitive-behavioral, educational approach the relationship is more explicitly defined as a teacher-student relationship, in which the student is the consumer who buys educational services. Once complaints and problems are transformed (with the help of the teacher) into goals for change, the student can evaluate whether the program can teach him the appropriate means and skills for reaching these goals. Similarly, it is the teacher's task to assess whether concrete, teachable methods are available to help the clients in their change projects. However, the consumer who buys the services will be the final judge of the appropriateness of the program. The teacher has to constantly be aware of this role: One is not to tell the students what they should do or in which direction they should change, but is to provide information and to teach them methods for changing interactions.

Thus, the therapeutic contract in an educational relationship is limited to selling information and skills training, appropriate to the consumer's goals, to a student who in turn is requested to determine these goals and to put in the effort of learning. The patient pays for expert information and help, not for friendliness, human warmth, empathic understanding, trust, and care. This does not mean that the relationship itself has to be cold, sober, and neutral. On the contrary, a consequence of the fact that human kindness and warmth are not seen as part of the direct therapeutic contract is that these qualities do not have to be pretended or to be therapeutically manipulated. The affiliation and encouragement which the client receives is personal, more genuine, free of charge, and therefore especially valuable and helpful. It is obvious that much of the educational approach is incorporated in the behavioral approach and that the difference between the two is a matter of the explicit label which the relationship carries, the overall context, emphasis on cognitive elements, and ranking of the acceptance of the clients' goals without evaluation as a primary value.

It is relatively easy to imagine the concrete form which the educational relationship takes in the classes and training groups. As already noted, the staff gears the group interactions toward teaching and learning, operationalizing conflicts and problems in a language of change and skill acquisition.

As for the marital therapy sessions, it seems that the educational relationship is usually established during critical moments early in the program. Actually, one attempts to set the tone of the relationship even before the couple enters the 3-week training. The preliminary intake interview is deliberately kept businesslike; the staff member explains the kinds of things the program has to offer and then asks the clients to say something about their problems, "so that we can discuss together whether we have for sale what you would like to buy."

A second critical moment in the establishment of the teacher-student relationship is the round-table conference. The act itself of giving a written report to the clients stresses the concrete nature of the therapeutic exchanges. Furthermore, the fact that the content of the report is defined as the therapist's subjective view, with the clients having ultimate say over the appropriateness and applicability of the explanatory hypotheses, confirms the consumer role. Finally, the report ends with a proposed strategy for change, emphasizing concrete skills to be learned in the next weeks. This teaching plan is discussed and only goes into effect when all parties come to an agreement.

A third critical moment sometimes arises when one of the clients seems uncooperative or even seems to boycott the sessions. Construing this as "resistance," i.e., as a neurotic characteristic of that client, would mean a relapse into the medical illness model. In accordance with the interactional and educational approach, the following procedure is used: Uncooperative behavior is defined as a teacher-student relationship conflict. The therapist who would be most sensitive to and irritated by a client's unwillingness to collaborate would announce this at the first opportunity and propose to work it out in a structured problem solving discussion. This discussion between therapist and client is coached by the cotherapist and the client's partner. The adversaries explain their points of view to each other and attempt to negotiate a solution. This procedure has proven to be extremely useful. Often the client's uncooperativeness proved to be based on misunderstandings, false expectations caused by unclear messages from the therapists, and ambiguities in the teaching contract. Dealt with at this level, the conflict has usually been clarified and resolved, with all partners being more committed to the change program.

Essential also for the educational therapeutic relationship is the fact that clients and therapists share the same language. The establishment of a student-teacher relationship is easily accepted by the couples and in fact seems to create confidence and trust. An indicator for this assumption is the fact that the program did not have one single dropout among the first 150 couples.

From a therapeutic point of view, it can be argued that the student-teacher relationship solves the following paradoxes inherent in the medical psychiatric model:

1. All therapy is geared toward fostering and increasing the patient's autonomy, since we know that people who develop a strong sense

of autonomy are less likely to need psychiatric help. At the same time, therapy often implies passivity on the part of the patient, the expectation being that some cure will be applied to her. (Actually, the term patient is derived from the Latin word *patiens,* which means passive and submissive. In this frame of thinking the surgery room represents the ideal of the medical model, since the patient is anesthetized so that she will not interfere with the treatment given to her.) The dilemma between wanting to foster autonomy and risking increased dependency through treatment has been one of the major points of critique formulated against psychiatry (Bakker, 1975; Goffman, 1961; Wing & Brown, 1970).

2. Next to a sense of autonomy, people need a sense of belonging, of participating in a social structure in order to function psychologically. Again, though therapy is often geared at fostering relationships, the therapeutic relationship in itself may seduce the patient into seeking permanent psychotherapy as a substitute for intimacy and for a social support system. The solution propagated by some therapists, namely, to make the relationship itself frustrating to the clients, seems hardly satisfactory. The solution of the educational model is to exclude the intimate and supportive aspects of the relationship from the direct treatment contract. In addition to already focusing on improving the intimate aspects of the couple's own relationship, the Communication Center's program, like other time-limited programs, seems to have an additional advantage in this regard, namely, the fact that treatment is limited to 3 weeks. Hence, the social support system of the groups and of the therapist team are clearly structured as temporary and transitional.

Research and Further Developments

The program was established with a built-in research design to evaluate its effectiveness. The Dutch revisions of the Relationship Inventory of Barrett-Lennard (1962) and the Personal Orientation Inventory of Shostrom (1963)[3] are given to the clients before, after, and at 6-month and 2-year intervals following the program. Couples on the waiting list

[3] The Dutch revision of the Relationship Inventory of Barrett-Lennard consists of a questionnaire of 50 items, which is presented in two forms: (a) the Self-form, which reflects one's own attitudes toward the partner and (b) the Partner-form, which reflects one's perception of the partner's attitudes. There are five subscales: (a) empathy, i.e., the ability to understand the partner's experience and point of view, (b) directivity, i.e., the tendency to deal with the other in an authoritarian way and on one's own terms only, (c) positive regard, i.e., affection, commitment, and

are used as a control group, assessed at intake and 6 months later. Subjective rating scales for the evaluation of marital change and of the program's impact are also filled out by therapists, as well as clients. As reported elsewhere (Vansteenwegen, 1978; Verhulst & Vansteenwegen, 1974), significant differences in progress after 6 months were found between the control and the treatment groups. In the treatment group significant improvement occurred after the program and after 6 months on most of the variables measured. The changes remained significant after 2 years for a number of the variables evaluated by the Dutch versions of the Relationship Inventory and the Personal Orientation Inventory (perception of the partner's empathy and directivity; perception of one's own positive regard, transparency, and directivity; autonomy and time competence). The results indicate that the program is highly efficient, not only in bringing about changes in the relationship but also in fostering the personal growth and autonomy of the partners.

In the course of the years, as insight into the specific training needs of different client populations grew, the standard therapeutic program of the Communication Center has been further differentiated. A first striking observation was that the success rate of a Masters and Johnson therapy for sexual dysfunction was lower than expected, when performed within the context of the overall program. Sex therapy proved to be more efficient if one exclusively focused on the sexual relationship rather than also confronting the couple with their more general problems and conflicts. Consequently, the Communication Center organized a separate sexual dysfunction clinic. In some cases, then, the sex therapy is done before or after the 3-week relationship training program, a format which is successful but expensive.

A second observation was that couples who came to the program primarily to find out whether to stay together or to get divorced had different educational needs than couples whose primary goal was to solve current problems and to plan for a better relationship. Similarly, couples who were concerned about one another and about their children, but who wanted to learn how to implement an already-made divorce decision, found themselves participating in a program which seemed to emphasize marital enhancement and growth of the relationship. Consequently,

---

belief in the other's potential, (d) transparency, i.e., the willingness to self-disclose, and (e) unconditionality, i.e., the steadiness of one's acceptance of the other as a person.

The Dutch version of Shostrom's Personal Orientation Inventory consists of a questionnaire of 150 items, with two major subscales: (a) autonomy, i.e., the capacity for self-reliance and assertiveness, and the ability to reach independent decisions and (b) time competence, i.e., the sense of not being controlled by one's past history and of being free from undue guilt and resentment.

an alternative 3-week training program was set up for problems with regard to decision making. In this program emphasis is put upon values clarification and upon the exploration of the alternatives considered, with cognitive rehearsal of various interactional possibilities. Couples who, after years of living together, are still ambivalent about making a definite commitment to one another are also encouraged to enroll in this decision-making program.

A major program differentiation was also made in order to accommodate more atypical couples, e.g., the physically handicapped and couples who are interested in an alternative lifestyle, such as commune living. A more individualized program for all these "relationship minorities" was developed. The question was also raised whether the program was appropriate and useful to client with low IQs or with poor educational backgrounds. First, the program was revised, eliminating a number of technical terms and psychological jargon from the content of the classes. An effort was made to illustrate the technical concepts with examples not chosen from the world and the situations in which the staff is living, but with ones derived from the life circumstances of the average client. With these changes, the program proved to be able to accommodate people from different sociocultural backgrounds. Factors such as the IQ or the social level were found not to be predictive for success or failure (Vansteenwegen, 1978). It is our impression that the already high success rate of the program has been improved by the foregoing differentiations, based upon clinical experience and empirical findings.

A possible future differentiation has to do with the realization that relationship problems can be divided into structural and stress-related ones. *Structural conflicts* are related to unrealistic expectations about one another, dysfunctional interaction patterns, and inefficient problem solving approaches. *Stress related conflicts* are the result of external or internal situational stress. External stress includes serious illness of one partner; physical handicaps or mental retardation in a child; important social, financial, or professional problems; and parental interference. Internal situational stress can be caused by role changes within the relationship, for instance, becoming parents (Ackerman, 1965), or by age-related life crises (Buhler, 1968). At the present time it does not seem that these different types of conflicts would need separately designed programs. However, additional classes and exercises in marital stress management are being considered.

SUMMARY

This chapter describes the basic philosophy, the treatment context, and the therapeutic approach of an experimental educational program for couples at the University of Leuven in Belgium. The program, which features built-in evaluation research, has proven to be highly effective in establishing relationship change, as well as individual change. We assume that this is to a large extent due to the consistent implementation of a cognitive-behavioral approach in an educational group setting.

References

Ackerman, N. The family approach to marital disorders. In B. Greene (Ed.), *The psychotherapies of marital disharmony.* New York: Free Press, 1965.

Bach, G., & Bernhard, Y. *Aggression lab: The fair fighting training manual.* Dubuque, Iowa: Kendall/Hunt, 1971.

Bakker, C. B. Why people don't change. *Psychotherapy: Theory, Research and Practice,* 1975, *12,* 164-172.

Bakker, C. B., & Bakker-Rabdau, M. K. *No trespassing: Explorations in human territoriality.* San Francisco: Chandler & Sharp, 1973.

Barrett-Lennard, G. T. Dimensions of therapist response as causal factors in therapeutic change. *Psychological Monographs,* 1962, *76,* 1-36.

Berman, E. M., & Lief, H. I. Marital therapy from a psychiatric perspective: An overview. *American Journal of Psychiatry,* 1975, *132,* 583-592.

Buhler, C. The course of human life as a psychological problem. *Human Development,* 1968, *2,* 184.

Ely, A. L., Guerney, B. G., & Stover, L. Efficacy of the training phase of conjugal therapy. *Psychotherapy: Theory, Research and Practice,* 1973, *10,* 201-207.

Goffman, E. *Asylums: Essays on the social situations of mental patients and other inmates.* Garden City, N.Y.: Doubleday, 1961.

Gordon, T. *Parent effectiveness training.* New York: Wyden, 1970.

Haley, J. *Problem solving therapy: New strategies for effective family therapy.* San Francisco: Jossey-Bass, 1977.

Jacobson, N. S. Problem solving and contingency contracting in the treatment of marital discord. *Journal of Consulting and Clinical Psychology,* 1977, *45,* 92-100.

Mahoney, M. J. *Cognition and behavior modification.* Cambridge, Mass.: Ballinger, 1974.

Masters, W. H., & Johnson, V. E. *Human sexual inadequacy.* Boston: Little, Brown, 1970.

Olson, D. H. Marital and family therapy: A critical overview. In A. S. Gurman & D. G. Rice (Eds.), *Couples in conflict: New directions in marital therapy.* New York: Aronson, 1975.

Patterson, G. *Living with children.* Champaign, Ill.: Research Press, 1976.

Rappaport, A. F., & Harrell, J. Behavior exchange model for marital counseling. *The Family Coordinator,* 1972, *21,* 203-212.

Rogers, C. R. *Carl Rogers on encounter groups.* New York: Harper & Row, 1970.

Sager, C. J. *Marriage contracts and couple therapy: Hidden forces in intimate relationships.* New York: Brunner/Mazel, 1976.

Shostrom, E. L. *Personal orientation inventory.* San Diego: Eslits, 1963.

Vansteenwegen, A. *Onderzoek naar de veranderingen in de relationele grondhoudingen en de zelfrealisatie bij intensieve partnerrelatietherapie.* Unpublished doctoral dissertation, University of Leuven, Belgium, 1978.

Verhulst, J. Marital change: An intensive short-term approach. *International Mental Health Research Newsletter,* 1975, *17,* 7-10.

Verhulst, J., & Vansteenwegen, A. Werking, ideologie en doelmatigheid van een experimenteel therapeutisch programma. *Tijdschrift voor Psychiatrie,* 1974, *16,* 134-151.

Verhulst, J., & Verstraeten, D. Territoriality and marital therapy. In World Psychiatric Association, *Abstracts of the VI World Congress of Psychiatry,* Basel, Ciba-Geigy, 1977.

Watzlawick, P., Weakland, J. H., & Fisch, R. *Change: Principles of problem formation and problem resolution.* New York: Norton, 1974.

Weiss, R. L., Birchler, G. R., & Vincent, J. P. Contractual models for negotiation training in marital dyads. *Journal of Marriage and the Family,* 1974, *36,* 321-333.

Wells, R. A., Figurel, J. A., & McNamee, P. Group facilitative training with conflicted marital couples. In A. S. Gurman & D. G. Rice (Eds.), *Couples in conflict: New directions in marital therapy.* New York: Aronson, 1975.

Wing, J. P., & Brown, G. W. *Institutionalism and schizophrenia.* London: Cambridge University Press, 1970.

Chapter 6

An Educational Approach to Psychiatric Day Care

Hubert E. Armstrong, Jr.

Abstract

The Life Skills Program (LSP) is a psychoeducational approach to psychiatric day care based upon social learning theory, behavioral group psychotherapy techniques, the cognitive psychology of George Kelly, and emerging concepts regarding the role incompetence plays in the problems of the mentally impaired. In this chapter the theoretical orientation of the LSP is articulated, its objectives and structure are described, and preliminary evaluation data are presented.

For several years the Department of Psychiatry and Behavioral Sciences at the University of Washington has been developing mental health delivery systems based upon educational theories and techniques. The program described in this chapter is the first to appear in a community health center. The Harborview Community Mental Health Center (HCMHC) is a university-affiliated comprehensive community mental health center attached to Harborview Medical Center, a teaching hospital of the University of Washington. From its establishment 9 years ago the HCMHC has implemented a psychiatric day treatment program originally called the Partial Hospitalization Program (PHP), now called the Life Skills Program (LSP). Over the years the facility gradually and tentatively moved in the direction suggested by its elder sibling, the Adult Development Program at the University of Washington.[1] In 1977 when the present

[1] The Adult Development Program is described in Chapter 7 of this volume.—Eds.

author took over the directorship of the program, the groundwork for a psychoeducational approach to psychiatric day services was well laid. Since that time the development of psychoeducational procedures as applied to the seriously behaviorally impaired, as well as poverty level and minority populations of an inner-city catchment area, has continued. The present chapter describes the implementation of psychoeducational procedures for mental health service delivery developed by the Life Skills Program.

The assumption of studentship fosters the realization of mental health objectives. This hypothesis suggested by the experience of the Adult Development Program provides the clinical and research base for the Life Skills Program at HCMHC. The notion is not new, and a brief summary of its history may be helpful. Freud thought of psychotherapy as being primarily a reeducational experience. Fifty-five years ago Witmer (1925) observed that clinical psychology has more in common with pedagogy and sociology than with medicine. Twenty-five years ago Sanford (1955) called for a "democratization or secularization of health and welfare professions" (p. 83). He saw a viable role for mental health professionals similar to that of a teacher. More recently a number of authors have suggested the viability of mental health delivery systems based jointly upon educational technology and social learning theory (Adkins, 1970, 1974; Armstrong & Bakker, 1971; Armstrong & Booth, 1979; Bakker & Armstrong, 1976; Carkhuff, 1971; Guerney, Stollak, & Guerney, 1971; Hobbs, 1967; Spiegler, 1977). Maxwell Jones (1974) is perhaps the most prominent voice from psychiatry to argue that the mental health field has been remiss by ignoring potential contributions from the field of education. Recently Jones has joined with Szasz (1961) in emphasizing the potential debilitating impact of the disease model upon the clientele served by psychiatry. Jones joined the foregoing list of authors in suggesting the potential advantages which could accrue to psychiatry from borrowing educational principles and procedures.

Overview and Field Setting

The objective of the Life Skills Program is the delivery of psychiatric day care services to a population of seriously psychiatrically impaired, primarily poverty-level, substantially third-world clientele. We have chosen to fulfill this mandate by means of an innovative delivery system consistent with the principles of social learning theory, the cognitive psychology of George Kelly, and emerging psychoeducational mental health theory. In this effort we propose to generate preliminary statements of the acceptability and effectiveness of this approach to psychiatric service delivery.

The clientele of the LSP are seriously psychiatrically impaired people similar to those participating in the day care programs of most comprehensive community mental health centers serving inner-city populations. The active caseload averages 70 clients, with an average daily attendance of 45. Fifty-five percent of LSP clients are diagnosed as psychotic, with the remainder diagnosed as character disorders and neurotics. Thirty percent of the LSP clientele carry secondary diagnoses of alcohol or drug abuse. Fifteen percent of our clientele are considered to be actively suicidal at the time of their admission to the unit. At admission the mean score of LSP clients on the Global Assessment Scale (GAS) (Endicott, Spitzer, Fleiss, & Cohen, 1976) is 37.4 (SD = 18.68), reflecting moderate psychiatric impairment (scores below 30 on the GAS indicate severe psychiatric impairment). One-fourth of LSP clients have had no previous psychiatric hospitalizations, yet the average number of previous hospitalizations for our clientele is 3.2. Eighty-three percent of LSP clients are on public assistance and are supported in the program by medical coupons or by Medicare. Thirty percent of LSP clients represent minorities of color.

The LSP staff includes a full-time psychologist director, a program coordinator (master's-level occupational therapist), a psychiatric nurse, three counselors (one counseling psychologist, one master's-level vocational counselor, and one master's-level anthropologist with training in clinical psychology), and one social work assistant. A psychiatrist serves as medications consultant 6 hours per week. A clinic clerk completes the regular LSP staff. Trainees from a number of disciplines including clinical psychology, nursing, occupational therapy, and vocational counseling augment the regular staff. All regular staff provide case management, course development, teaching, and counseling services. These functions will be described later.

The LSP occupies 5,000 sq. ft. In addition to office space and a reception area, there are a student lounge which doubles as a classroom, two additional small classrooms, and one large classroom which doubles as an occupational therapy room supplied with storage areas, work tables, and kiln. The workshop is equipped with hand woodworking tools, work tables, and storage cabinets. The LSP shares a kitchen, dining area, and adjacent large group meeting room with the rest of the Community Mental Health Center. In addition to these facilities, the LSP occasionally uses classroom space elsewhere in the Community Mental Health Center.

Theoretical Orientation

The LSP is described as a "learning place, not a get-well place." This statement is meant to imply that the behavioral phenomena for which the LSP assumes primary responsibility are not manifestations of disease but are learned behavioral patterns or represent the absence of *learned* desirable life skills. The format of our program is educational: a curriculum of seminars, laboratory classes, and individual study, all of which are designed to assist individuals in the learning of new behaviors and the unlearning of old maladaptive habits. A central element of our educational model is that the assumption of *studentship* entails role behaviors with advantages over those behaviors elicited by the *patient* role. The educational model, as we implement it, does not ignore the appropriateness of psychotropic medications. Rather, our model asserts that a learning atmosphere is ideally suited to many critical needs of mental patients. Medications are viewed as a necessary adjunct which enable a substantial number of people to benefit from the psychoeducational opportunities offered, just as the case would be in an academic educational setting. We anticipate that the demand characteristics of the term *student* (an active, searching orientation) will carry advantages over those fostered by the term *patient* (a passive and "patient" orientation). Like Hobbs, we have sought to implement a program in which "goals and procedures are couched in the idiom of competence and self-fulfillment rather than illness or pathology" (Hobbs, 1968, p. 17).

Clearly, social learning theory provides a solid theoretical foundation for the implementation of an educational model within a mental health context (Bandura, 1961). A view of problematic behavior as a manifestation of learned maladaptive skills or the failure to have developed appropriate behavioral repertoires leads comfortably to a delivery system described as a *learning place,* where studentship is fostered and skill building is viewed as the primary objective.

A second theoretical underpinning of our approach is the cognitive psychology of George Kelly (1955), who offered intriguing suggestions for the structuring of psychological and psychiatric delivery systems. Since Kelly's notions are not as familiar to mental health professionals as are concepts of social learning theory, a summary of his views of the treatment process should be helpful.

In Kelly's view those conditions unfavorable to fostering new development and acquiring new constructs and behaviors were: (a) threat, (b) preoccupation with old material, and (c) unavailability of a laboratory for testing out new constructs and new behaviors. The first two of these, and to a degree, the third, are conditions which exist to a substantial extent in most conventional psychiatric treatment programs.

One of the most significant threats posed by such programs is the statement about the patient implicit in the disease view of mental illness and the stigmatization associated with the role of the mental patient (Goffman, 1961; Stuart, 1970). The treatment by a psychiatric service and the diagnosis which is carried as an admission ticket are often taken by the patient as uncomplimentary statements about her. Though the belief that they are receiving help may be comforting, many patients experience intense threat from the idea that their behavior is a result of mental illness and thereby less amenable to self-control than other types of behavior. A learning explanation for such behavior and avoidance of the disease model can be a comforting alternative.

A preoccupation with old material can be avoided. There should be emphasis on eliminating behavioral deficits and on taking the behavioral view that the best way to acquire new behavior is to engage in it (as opposed to searching for its antecedents).

Finally, a psychiatric delivery system based upon learning principles provides a laboratory for experimentation with new behavior described and prescribed by Kelly. He referred to the "stabilizing effect that accrues from seeing oneself in a functional relationship to one's surroundings, whether these surroundings be things or people" (Kelly, 1955, p. 365). This observation is consonant with our efforts to teach a *behaviorist view of behavior,* which we define as the functional relationship between a person and the world around him.

THE EDUCATIONAL MILIEU

Social learning theory implies the feasibility of educational approaches to mental health, but the gap between this implication and the practical demands of mental health service delivery is enormous. To bridge a similar gap in the field of special education, Hewett (1964) described a hierarchy of educational tasks, articulating the problems and tasks facing student and teacher at various stages of the educational relationship. These tasks, beginning with the most rudimentary and moving to the more complex, are: primary level, acceptance level, order level, exploratory level, relationship level, mastery level, and achievement level. As the concept of hierarchy implies, a person usually has to acquire lower-level skills before he is able to accomplish higher-level skills. We have found Hewett's task hierarchy a most useful framework for conceptualizing the various tasks faced by both client and teacher at the LSP. An examination of each step and our methods of approaching the issues raised at that task level should provide a helpful introduction to LSP programming.

The Primary Task Level

At the primary task level the teacher approaches an inaccessible student on the latter's own terms in an effort to establish contact and a base of trust and mutual understanding upon which future goals and relationships can be based. Most LSP clients are functioning well above this level most of the time. Some clients' needs are primarily at this level, however, while a larger group periodically displays needs of this type. Many clients referred by the courts or family feel compelled to attend the program. They face primary educational tasks, as do clients with limited intellectual resources or with paranoid ideation. For quite different reasons, clients in these classifications are inclined to be either apathetic about involvement or to actually resist it. At this level the responsibility of program staff exceeds that of the client. Often the task of staff at this level involves the establishment of special incentive systems employing either social or tangible rewards.

One LSP client with severe impulse control problems, minimal social skills, and borderline intelligence was both unreliable in attendance at the LSP and disruptive in class. A program involving her guardian was established. Daily attendance cards were signed by the student's instructors if she was present and not disruptive. Cards were submitted to the guardian at the end of the week, with each being worth a prescribed portion of the client's weekly allowance. Regular attendance with minimal disruptive behavior resulted in more allowance money, while poor attendance and/or unacceptable disruptive behaviors resulted in diminished amounts. Case management arrangements of this sort, as well as the personal contact among staff, client, and significant others required to carry them out, are a fundamental aspect of LSP educational efforts. In our experience such programming fosters the trust, mutual respect, and optimism necessary for skill development at higher levels of educational tasks.

The Acceptance Task Level

At the acceptance level an interpersonal relationship with the teacher has begun, allowing the establishment of educational goals and plans based upon individual contact between client and staff person. Classroom demands are still minimal, with the primary educational objectives being to foster the client's comfort and mutual trust with professionals and to enhance his capacity to relate to them. Many LSP clients who have had unfortunate experiences either with mental health professionals or with schooling spend considerable time and energy at this level of the educational hierarchy. Clients with paranoid symptomatology and/or with problems in taboo areas (such as sexual identity) often enter the program at this level or become involved at this level as they begin to come to terms with these problems. Field trips, lunchroom encounters, and between-class informal contacts, as well as individual sessions between staff person and client, provide important opportunities for skill building at this level.

The Order Task Level

Mastering the subject matter of an LSP class is not the student's primary task at this or other basic levels of the educational task hierarchy.

Instead, the student at the order task level increasingly is held responsible for meeting expectations in terms of classroom structure and routine. In the LSP the emphasis at this level is regular attendance, promptness, self-control (decreased frequency of temper outbursts, off-target verbalizations, hallucinatory behaviors, etc.), and participation in classroom activity. Each LSP student is issued a folder containing his class schedule, notepaper, pencil, etc. The importance of bringing folders to class, taking notes, and keeping records of homework assignments is emphasized. The concept of studentship is stressed in this way. These skills are fundamental to the more advanced steps in the task hierarchy, which cannot be handled effectively unless skills at more rudimentary levels have been acquired.

One LSP client was not only unable to track class content and was, therefore, unable to benefit much from it, but was also highly disruptive in class. She would typically enter class late, leave early, and when present, disrupt class activities with her temper tantrums and frequent outbursts that "this isn't doing me any good." The following educational plan was implemented. Teachers allowed the client to come late, but they urged prompt arrival and insisted upon her leaving class when she became disruptive. Leaving was cued by a preestablished signal given by the teacher. The student's counselor introduced a behavior rehearsal project which involved the client's pretending to be disruptive (she was to act mildly disruptive at prearranged times) followed by the teacher's signaling her to leave. Finally, the client was to pretend disruption at unscheduled times, thus eliciting the exit cue from a teacher who was unsure if it was a genuine disruptive event or a practice one.

This procedure of having a student pretend disruption and practice it first on schedule (followed by appropriate management procedures) and then at unscheduled, extemporaneous times has been a most helpful device for teaching self-control skills at the LSP. Planning for the final step—extemporaneous practice—among other things, appears to give the client a face-saving way of extricating herself from a genuine loss-of-control situation. She can call it a practice session and thereby withdraw from the situation much more readily than she could without such a device.

In addition to the tailoring of individually oriented case management projects, LSP staff at this level have the task of engineering those aspects of an educational milieu which actively foster the assumption of studentship. The majority of our clients have been well-trained by previous experiences to be patient, to function in the passive role of a patient, and to view their behavioral problems as something they catch. The implication of these beliefs and attributions produces behaviors counterproductive to the aims of mental health education, as we offer it at the LSP. Two examples of the milieu management of studentship will suffice—fostering attendance and bringing folders to class. Average class attendance in the LSP was 48% in 1976, a figure very close to that reported by most day treatment programs having comparable populations but far below a desirable attendance level in a program based on continuity, with the activities of one day building upon those of the previous day. The problem of low attendance was approached in two ways. The first step was to emphasize the importance of regular attendance in GATE groups (e.g., homeroom classes; see the description of GATE groups in another section) and in orientation sessions, as well as periodically in student body meetings. The second step was to compute attendance figures for each client and to send a letter each month providing feedback to each client regarding his attendance record. The client was complimented if attendance was high (over 85% for the month) or requested to give special attention to attendance if it was low (below 85% for the month). Subsequent to these two efforts, attendance has stabilized at 73%.

Folder bringing was approached initially in the same manner as attendance. Discussions of its importance were presented in GATE group and student body meetings. These efforts produced compliance in folder bringing 45% of the time, a figure considered unsatisfactory by LSP staff. A social reinforcement strategy was initiated, whereby clients were told that teachers would initial the students' calendars on those days they brought their folders to class (initialed calendars were not redeemable; this was simply social reinforcement). Compliance jumped to 90%. Reversal to the baseline condition followed by a second introduction of the social reinforcement condition demonstrated the remarkable influence of the intervention procedure on this aspect of studentship (Armstrong, Rainwater, & Smith, Note 1).

The Exploratory Task Level

The LSP counselor's or teacher's educational objective at the exploratory level is to assist in establishing a goal of genuine interest to the client. The client's task is to get "turned on," to gain the realization that the LSP is his advocate and therefore will assist him in doing what he wants. Our clients often enter the program with a general mistrust of mental health agencies, believing them to be agents for family or courts interested in changing them against their wills. The exploratory level differs from the acceptance level in that it is task or goal oriented, whereas the acceptance level is more directed toward building interpersonal rapport. For example, one LSP client entered the program diagnosed as a paranoid schizophrenic and displayed an elaborate delusional system. He expressed satisfaction with his delusions but showed great interest in the physical education curriculum of the LSP. He got deeply involved in dance, jogging, and relaxation training courses. The basis of involvement was acceptable to staff as a manifestation of the exploratory phase of the hierarchy.

The Relationship Task Level

For the client the educational task at the relationship level is the establishment of social relationships with both staff and fellow clients which will prove motivating. More than their acceptance is involved here. LSP students (and schizophrenic patients in general) have often experienced minimal social involvement and present substantial social skills deficits (Hersen & Bellack, 1976). Problems at this level occasionally become evident in disruptive classroom behavior, but more frequently they involve serious disengagement from classroom activity. Progress at this level is observed as clients begin to receive, accept, and return feedback from fellow students and teachers. Often progress at this level involves peer relationships which result in a preference for conversations in the hall and skipping classes rather than the kind of responsible studentship viewed as the task of other levels of the hierarchy. This trade-off must be understood and sometimes accepted as a desirable sign of accomplishment for a given client's status on the educational task hierarchy.

The Mastery Task Level

We consider the five levels of the hierarchy of educational tasks already discussed to be what educational psychologists have termed the "learning-how-to-learn" process. Students are unable to gain mastery of skills (whether they be academic or mental health related) without first learning those skills related to basic steps of the educational hierarchy. The belated appearance of educational approaches in mental health service delivery appears to us to be due in significant part to the ignorance of mental health professionals regarding those educational tasks and procedures which provide the foundation for mastery learning. Noneducators often equate education with the mastery of subject matter and consider the acquisition of skills basic to that process to be the province of some other enterprise (child rearing or psychotherapy, for example) from which educators and educational technology should be excluded. The corollary to this is that mastery level learning is considered to be the province of the educator and other disciplines (including mental health professionals) often choose not to become involved. Services to the mentally ill have suffered from this dichotomizing of tasks and professional responsibilities. The mentally ill have things to learn which, though nonacademic, are often appropriately viewed as mastery level activities: self-control techniques, relaxation skills, social skills, home management skills, and skills in dealing with social agencies, to name but a few. A psychiatric treatment program must offer learning opportunities at this level at the time when the client is prepared to receive them. The content of the LSP curriculum is designed to impact at this level. It will be described after a comment about the final level of the task hierarchy.

The Achievement Task Level

The achievement task level represents the final step in the hierarchy of educational tasks. The learning process here is primarily managed by the client/student who has substantially accomplished tasks at all lower levels of the hierarchy and is in the process of moving from reliance on the LSP toward an autonomous role in the community. Movement up the educational task hierarchy entails a shifting of degree (as well as type) of responsibility between staff and client, with staff responsibility and initiative being relatively high at lower levels and diminishing as student involvement increases at the higher levels of the hierarchy. At the achievement level the teacher is essentially uninvolved, except as an interested bystander and source of friendly support. At this point the client has the skills to do it herself.

THE CURRICULUM

The structure upon which educational tasks at all levels of the hierarchy are implemented in the LSP is the curriculum. The LSP curriculum is composed primarily of three elements: GATE groups, classes providing structured learning experiences, and individual study. These activities take place in 3 hours of student involvement each day. Figure 6.1 presents an example of the morning program which runs from 9:30 a.m. until 12:30 p.m. (ending with lunch). The afternoon curriculum is similar, running from 12:00 (beginning with lunch) until 3:00 p.m. The morning and afternoon programs serve primarily separate client groups, although cross-registration is encouraged for those needing more than 3 hours per day of involvement or desiring courses which are available only in the alternative session.

GATE Groups

GATE is an acronym for Goal Attainment Through Education. Each client of the LSP is assigned a GATE group which serves as homeroom. Counselors/teachers rotate the leadership of GATE groups each month so that they have the opportunity to become familiar with as many clients as possible. The groups meet every day of the week and assure counselors of daily personal contact with each client. GATE groups provide a variety of functions and educational opportunities in addition to staff/client contact for case management purposes. The groups are relatively unstructured (in contrast to the structured learning of all other classes) and thus allow for maximum peer and staff/client interaction attentive

Figure 6.1. Life Skills Curriculum

Time	*Monday*	*Tuesday*	*Wednesday*	*Thursday*	*Friday*
9:30 to 10:15 A.M.	Assertiveness Training Ceramics Lab Relaxation Exercises Vocational Exploration	Substance Abuse Nonverbal Communication Skills Physical Education	Assertiveness Training Viewing Behavior through T.A. Relaxation Exercises Vocational Exploration	Substance Abuse Nonverbal Communication Skills Physical Education	Assertiveness Training Viewing Behavior through T.A. Creative Expression Socialization Group
10:20 to 11:05 A.M.	Ceramics Lab (continued) Changing Behavior Vocational Assessment Weight Control	Dance Group Human Sexuality Practical Living Skills Stress Management	Group Interactions Changing Behavior Depression Weight Control	Mental Health Myths Human Sexuality Practical Living Skills Stress Management	Group Interactions Creative Expression (continued) Depression Socialization Group
11:10 to 11:50 A.M.	GATE Groups	GATE Groups	GATE Groups	GATE Groups	GATE Groups
12:00 to 12:30 P.M.	Lunch	Lunch	Lunch	Lunch	Lunch

to issues as they arise, as well as for social support and staff and peer feedback. The homeroom function of GATE groups is significant. Announcements are made, and the educational philosophy and procedures of the LSP are reinforced in this class. GATE groups are attentive to educational tasks at stages of the educational hierarchy up through the relationship level, and they even occasionally offer specific structured experiences directed at the mastery level.

Structured Learning Classes

Structured learning experiences constitute the second major element of the LSP curriculum and represent the content areas within which mastery learning opportunities are offered. Examples of structured learning approaches have increasingly appeared in the mental health literature (Goldstein, 1973; Hersen & Bellack, 1976), and their use and implementation are widely accepted, particularly in the area of social skills training. Our curriculum reflects these developments but extends into a number of related areas as well. Our curriculum development has been guided by Lazarus (1976), who has offered the acronym BASIC ID to represent the seven zones of human functioning he has considered significant to psychotherapeutic activity: behavior, affect, sensation, imagination, cognition, interpersonal, and drug or physiological. Lazarus' thesis is that significant and enduring personality change occurs when three or more zones are involved in the learning experience. Each zone is represented in the LSP curriculum, which offers several courses providing skills training relevant to activities of daily living. Examples are courses by such names as Getting Around (instruction in the use of public transportation), Cooking, Home Management, Vocational Reentry, and Personal Health. Other courses are oriented toward specific problematic behaviors. Examples include: Changing Food Habits, Child-Rearing, Substance Abuse, Fear and Anxiety Reduction, Memory Training, and Depression. These courses, along with Changing Behavior, have in common their semididactic approach and their attempt to convey to the student information about techniques and, to a degree, theory that will enable him to assume a cognitive control over his own behavioral destiny.

A second major division of our curriculum emphasizes learning by doing. This is the laboratory portion of the LSP. Included here are classes by such names as Effective Communication, Couples' Communication, Human Relationships Laboratory, Behavioral Rehearsal, Relaxation, and Assertiveness Training. The common feature of these courses is their emphasis on action, or learning by doing, and their utilization of behavior rehearsal techniques. Each course has a syllabus outlining overall goals, session objectives, daily lesson plans, and homework assignments, as well as the method of evaluating course objectives.

Individual Study

The third primary aspect of the LSP curriculum is individual study. Clearly, classroom experiences and structured learning, as significant as they are in most educational enterprises, are not the only means to skill acquisition. Classroom experiences present substantial limitations for many tasks, especially at basic levels of the educational task hierarchy. Even at the mastery level, skill acquisition in certain sensitive (especially to the client) or taboo (to either the client or her peers) areas is often most readily accomplished in individual study or coaching sessions. The principles by which individual study is conducted at the LSP are similar to those employed in classroom instruction. Sessions, though tailored to the individual needs of a client's present project, are structured, goal oriented, and time limited. Structuring of these projects generally falls within the framework taught in the Changing Behavior class and has been described as behavioral change by McFall (1976). In addition to those interventions which lend themselves to use as self-control measures by the clients themselves, individual study allows the use of virtually the full range of behavior modification strategies as intervention techniques. In our experience individual study offers an important potentiation effect for classroom activity at all levels of the educational task hierarchy. Staff also seem to gain an important additional element of job satisfaction from this aspect of the educational enterprise.

MEDICATION MANAGEMENT AND THE ROLE OF THE SCHOOL NURSE

The LSP is a school for behavioral change, an opportunity to learn those personal, interpersonal, and community survival skills which persons labeled mentally ill often lack. We do not maintain that LSP clients lack illness, only that LSP programming is primarily attentive to the non-illness aspects of behavioral disability. We acknowledge the critical importance of psychotropic medications for many of our clients who would be unable to benefit from the educational opportunities if they did not have these medications. Seventy-eight percent of LSP clients receive some form of psychotropic medication during their involvement with LSP. Forty-seven percent of LSP clients receive neuroleptics, 17% receive antidepressant medication, and 10% receive Lithium. Clients are initially evaluated for medications (upon referral from teachers and counselors), and medications are prescribed by the LSP consulting psychiatrist, who provides 6 hours per week of evaluation time. The bulk of medication management is the responsibility of the LSP school nurse, who monitors client reaction to medications by means of teacher/counselor feedback and staff case conferences, individual medication evaluation appointments, and weekly attendance for this purpose at each GATE group. The school nurse also provides clients with general information regarding medications and their use and abuse.

Obviously, the availability of psychotropic medications in a psychoeducational mental health unit produces mixed effects. There is no question of the need for medication demonstrated by a sizable portion of LSP clients. Medication, however, can be delivered in a variety of ways and in a variety of settings (the HCMHC Medications Clinic, for example). Clients do not enroll in a school for behavior change because of their need for medication, but in spite of it. At the LSP the chemistry of behavior change is viewed as a potentiator of our approach to the psychology of behavior change. Predictably, our clients don't always understand this. Many LSP students have gained considerable dependence on medications and have come to view them as the major contribution mental health care has to offer. The administration of medicine tends to undercut the message that behavioral problems are learned, not caught. Further, if not carefully managed, medication coverage tends to reinforce the sick role and to seduce the LSP student away from taking an active responsibility for skill acquisition. One method of handling this dilemma would be to eliminate medication coverage from the program and to subcontract this task to the Medications Clinic, thus attempting through separation of functions to avoid the conflicting messages provided by a simultaneous presentation of the chemistry and the psychology of change. We have considered this to be both unnecessary and undesirable, and we believe that coming to terms with the paradigm clash is a skill our students must gain.

Experience has supported the contention that the disease and behavioral models of problematic behavior can indeed coexist in a fashion where the one complements, rather than detracts from, the other. Considerable education regarding the nature of the two models and their interplay is necessary. This takes place in GATE groups and is done both by the group leader and by the school nurse during her weekly visits. Other courses take every opportunity to reinforce the message that behavior is learned but that medications alter conditions which either impair or facilitate the learning process. The consulting psychiatrist and the school nurse, along with other LSP staff, watch their language, carefully endeavoring to avoid inadvertent reinforcement of *patientship* through references to *patient, therapist, treatment, cure,* and *sickness.* Both the nurse and the psychiatrist assess mental status and evaluate current adjustment through a questioning of what learning has taken place in school, what problems are interfering with educational tasks, and what environmental conditions at home are supporting or interfering with school attendance, homework assignments, or other learning activities. In our experience studentship need not be weakened by the use of medicine in a psychoeducational mental health program substantially more than it is in an academic setting. Certainly the issue is critical, however, and vigilance to the problem must be maintained.

In addition to having primary responsibility for medication coverage, the school nurse provides a number of functions including study, crisis management, the development and teaching of structured classes such as Illness Prevention and Health Maintenance and Women's Health Problems, and the supervision of trainees from the nursing profession.

CLIENT FLOW

Upon referral to the LSP, potential clients are given an orientation interview consisting of an overview of the LSP, its philosophy and curriculum, initial goal-setting, and tentative curriculum selection. Criteria for involvement are stressed in this session: (a) the client must want to come to the LSP (LSP staff rely on the line, "You can't come here if you don't want to"), (b) the client must have some identifiable mental health objective, and (c) the client must be able to refrain from interfering with the education of other students. If clients have been referred by an outpatient service of HCMHC (including Entry, the HCMHC evaluation and referral unit), clients have been administered the Psychiatric Status Schedule (Spitzer, Endicott, Fleiss, & Cohen, 1970), the Internal-External Locus of Control Scale (Rotter, 1966), and the Hopkins Symptom Check List (Derogatis, Lipman, Rickels, Uhlenhuth, & Covi, 1974). If the referral comes from the HCMHC inpatient service, these instruments have not been administered and referral is made to the Entry program for this purpose. If a client enrolls in the LSP, he is assigned to a case manager who assists in the elaboration of goals and curriculum selection. Referral for a medication evaluation is made at this time if considered appropriate by the case manager. Assignment to a GATE group is also made at this time, and further orientation to the LSP, goal articulation, and curriculum development occur there. The initial curriculum continues for 1 month, at the end of which the case managers assist clients in evaluating progress toward goals, establishing goals for the coming month, and selecting the next month's curriculum. The average duration of client involvement in LSP is 5.7 months, with an average enrollment in 88 separate classes (including GATE groups and repeat enrollment in selected classes).

CASE MANAGEMENT

Case managers have a number of responsibilities, including charting (this includes updating of the problem list), client advocacy (including interaction with other agencies), family conferences, facilitation of client goal articulation, and curriculum selection. Of these, only the charting function needs elaboration here.

Charting

Case managers are responsible for one comprehensive chart note per week for each client in their caseloads (which average 15 clients). This note, labeled *case review*, summarizes the discussions about the client at one of that week's three case review staff conferences. *Charting* follows *DIG-A-Plan format*, the acronym for data, interpretation, goals, and plan. Entries in each section of the DIG-A-Plan note are referenced to the number of the relevant item on the problem list of the problem-oriented medical record. Problems described by the list conform to the terminology of the Psychiatric Status Schedule (Spitzer et al., 1970). The last case review of each month presents the curriculum plan for the coming month.

At the end of the month the case manager rates the severity of each item on the client's problem list according to a 5-point scale (from 0–4) that reflects the number of basic living patterns influenced by the problem at that time. Basic living patterns reviewed are the following: intrapersonal functioning; self-management; social and interpersonal functioning; and vocational, avocational, and educational functioning. Table 6.1 presents examples of behavioral excesses and deficits related to each of the four living patterns which are used to evaluate problem severity. Table 6.2 presents a rating system for estimating the severity of each problem. The system is used in the record keeping.

PROGRAM EVALUATION

Several evaluation projects have been implemented to assess both process and outcome variables related to LSP procedures. Initial evaluation efforts sought to identify those LSP student characteristics predictive of attrition, active and appropriate participation in the program, and problem change (Armstrong, Tracy, Rock, & Hays, Note 2). The ultimate objective of these efforts is the generation of prescriptive statements characterizing those clients best served by various aspects of psychoeducational service delivery. Needless to say, present efforts leave us far from having attained that objective. Toward this end, however, 44 LSP subjects completed pretreatment measures including the Minnesota Multiphasic Personality Inventory (MMPI) (Hathaway & McKinley, 1967), the Psychiatric Status Schedule (PSS), and the Internal-External Locus of Control Scale (IELCS). This intake battery, plus the four demographic variables of age, education, income, and number of psychiatric hospitalizations, provided a total of 42 independent variables. Dependent measures were termination status, participation in the program, and problem change.

Participation is monitored in each LSP class by the instructor, who uses the participation ratio, an index of both quantity and quality of a client's participation in class. Teachers rate the participation of each student for each session on the following scale: 0 = no participation and/or inappropriate or disruptive verbalizations; 1 = only solicited, though appropriate, verbalizations; 2 = one or two unsolicited and appropriate verbalizations; and 3 = multiple appropriate verbalizations. Total points obtained by a student are divided by the possible points scored for the number of class sessions the student actually attends. This figure multiplied by 100 yields the participation ratio. Problem change is rated by counselors in the manner described in the preceding case management section. The *problem change score* is the severity rating at the end of the first month, minus the severity rating at the end of the second month.

The results of this study can be summarized as follows: only locus of control predicted termination. Early terminators were significantly

Table 6.1
Basic Living Patterns Used to Judge Problem Severity

INTRAPERSONAL FUNCTIONING	
Hallucinations	Self-concept deficits
Delusions	Paranoid ideation
Self-depreciating thoughts	Ambivalence
Suicidal ideation	Fear/anxiety
Denial of mental health problems	Depression
Insight deficits	Self-blaming
Judgment deficits	Unrealistic expectations
SELF-MANAGEMENT	
Drug abuse	Poor impulse control
Alcohol abuse	Bizarre behavior
Personal hygiene deficits	Lack of daily routine
Suicidal behavior	Food and other substance abuse
Activities of daily living deficits	
SOCIAL–INTERPERSONAL	
Aggressive behavior	Poor communication skills
Verbal aggression	Demanding/manipulative behavior
Social isolation	Marital problems
Dependency	Lying
Child-rearing deficits	
VOCATIONAL, AVOCATIONAL, EDUCATIONAL	
Interviewing skills deficits	Concentration deficits
Job search deficits	Interest deficits
Attendance and promptness deficits	Knowledge deficits

Table 6.2
Rating System for the Problem Progress Record

0 = no disorder

This rating indicates data relevant to this problem have been volunteered by the student or elicited by the case manager. On the basis of this data, no problem is judged to exist at this time.

1 = borderline

This rating indicates that there is behavioral evidence for this problem, but not to the point where it interferes with any basic living pattern (intrapersonal/self-management/social-interpersonal/vocational, avocational, educational).

2 = mild

This rating indicates there is behavioral evidence that this problem interferes with at least one basic living pattern.

3 = moderate

This rating indicates there is behavioral evidence that this problem interferes with two or three basic living patterns.

4 = severe

This rating indicates there is behavioral evidence that this problem interferes with all four basic living patterns.

X = problem not discussed

This rating indicates that no data have been volunteered or elicited since the last rating. This rating should not be confused with a rating of no disorder.

more internal than were those remaining in the program (t (42) = 2.85, $p < .01$); terminators dropped out of the LSP less than 10 days after enrollment. Six variables from the PSS (overt anger, disorientation, agitation, denial of psychiatric symptoms, student role impairment, and speech disorganization) were inversely related to participation ratio ratings (F values for stepwise regression analyses exceeded .001 probability levels for each variable). Only interview belligerence (also a PSS score) predicted problem change (r = .55, $p < .001$). Interestingly, no MMPI scores predicted dependent measures. In summary, this study suggests that clients presenting as belligerent, agitated, disorganized both in speech and action, displaying overt anger, reporting student or trainee role impairment, and/or denying psychiatric symptoms are significantly less likely to benefit from LSP involvement. In addition, persons with higher external control scores are more apt to remain enrolled in the program than those with more internal orientations.

Ongoing evaluation efforts seek to demonstrate the LSP program's capacity to foster studentship and relate this to outcome. The results of one such investigation (Armstrong, Rainwater, & Smith, Note 1) attempting to enhance compliance on folder bringing were summarized earlier. Ongoing attempts are being made to evaluate the effects of increasing one student behavior upon other behaviors of the same response class (such as note taking, participation, attendance, promptness, etc.) and upon student problem ratings.

At this time follow-up data using the PSS, Hopkins Symptom Check List, and IELCS are being collected on all outpatient units of the HCMHC. Hopefully, this project will yield more meaningful information regarding prescriptiveness (i.e., for whom is the program effective?), long-range outcome, and the impact of LSP procedures as compared with programs employing other approaches.

Discussion

A number of problems can be expected by mental health centers attempting to implement psychoeducational approaches to mental health service delivery. Perhaps the most notable of these is the suspicion generated by any innovation, especially when based upon alternative theoretical models. Innovation implies criticism of convention, and no one likes to be criticized. At the LSP we handle this problem by reliance on extensive public relation efforts stressing that we view our approach as being employed in addition to, not instead of, alternative programmatic approaches to mental health problems. We stress that there is strength in using diverse strategies to mental health problems, that no program can be all things to all people, and that our educational approach is clearly not appropriate for everyone. Most professionals seem satisfied with this soft-sell and are able to evaluate our procedures on their merits.

Another related problem is third-party reimbursement for services. The majority of our clients are supported by Title XIX funding. The Department of Institutions (for the state of Washington) site visitors were at first frankly skeptical of our educational rhetoric and seemed concerned that standards of psychiatric care might be ignored by classroom approaches. On closer evaluation, however, site visitors have consistently lauded the LSP for setting a high standard of excellence in the delivery of psychiatric day care by any standard. What has been critical in this regard is the excellence of our record-keeping procedures. In fact, our charting procedures (described earlier) exceed state standards in some respects. The price of innovation is the necessity for impeccable procedures in those areas not in direct conflict with innovative principles. Conventional treatment programs can survive shoddy charting and other procedures which would jeopardize the existence of an innovative approach. As a result of our observance of this principle, we have had no problem with third-party reimbursement.

A significant problem for a psychoeducational mental health treatment unit is the contrast it poses to the expectancies of its clients. Goldstein (1962) has articulated the critical role of consistency between what a client expects and the nature of the psychotherapy that is offered. Clearly people are referred to the LSP expecting to be treated in basically the same manner that they have been treated in the past (i.e., as patients). The initial task of LSP staff, therefore, is rapid indoctrination designed to lessen the gap between the expected and the observed and preparation of the customer for his new role. Basically, this is handled in the LSP in the same fashion that many behavioral psychotherapists cope with the problem, i.e., through education regarding the nature of problematic behavior and its amenability to influence, and through normalization of clients' self-perceptions by educational efforts.

The problems faced by programs such as the LSP seem far outweighed by the advantages of the educational approach. In our experience the "idiom of competence" (Hobbs, 1968, p. 17) and education has easily as much to offer the customer of mental health services as does the "idiom of illness or pathology." A major advantage of the approach is its potential for the destigmatization of the mental patient. LSP clients frequently describe relief that they can tell friends and relatives (and the police!) that they are going to school. An additional advantage is the enthusiasm of referring professionals for a "learning place," a "school of behavioral change," a mental health delivery system attentive to the strengthening of competence. Judging from the response of clients, referring professionals, and enthusiastic staff of the LSP, our educational, skill-building, competence-enhancing approach to some of the problems of the mentally ill is an idea whose time has come.

Reference Notes

1. Armstrong, H. E., Jr., Rainwater, G., & Smith, W. R. *Studentship as a function of contingent social reinforcement.* Unpublished manuscript, University of Washington, 1979.
2. Armstrong, H. E., Jr., Tracy, J. J., Rock, D., & Hays, V. *The life skills program: A psychoeducational approach to psychiatric day treatment.* Unpublished manuscript, University of Washington, 1979.

References

Adkins, W. R. Life skills: Structured counseling for the disadvantaged. *Personnel and Guidance Journal,* 1970, *49,* 108-116.

Adkins, W. R. Life coping skills: A fifth curriculum. *Teachers College Record,* 1974, *75,* 507-526.

Armstrong, H. E., Jr., & Bakker, C. B. The adult development program: An educational approach to behavioral change. In J. H. Straughan

(Ed.), Proceedings of the Fall Conference, Association for the Advancement of Behavior Therapy-Northwest. *Behavior Therapy,* 1971, *2,* 431-432. (Summary)

Armstrong, H. E., Jr., & Booth, C. An educational program for obesity management: Preliminary report. *Psychotherapy: Theory, Research and Practice,* 1979, *16,* 280-285.

Bakker, C. B., & Armstrong, H. E., Jr. The adult development program: An educational approach to the delivery of mental health services. *Hospital and Community Psychiatry,* 1976, *27,* 330-334.

Bandura, A. Psychotherapy as a learning process. *Psychological Bulletin,* 1961, *58,* 143-159.

Carkhuff, R. R. Training as a preferred mode of treatment. *Journal of Counseling Psychology,* 1971, *18,* 123-131.

Derogatis, L. R., Lipman, R. S., Rickels, K., Uhlenhuth, E. H., & Covi, L. The Hopkins symptom check list (HSCL): A self-report symptom inventory. *Behavioral Science,* 1974, *19,* 1-15.

Endicott, J., Spitzer, R. L., Fleiss, J. L., & Cohen, J. The global assessment scale: A procedure for measuring overall severity of psychiatric disturbance. *Archives of General Psychiatry,* 1976, *33,* 766-771.

Goffman, E. *Asylums: Essays on the social situations of mental patients and other inmates.* Garden City, N.Y.: Doubleday, 1961.

Goldstein, A. P. *Therapist-patient expectancies in psychotherapy.* New York: Pergamon Press, 1962.

Goldstein, A. P. *Structured learning therapy: Toward a psychotherapy for the poor.* New York: Academic Press, 1973.

Guerney, A. G., Stollak, G. E., & Guerney, L. The practicing psychologist as an educator—an alternative to the medical practitioner model. *Professional Psychology,* 1971, *2,* 276-282.

Hathaway, S. R., & McKinley, J. C. *Minnesota multiphasic personality inventory manual* (Rev. ed.). New York: Psychological Corp., 1967.

Hersen, M., & Bellack, A. S. Social skills training for chronic psychiatric patients: Rationale, research findings, and future directions. *Comprehensive Psychiatry,* 1976, *17,* 559-580.

Hewett, F. M. A hierarchy of educational tasks for children with learning disorders. *Exceptional Children,* 1964, *31,* 207-214.

Hobbs, N. The reeducation of emotionally disturbed children. In E. M. Bower & W. G. Hollister (Eds.), *Behavioral science frontiers in education.* New York: Wiley, 1967.

Hobbs, N. Reeducation, reality and community responsibility. In J. W. Carter (Ed.), *Research contributions from psychology to community mental health.* New York: Behavioral Publications, 1968.

Jones, M. Psychiatry, systems theory, education, and change. *British Journal of Psychiatry,* 1974, *124,* 75-80.

Kelly, G. *The psychology of personal constructs.* New York: Norton, 1955.

Lazarus, A. A. *Multimodal behavior therapy.* New York: Springer, 1976.

McFall, R. M. Behavioral training: A skill-acquisition approach to clinical problems. In J. T. Spence, R. C. Carson, & J. W. Thibaut (Eds.), *Behavioral approaches to therapy.* Morristown, N.J.: General Learning Press, 1976.

Rotter, J. B. Generalized expectancies for internal versus external control of reinforcement. *Psychological Monographs,* 1966, *80* (1, Whole No. 609).

Sanford, F. H. Creative help and the principle of habeas mentum. *American Psychologist,* 1955, *10,* 829-835.

Spiegler, M. D. School days—creditable treatment. *Psychiatric Quarterly,* 1977, *49,* 221-229.

Spitzer, R. L., Endicott, T. J., Fleiss, J. L., & Cohen, J. The psychiatric status schedule: A technique for evaluating psychopathology and impairment in role functioning. *Archives of General Psychiatry,* 1970, *23,* 41-55.

Stuart, R. B. *Trick or treatment.* Champaign, Ill.: Research Press, 1970.

Szasz, T. S. *The myth of mental illness.* New York: Wilbur & Hoeber, 1961.

Witmer, L. Psychological diagnosis and the psychonomic orientation of analytic science. *Psychological Clinic,* 1925, *16,* 1-18.

Chapter 7

Learning to Cope: Content and Context of an Educational Program

Cornelius B. Bakker
Marianne K. Bakker-Rabdau
Johan Verhulst

Abstract

The autonomy and personal effectiveness of individuals depends largely on their coping skills. The Adult Development Program was designed to provide learning opportunities to overcome coping skill deficits. The content of these classes, combining a specific conceptualization with behavioral techniques, is exemplified. The context which enhances the effectiveness and the consistency of an educational approach is discussed in terms of the character of the customer-provider relationship and the setting in which the services are rendered.

In January, 1970, the Adult Development Program (ADP) began its services in the Department of Psychiatry of the University of Washington (Bakker & Armstrong, 1976). The mandate given by the department chairman to the initiator was to organize a set of intensive, ambulatory psychiatric services akin to those rendered in day care programs. Site visits by the senior author to major day care programs in the United States and Europe extinguished all desire to emulate existing models. Most such units had started with therapeutic enthusiasm, but as an increasing percentage of the available time was taken up by chronic patients who showed little evidence of change, the mood of the professional staff shifted toward resignation, self-protective indifference, or despair.

To prevent a similar fate, we developed a set of guidelines for our program which has served well and is now, almost 10 years later, still in effect without change. A brief description of these principles follows.

GUIDELINES FOR THE PROGRAM

Objectives

The services to be rendered by the ADP were designed to enhance the clients' personal autonomy; in practical terms, to teach coping skills. This was and still is the primary objective; all other aspects of the program are subservient to this objective. Since the literature contained no ready blueprints for such a program, we had to systematically develop both a relevant curriculum and effective teaching methods.

Model

To render a specific service by means of an internally consistent set of approaches requires a careful articulation of the way in which one views human behavior. The perspectives which have served as working hypotheses in the ADP are:

1. An individual's problems, dissatisfactions, and limitations are a function of the ability to cope with the perceived challenges of the situation.
2. This coping ability depends to the largest extent on the *skills* and the *information* which the individual has learned in the past.
3. This ability can be enhanced by additional training.
4. The manner in which a person perceives a situation (the cognitive set) is primarily the result of past social learning experiences.
5. This set can be altered by new learning (cognitive restructuring).

These assumptions are implied in the term *educational model* whenever it is used in the remainder of this discussion.

Customers to Be Served

The objectives of the program and the model on which it was based have a sobering consequence: only insofar as the program is able to train individuals in more effective skills is it justified to sell its services. The history of psychiatry shows that the practitioner's ardent belief in the logic or profundity of a theory is hardly a guarantee of favorable results. In order to minimize the customer's risk of paying for useless services, we have placed several limitations on admission to the program.

1. Only individuals capable of participation in the learning process provided by the ADP are allowed to enter the program.
2. Individuals complaining of problems for which known effective and efficient remedies are available elsewhere are referred to the appropriate service.
3. No person coming to the ADP as a result of coercion is accepted.
4. Intensive enrollment in the program is limited to 6 months, unless special compelling reasons for an extension are presented to the staff.

Evolution of the Educational Approach

The educational model, as it was articulated at the inception of the ADP, drew from the work of Hobbs (1966) with children, the clinical and theoretical advances of social learning theory, the psychology of personal constructs (Kelly, 1955), human ecology (Bateson, 1972), and systems theory (Miller, 1978).

The climate for the development of educational approaches was obviously favorable in the 1970s, as the literature reflects (Authier, Gustafson, Guerney, & Kasdorf, 1975; Ivey, 1971). A good deal of this development occurred in a rather haphazard fashion. Community mental health centers and psychiatric hospitals began to organize various courses as an adjunct to treatment. Assertiveness training, in particular, gained widespread popularity (Heimberg, Montgomery, Madsen, & Heimberg, 1977), but many other programs, some of them dealing with social skills (Argyle, Trower, & Bryant, 1974) and others with self-management (Kanfer & Goldstein, 1975), were created. The movement is now spreading beyond the mental health field. Courses related to self-management and interpersonal skills are becoming part of the curriculum of many colleges and universities, and even some secondary schools are beginning to experiment with such courses. The medical community has become increasingly interested in this development because of its potential for illness prevention and health maintenance (Benson, Shapiro, Tursky, & Schwartz, 1975; Pomerleau, Bass, & Crown, 1975; Williams, 1978). Beyond that, the general public has been exposed to a nearly limitless array of workshops, purported to enhance personal growth, which range from the apparently legitimate to the esoteric and charlatanic.

In the decade of the ADP's existence broad and significant changes have taken place in the mental health arena. Most members of the mental health professions no longer view the behavior therapies with suspicion. At the same time, the behaviorists themselves have broadened their repertoire; the variety of interventions, as well as the range of application, has been vastly expanded. Modeling, role playing, and behavior rehearsal have taken their place beside desensitization, contingency management, and other operant procedures. An increasing number of behavior therapists are using educational techniques, and many, if not most, behaviorists recognize the importance of cognitive methods for the facilitation of change.

Clearly these 10 years of evolution have resulted in an increasing acceptance of a cognitive-behavioral perspective on human behavior (Ellis & Harper, 1971; Meichenbaum, 1977). A wide variety of practical clinical interventions can be placed within this framework, and the educational approaches, be they individual or group-directed, form an important segment of them.

At this point of evolution and development, it appears fruitful to make explicit what has been learned along the way. In the following sections several aspects of the content and the context of the ADP, especially those which have been of practical value for the implementation of the program, will be described.

PROGRAM CONTENT

In an educational approach it is characteristic that the objectives be formulated in operational terms. The educational approach has this in common with other behavioral therapies. However, when one sets out to develop a specific curriculum and wishes to define the content of the courses which are to train the customers in coping skills, it becomes apparent that these coping skills are rather difficult to define. Are these skills recognized by the behavior involved or by their results? Can they be grouped and classified, or are they a conglomeration of independent, learned behaviors? Are the skills behavioral patterns which are best taught together? Or, are they separate behavioral units which must be acquired in sequence?

Equally important for the formulation of course content were the considerations: how much and what kind of information it should contain and, especially, whether the students required a rationale for the form and the content of the classes. Because little useful guidance was available in either literature or tradition, the program went through an extended trial-and-error phase of approximately 3 years. It serves little purpose to review these experiences in detail. It may suffice to say that the classes were divided into those dealing with interpersonal problems and those focusing on self-management and that initially they were best described as behaviorally oriented, open-ended group discussions. It rapidly became apparent that the types of problems brought up by the clients fell into different categories, and as the program's volume expanded (from an average of 5 new enrollments a month during the first year to 150 a month at present), separate groups were created for each type of problem. Simultaneously, the staff gained skill in assisting the students with goal definition and advising them about the appropriate classes to attend.

As a result of these experiences, a curriculum has evolved in which each class has a stable content that is either related to a specific set of coping skills or to a well-defined set of problems. For instance, Affiliation Skills Training exemplifies a skills-related class, while the course on dealing with depression (for individuals on renal dialysis) represents a class oriented toward a specific problem. In practice, it frequently occurs that a person signs up for a problem-related class, such as Dealing with Depression, but after data gathering and assessment she detects that, for instance, inadequate aggressive skills are the primary determinant of the episodes of depression. The student then may take Aggressiveness Training in the subsequent month, with a clear understanding of the objectives and an articulated rationale for participating in the classes.

This brief discussion underscores the fact that the backbone of the program is formed by the skill-oriented classes and that the recognition of specific skill deficits is a necessary element of the problem-focused classes. An internally consistent, readily understandable conceptual basis for all the courses seemed necessary. This seemed especially important since most students were not inclined to participate actively in the classes until they understood the reasons for the class content and after they had experienced some type of an "Aha!" effect.

The conceptual framework developed at the ADP was derived from two sources: (a) the way in which the problem areas naturally separated out and (b) the interest of the first two authors in *ethology,* which is the study of behavior from an evolutionary perspective. Ethology looks at behavior as adaptation to environmental challenges. It recognizes that such adaptations have slowly evolved and represent a species' ability to cope with the environment enough to ensure offspring survival.

The language of ethology, in which coping with environmental challenges stands central, provided a ready example for the language we sought to develop for a program dealing with coping skills. Recent developments in ethology and sociobiology (Wilson, 1975) have generated an extensive classification of social behaviors observed in higher primates. The function of these behavior patterns for the survival of the individual has been studied intensively in nonhuman species, and similar studies in humans are beginning to be carried out (Austin & Bates, 1974). A comparison of the classes which naturally evolved from the problems presented by ADP students and the classifications of behavior patterns observed by ethologists showed such close similarity that we decided to explore the utility of placing the categorization of skill areas on an ethological foundation. The results of this exploration have been published in detail elsewhere (Bakker & Bakker-Rabdau, 1973). For the present, it suffices to mention a few categorizations that have been particularly useful.

In the area of interpersonal behaviors a distinction was made between *affiliative behaviors*—those seeking to establish or strengthen interpersonal contact—and *competitive behaviors.* The competitive interactions in turn were divided into those employed for acquisition and those used for defense. Some of the other useful distinctions included courting, parenting, and sexual behavior. Thus we could scrutinize the behaviors related to each of these areas to decide which adaptive skills were used and what purpose they served. This perspective continues to provide a frame of reference that greatly facilitates the systematic observations of coping behavior in order to isolate specific components of skills that subsequently can be taught to the students. As a consequence of this process, the number of classes at the ADP has steadily grown, with the content of each class becoming increasingly specific.

We have learned from the ethologists that each species, including our own, has similar environmental challenges with which to cope and uses various strategies to do so. However, for the higher primates, and especially for the human species, such strategies clearly include the transfer of coping skills through the social learning process. The purpose of the ADP classes is to provide learning opportunities to enhance this process and to make up for deficits in the individual's past skills training.

The manner in which ethological concepts are integrated in the program can best be exemplified through a more detailed discussion of several of the skills training classes. The Assertiveness Training and Aggressiveness Training classes (Bakker-Rabdau, 1977) will serve as examples. Both classes are part of the Competitive Skills Pathway and represent the second step in a three-part program developed to teach competency in handling interpersonal situations involving conflict—territorial as well as hierarchical—in a way that minimizes interpersonal hostility and promotes cooperation. The first segment of the Competitive Skills Pathway introduces the student to a territorial model of interpersonal conflict (Bakker & Bakker-Rabdau, 1973), with explicit definitions of aggression. assertion, and hostility, together with methods of data gathering, goal setting, and other problem solving techniques. During this first month the student learns to assess the adequacy of his own competitive skills relative to future goals and determines which skills to acquire in subsequent months (Bakker, Bakker-Rabdau, & Breit, 1978). Depending on the skill area(s) to be mastered, the student may sign up for Assertiveness Training, Aggressiveness Training, or both. The student has the option to take these classes concurrently or in sequence. Classes may be repeated as necessary to acquire the desired competence. The Competitive Skills Pathway has a third segment, conflict resolution, which deals with more complex interactions, such as managing long-standing disputes, multiple issues, problems with reference groups, etc.

A few basic concepts will be elucidated to provide the reader with a clearer understanding of the classes taught in the second segment of the pathway. In the territorial model used, *assertion* is defined as behavior which defends a territorial holding against acquisition by others. Conversely, *aggression* is defined as behavior related to the acquisition of additional territory. By defining aggression (acquisition) and assertion (defense) in this manner, the students can readily understand that behaviors appropriate to keeping territory are quite different from those appropriate to gaining new territory. This becomes especially clear when one provides examples and scrutinizes the adaptiveness of the behaviors in the overall social context in which they necessarily take place.

The meaning of assertion and aggression is further clarified by defining *hostility* as behavior which is retaliatory, designed to damage the person or the territory of the victor in a set of territorial transactions. Such hostile behavior can be expected from a previous owner who loses holdings she wants to keep or from individuals who failed to obtain areas to which they feel they have a definite right.

The territories involved in human disputes are frequently intangible, such as attention, areas of functioning, privacy, and status. Nevertheless, the feelings of ownership relative to such areas are no less strong than those relative to more concrete property. Moreover, techniques of aggression for such areas are largely subtle verbal attacks, i.e., weapons. These weapons are designed to undermine the opponent's defensive abilities and thus acquire ownership by what could be called psychological warfare. Unfortunately, the individual deprived of territory by such trickery retains residual feelings of ownership and develops resentment resulting in hostile behavior toward the aggressor. Such hostility may lead to actual violence, to a withdrawal from the relationship with the aggressor, or to depression and self-destructive behavior. As these theoretical notions are taught, it is repeatedly emphasized that such knowledge is useful only if it leads to the student's carefully keeping data on what actually transpires in the conflict situations he encounters.

Assertiveness Training Class

The goal of the Assertiveness Training class is to teach the student techniques to courteously but firmly defend territorial holdings against the unwanted aggression of others. It is emphasized that these techniques are for the defense of territory already owned and that using them for the acquisition of new territory is maladaptive. Unfortunately, many of the so-called assertiveness programs are taught as if defense and acquisition (and for that matter, affiliation) all involve the same skill. The result is that the student frequently pays a heavy price of increased conflict and reduced interpersonal harmony. The skill taught in the ADP Assertiveness Training class is divided into several areas.

Data collection. The students are asked to keep daily logs of their interactions with others in order to identify conflict situations. In this log keeping emphasis is placed on the precise rating of feelings related to territorial conflict: (a) irritation and anger, which point to a transgression by someone else; (b) guilt, which is related to being a trespasser on another's territory (the "hand in the cookie jar feeling" which must be differentiated from shame, a feeling of "I am no good, I have failed"); and (c) resentment, the desire to engage in hostile, destructive behavior. Students are taught to rate these feelings on a 10-point scale and to record them, together with the situations in which they occurred. This includes a detailed account of interpersonal transactions and especially conversations. This procedure is very useful in the identification of the particular weapons used in the conflict.

Identifying the issue. Once the student has identified a competitive interaction (by log keeping), he learns how to analyze the components of the conflict in order to identify which of his own territories is under attack and to formulate a specific defense statement.

Keeping to the issue. The element of keeping to the issue involves stating one's defense in a courteous, nonaccusatory manner (e.g., an *I* statement) and refusing to be distracted from the objective, i.e., retaining ownership of the territory disputed by behavior or verbalizations of the opponent. By means of role playing and behavioral rehearsal, the student then learns specific verbal defense techniques.

Body language. Appropriate body language with emphasis on facial expression and eye contact is demonstrated and practiced in the class. The student learns to use the confrontive body position, in which the plane of the body of the asserter is kept parallel to that of the aggressor. This stance includes an open, relaxed body posture and an impassive face. A steady, level, unblinking gaze is directed into the eyes of the aggressor, holding the attention of the opponent while the assertive message is delivered in a calm, quiet voice.

Identifying weaponry. A major segment of the class is devoted to identifying the weaponry of the opponent and learning specific means of defense against each weapon. The old rhyme "Sticks and stones may break my bones, but words will never hurt me" misses the mark. Words are the most powerful weapon of psychological warfare. The students learn therefore to identify the various types of weapons by attending to certain key words and phrases used by others, as well as themselves.

Should, for instance, is a key word which implies obligation and is used in a weapon called *guilt induction.* "You should go to see your mother" instead of "I want you to go see your mother" creates quite a different set in the listener. While the aggression is direct in the latter statement, and therefore easily recognized, it is hidden, wrapped in righteousness, in the former one. Similarly, the weapon of *helplessness* is identified by words such as *can't, need to, have to,* e.g., "I need to have your car" as opposed to "I would like to have your car." *Definition,* another weapon, can often be identified by words such as *always, never,* e.g., "But you never let me borrow anything of yours." In this manner the students are taught to recognize and effectively respond to the 15 major types of weaponry and their variations. The class also clarifies that it is important to avoid the use of weapons in both assertive and aggressive situations because such methods tend to increase conflict rather than solve it.

Generalization. To facilitate application of the newly gained skills in the natural situation, the teacher helps the students to analyze a number of their own specific conflicts and to role play the preferred assertive responses. For homework they are asked to try out these responses in the real situation and report to the class the results, including the impact on their feelings as registered in their log books. Frequently the students report positive results of the application of their newly learned techniques to a previously unresolved conflict and at the same time confirm that the previous hostility toward the opponent also has been appreciably reduced. This, on the one hand, reinforces the application of assertive responses and, on the other hand, strengthens the sense of understanding of the situation, thus increasing the probability that future situations will be analyzed and responded to according to the same model. Reported failures are also useful in the class for, when analyzed, they usually show that either the asserter used aggressive weapons to aid her defense or used assertiveness to gain territory, a situation where aggressiveness techniques are in order.

Aggressiveness Training Class

As already noted, in the territorial model used at ADP aggression means expanding one's territorial holdings (Bakker & Bakker-Rabdau, 1973). However, such aggression can be carried out in many different ways. To obtain a new car, a person might strong arm the car dealer and take one by force. Or, more commonly, he may barter and become the owner by paying the agreed-upon price. Both methods are aggressive, but unfortunately, the aggressees are not equally satisfied with each of these transactions. One is left with a lump on the head and resentment in the heart, while the other feels joy at having made a sale. The goal of the class, then, is not to prevent aggression as defined, but rather, to manage aggression by teaching the methods of acquiring territory in a way that will minimize interpersonal hostility and conflict.

Introduction. In the beginning of this class the student is desensitized to the words *aggression, aggressive,* and *aggressiveness.* This is done by carefully defining terms and looking specifically at the positive and negative connotations of such words as *aggressive* and *nonaggressive.* The word *aggressive,* as in aggressive salesman, connotes motivation, strength, forcefulness. Conversely, aggressive woman has, at least for some, meant bitchy, hostile, unpleasant. The positive meanings of nonaggressive may imply peaceful, friendly, agreeable, while the negative meanings of the same word are things like lazy, passive, Milquetoast-type, and so forth. Thus the word, depending on the circumstances surrounding it and the intentions of the speaker, may be used as either a compliment or a pejorative.

Students are then asked to pay close attention to their own, as well as others', behaviors in all types of transactions which represent a transfer of ownership rights from one person to another, identifying those types of interactions that result in high levels of hostility and those which result in goodwill. *Claiming behaviors* (those involved in seeking to expand one's turf) are described, as well as the relevant aggressive feelings (curiosity, interest, tension, guilt, excitement, joy). Students become aware of the signals which alert them to their own aggressive plans so that actions may be properly managed, thus increasing the chances that they will obtain what they want with a minimum of hostility from others.

Body language and cognitive set. Whereas in the Assertiveness Training class the student learns to recognize his own territorial rights and act appropriately assertive in verbal and postural behaviors, in Aggressiveness Training class the student learns to recognize the territorial rights of others and to in no way convey—either verbally or physically—a prior claim to such rights. The *confrontive stance* is vital to the asserter's defense; the *petitioner's stance* is vital to the aggressor's. In the petitioner's stance the plane of the aggressor's body is placed at an angle to that of the territorial owner. An open body position with submissive palm display and exaggerated play movements is used to put the aggressee at ease. The face is *affiliative*—smiling with chin slightly lowered and eyes widened—a submissive, courtly, but not groveling display. The purpose is to charm and disarm. While the appropriate assertive verbalizations are "I want"-type *statements* ("I do not want you to have my car tonight"), aggression is signaled with "May I?"-type *questions* ("May I borrow your car tonight?"). The attention to this simple distinction—*statements are assertive; questions are aggressive*—is vitally important to aggressive success.

By asking for something, the petitioner necessarily recognizes that ownership rights reside with the person who is asked. It would be unwise, in fact, to ask another for things that one already owns, as such requests signal nonownership. The student is taught that it is precisely the recognition of another's absolute rights to territorial holdings that works in favor of peacefully obtaining what is desired. The individual who can say no freely and without feelings of guilt can just as freely say yes and obtain the exceedingly reinforcing feelings of generosity which accompany the act of gift making. The less free the act of territorial transfer becomes (think of paying taxes), the fewer generous feelings accrue to the giver. The student is taught that the easier he makes it for the territorial owner to say no to requests, the easier it becomes for the same person to react generously.

Aggressive strategies. Once the students understood the rationale behind eschewing violence or verbal pressure (i.e., weapons) as methods of obtaining new territory—they may, indeed, be quite successful but cause interpersonal conflict and hostility—they are ready to learn various alternative strategies of aggression. The following strategies are taught:

1. *Something for nothing* is the act of simply asking for that which is desired in a straightforward manner: "Mom, I would like to use your car this afternoon from 3 to 5 o'clock to drive downtown to pick up some packages. May I?"; "May I have 10 minutes of your time to discuss plans for the weekend?"
2. *Something for something* (bargaining) means bartering one territory for another: "I would like to borrow your car tonight from 6 to 10. I would be willing to clean the garage and rake the leaves for you this afternoon before I go. Would you be willing to make such an exchange?"
3. *How can I make you an offer you can't resist?* involves the aggressor's asking the aggressee to name the price: "What would you take for the use of your car this evening from 6 to 10?"
4. *Establishing coterritories or becoming partners* is the stepwise negotiation technique applicable when two or more individuals with limited resources join to acquire an otherwise unobtainable holding. The student is repeatedly cautioned to manage her aggressiveness so that she not only applies no pressure for an affirmative answer but also unquestioningly accepts a no reply. One may, after all, ask again at another time. The aggressor has the right to ask; the aggressee has the right to refuse.

Proper management of aggressiveness is a vitally important coping skill. The person with inappropriate aggressive skills pays a terrible price in interpersonal relations, often not only failing to achieve his goals but also suffering social rejection and punishment. Properly managed aggressiveness, on the other hand, frequently achieves its ends and in many cases actually enhances interpersonal harmony and affiliation.

This discussion of the Competitive Skills Pathway and the somewhat more detailed description of the Aggressiveness and Assertiveness Training classes exemplify how a particular conceptualization is integrated with modern behavioral techniques to provide training in coping skills. The conceptualization is used in a manner which is very similar to Meichenbaum's (1976) use of the work of Schachter and Singer (1962) to provide a conceptual framework for his stress inoculation classes. The utility of the conceptualization depends on its being easily understood and readily accepted by the clientele.

Verhulst and Verstraeten (1977) investigated this in their marital program which used the territorial model of interaction and found that their concepts were indeed readily understood and applied by the participating couples. Obviously, this supports the utility of these ideas for training purposes, not their scientific validity.

Program Context

Interventions aimed at facilitating change do not operate in a vacuum. The power of the intervention is not a function of its form and content alone. On the contrary, the context in which the intervention occurs plays a role which is at least as important as the intervention itself. The nonverbal behavior of the therapist, the character of the building in which the intervention takes place, the reputation of the program, and the expectations of the client can make a good method even better or render it less effective. Different approaches require different contexts. Using behavioral techniques in a behaviorally oriented clinic is altogether different from applying the same methodology in a psychoanalytic service.

For the sake of clarity, a distinction will be made between two contextual aspects of an educational program, viz., the quality of *the relationship* between the consumer and the provider and the characteristics of *the field setting,* including the physical aspects of the facilities, as well as the administrative organization.

The Relationship

The central aspect of an educational program reveals itself in the quality of the relationship between the consumer (the student) and the provider (the teacher or the adviser). Students frequently have a history which includes extensive contacts with other mental health facilities and professionals. This implies an intensive training in the patient role, a gratuitous by-product of the therapeutic process. Avoiding reactivation of the therapist-patient interaction requires that the responsibilities of the student, as well as the teacher, be made explicit.

Respecting the students' goals. If the student wishes to participate in the program, he must pursue his own goals. It does not matter whether at one point or another these goals were suggested by others; what is essential is that the student scrutinizes the options, determines the direction in which he wishes to proceed, and makes a commitment to this pursuit. It is the adviser's task to assist the student in articulating the goals and clarifying their consequences, as will be discussed in the next section. However, it must be conveyed to the student that the teacher lacks the superior wisdom to tell him how best to conduct his life. (After all, there is little evidence to show that mental health professionals are more satisfied or competent in handling their lives than any other group of human beings.) Program personnel further emphasize this character of the relationship by warning the student to watch for staff breaches of this basic rule.

Assisting with goal definitions. Many students enter a program far more aware of their dissatisfactions than of the specific changes they wish to make and the ultimate goals they want to pursue. Goal setting and planning for the future are themselves skills which are not equally developed in all human beings. An adviser must be able, therefore, to train the students in goal setting, strategy development, and goal commitment. This process of assisting the student in developing concise goals must be truly heuristic in nature if the autonomy of the student is to be increased at the same time. In a *heuristic process* the adviser raises questions, offers alternatives, elicits comparisons, and asks the student to assess the relative personal value of each option. This joint effort is carried out in a relaxed, unhurried fashion and may occasionally take as many as eight 1-hour sessions. It is essential that the student eventually gain a personal sense of what she wants and not base any goals only on the trust that the adviser is a good person and probably knows best.

The adviser underscores the need for such self-reliance when it is made explicit that every course of action has a cost attached to it and that it is the student who pays the price, not the adviser. Suggestions, be they from a teacher or anyone else, deserve to be viewed with skepticism, for it is always the student who bears the consequences.

Establishing a contract. Once the student is committed to action, an informal time-limited (usually 1 month or 6 weeks) contract covering the responsibilities of the student, as well as the staff, is agreed upon. This contract is based on a customer-provider model which implies that the customer has the prerogative to decide whether or not he likes the services he is receiving. The student is not only free to reject the services and seek assistance elsewhere but he also has the privilege to switch advisers within the program. To safeguard this freedom, the staff must avoid subtle coercion by means of interpreting why the student desires to change advisers. Students are told, therefore, that they do not have to give reasons for their requests.

The counterpart of this relationship is that the teacher retains the freedom to *not* render services for which he has not contracted. Most fundamental in this respect is that the provider does not assume responsibility for the student's life by becoming his teacher. The provider is responsible for the quality of the teaching, but not for the welfare of the student. This refusal to assume the responsibility for the student's welfare distinguishes the educational approach sharply from the psychotherapeutic interventions that originated in the field of medicine. The training of physicians and other health professionals strongly emphasizes the ongoing responsibility which one assumes for the patient. While this responsibility pertains only to health matters, it automatically becomes a responsibility for almost any aspect of the patient's life once the term *health* is expanded to *mental health.* It is not surprising, therefore, that the rejection of this responsibility is the aspect of the educational model which is most frequently criticized by other mental health professionals.

The teacher-student relationship implies also that the teacher does not have to be able to teach everything. A teacher has only a limited amount of skill and information to offer to the student. This obvious fact underscores the contrast between the student-teacher relationship and the therapist-patient interaction, for in the latter the therapist commonly feels obligated to do something for the patient, even if she has no idea how to provide a service of consequence.

A specific time-limited contract in which both student and teacher define what their respective obligations and responsibilities are keeps both parties from attempting to do the impossible. The teacher will not expect the student to acquire the desired skills without completing the homework tasks or participating in the practice sessions, while the student soon learns not to pressure the teacher for what goes beyond the contract. The clarity of this process helps a great deal in preventing prolonged, unsuccessful therapies.

Creating enthusiasm. The foregoing description of the student-teacher relationship may create the impression that such an interaction tends to be cold and mechanical; nothing, however, is farther from the truth. None of the outlined aspects of the relationship stand in the way of human warmth, friendship, and encouragement. In fact, a good teacher is an enthusiastic teacher. An outstanding teacher creates a joyful excitement in the student, for if the teaching is effective, the student experiences the liberating awareness of new possibilities, as well as the enormous pleasure of grasping that which previously seemed not understandable.

It is the mark of a good teacher that his students gain a growing sense of power (McClelland, 1975), together with the extension of their mastery. In fact, this sense of increased power and mastery may well be the most important intrinsic reinforcer of the student's learning process.

Classroom management. At the beginning of the course each instructor explains to the students what purposes the classes can serve and the ground rules for classroom conduct. The discussion of the course objectives includes examples of the kind of difficulties for which the class will have relevance, as well as of the type of problems which lie outside its scope. In the Couples Negotiations class, for example, it is clarified that the participating couples will have the opportunity to learn a specific way of settling conflict by means of negotiation and that they will be given the necessary theoretical foundation for this method. However, the emphasis will be on practical training so that everyone can become skilled at applying this negotiation method. It is also explained that negotiation requires two people who are both interested in resolving the conflicts; methods for dealing with a partner who refuses to negotiate can better be obtained in the Assertiveness Training and Aggressiveness Training classes.

The explanation of the ground rules for class conduct includes a delineation of the responsibilities of the student and the teacher. The instructor is responsible for providing the course material in an understandable fashion. If the student does not grasp the material, she may assume that this is due to the instructor's lack of clarity. However, it is the student's task to ask questions when anything is unclear. If the student wishes to gain the maximum from the course, active participation and completion of the homework tasks are essential; but it is entirely acceptable to the instructor if the student remains passive and omits the homework: "Don't expect to be pulled into action against your will or nagged about the homework . . . but on the other hand, you may not learn much that way." It is pointed out at the same time that the student can always repeat the course and that many people have found it useful to do so two or three times. There is a lot of material to be covered in each class, and the instructor will feel free, therefore, to cut short any discussion that carries the class too far afield, regardless of how interesting the topic may be. By the same token, it is the objective of the instructor to give the students ample opportunity to learn how to apply the material in their own lives, and it is therefore important that they bring their own situations as examples and raise all the questions they have. If there is not enough time during the class period itself to handle all these situations, the student is encouraged to make an individual appointment with the instructor. In a similar fashion, the students are requested to arrive on time for the classes because the instructor plans to start each session promptly at the scheduled hour. If, however, anyone happens to be late for one reason or another, it is perfectly acceptable to join the group as long as this is done with minimal disruption.

This set of ground rules gives the instructor the prerogative to manage the attention structure of the class. If student-student interaction becomes diffuse and loses its relevance to the class objectives, the instructor can intervene and shift to an instructor-student mode of interaction. If, on the other hand, the student-instructor interaction is monopolized by one student or polarizes into total passivity vis-à-vis an instructor's monologue, he can choose to switch to an increased student-student interaction, for instance, by means of action techniques.

Every teacher also functions as a model (McFall & Lillesand, 1971; McFall & Twentyman, 1973) and frequently provides personal examples of the way in which problem situations can be handled. He is not required to be aloof, a reflecting mirror, or an unknown stranger and does not have to be paralyzed by the fear of countertransference. The classroom situation itself often provides excellent opportunities for modeling interpersonal skills. When a student monopolizes the class or disrupts it with inappropriate behavior, the instructor is expected to manage this situation. After all, in an educational format the instructor has the primary attention of the students. He has the prerogative to relinquish the attention temporarily to a student but remains responsible for taking it back if the effectiveness of the educational process requires this. The monopolizing student interferes with the prerogatives of the instructor and the learning of the other students. An assertive, although courteous, response is called for. Ordinarily, this is accomplished by acknowledging that the student is raising some very interesting points but explaining at the same time that (a) this issue will come up at a later time in the course, and it would be appreciated if the student would raise the issue at that time; or (b) several of the student's situations have been discussed, and the instructor would like to focus on some examples from other students; or (c) this material lies outside the scope of the class, but the instructor would be delighted to take it up with the student individually after class.

Such responses model for the students a variety of assertive behaviors and at the same time demonstrate their utility, because ordinarily they solve the problem without doing damage to either the student or the instructor. Occasionally, such responses fail to get the desired results. This occurs usually vis-à-vis students with bizarre, inappropriate, or unusual stereotypical behaviors. An example of the latter was a blind student who would repeatedly distract and annoy the class with blindisms (rhythmic rocking of the body and waving motions of arms and head). In such cases the student is asked to stay a few minutes after class, at which time she is asked to drop out of the class, and it is suggested that a special project be started with one of the staff members to bring the disruptive behaviors under control. It is emphasized that the suggestion is only relevant if the student does, indeed, eventually wish to participate in the classes. Frequently, as for instance in the case of the student with blindisms, it has been possible to let the student return to the classes within a relatively short time. Making such return contingent on specific behavior changes apparently serves as a powerful reinforcer by itself.

The teacher-student relationship is part of a larger topic, i.e., teaching style. In his intensive study of teaching styles and pupil progress, Bennett (1976) found that teaching style has a more powerful effect on learning progress than the personality of the students. Well-structured teaching approaches are most effective with highly anxious students. Formal classroom styles work well for all students except low achievers, who benefit more from a casual setting which encourages self-expression. This extensive topic goes well beyond the scope of this discussion, and it is mentioned only to emphasize that the teaching process itself is in need of much systematic study to establish which teaching styles and techniques are most effective for training students in coping skills.

The Field Setting

The field setting includes the physical environment in which the training is offered, as well as the procedures and mechanics of the program. This aspect of the context begins to develop at the moment that the prospective student hears about the program and decides to explore the possibility of enrollment.

Repeated contacts with hospitals and clinics have taught individuals to behave as patients in such settings. Schools and other training institutions invite behaviors which are altogether different. While it seems entirely normal for a patient to spend hours in a crowded waiting room to be seen by a doctor, it is certain that few students would remain in a classroom if the teacher was more than 15 minutes late. In facilitating appropriate student and staff behavior in an educational program, it is important to eliminate from the field setting any similarities to hospitals and replace them with characteristic aspects of teaching facilities. The development of an educational program in a hospital or a mental health clinic creates some obvious problems; therefore, if feasible, it is preferable to use a facility which neither by name nor appearance suggests a medical service. It is for this reason that the Adult Development Program was housed in a remodeled apartment building, rather than in the hospital.

Information and enrollment. Since the creation of the context begins when the student first hears about the program, it is important to carefully manage the form of this information. In the Adult Development Program the name itself suggests education rather than therapy. Brochures, announcements, and news releases are verbalized in terms of education, rather than therapy. The University includes the class announcements in its quarterly publication on continuing education. In the ADP an individual becomes a student (not a patient) when he registers for classes (rather than joins groups) which are given by instructors whose clinical titles (psychiatrists, nurses, psychologists, social workers) are not mentioned; individual work is done with advisers, rather than therapists. Classes are announced on a monthly schedule (see Figure 7.1) to emphasize the structured character of the program.

Introduction into the program. An individual who wishes to enroll in the program can either directly sign up for specific classes if she feels sufficiently informed about their content to decide which are relevant to her goals or may meet first with an adviser. The purpose of the session with the adviser is to provide the student with the information needed to assess whether the program is appropriate to her goals, and which classes, possibly in combination with individual sessions, would be most useful. If the student is not able to make such a decision in one session, additional appointments with an adviser are offered to assist the student in determining her objectives (using the previously described heuristic method) and developing a strategy that effectively uses the program offerings.

Medical-legal realities make it the staff's responsibility to prevent individuals from enrolling in an educational program for problems which are caused by recognizable medical conditions, especially diseases of the central nervous system. To accomplish this with minimal dissonance, the staff is trained in observing unobtrusively for subtle behavioral indications of possible organic brain disease. A simple check list, designed for this purpose, allows the adviser to make the necessary observations without disrupting the flow of the interview. This check list is completed afterward and becomes part of the student's record. In the extremely rare case where central nervous system pathology is suspected, the student is told that his difficulties could be caused by neurological problems and that staff would want him to consult a physician about this before he enrolled in the ADP.

Training and maintenance of staff behavior. Staff behaviors are closely associated with the model on which the program is based. In psychodynamically oriented clinics the therapist approaches the client, hoping to uncover the hidden motivations behind the overt behaviors—insights are highly regarded and clever interpretations are reinforced. Humanistic programs tend to emphasize the expression of feelings, the development of close relationships, and the reduction of competitiveness. Staff members are reinforced when they successfully persuade their clients to shift their lifestyle in the direction of this ideal type.

Figure 7.1. Monthly Schedule of Classes at the Adult Development Program

ADULT DEVELOPMENT PROGRAM

MAY 1979

Class/Instructor	*Monday*	*Tuesday*	*Wednesday*	*Thursday*	*Friday*
Making Major Changes in Your Life – An Introduction to ADP C. Bakker, Staff (4 weeks; starts 5/11)					10:00 a.m.-12:00 noon Conference Room
Interpersonal Relationships (Territoriality) E. Goldsmith, Staff (4 weeks; starts 5/7)	9:30-11:00 a.m. Lounge		9:30-11:00 a.m. Lounge		
Assertiveness Training M. K. Bakker (6 weeks; starts 5/22)		9:30-11:30 a.m. Mural Room		9:30-11:30 a.m. Mural Room	
Aggressiveness Training (Initiating Change) M. K. Bakker (6 weeks; starts 5/22)		11:30 a.m.-12:30 p.m. Mural Room		11:30 a.m.-12:30 p.m. Mural Room	
Feelings and Emotions Management A. Ivanoff (4 weeks; starts 5/7)	11:00 a.m.-12:00 noon Mural Room		10:00 a.m.-12:00 noon Mural Room		

Figure 7.1 *(continued)*

Class/Instructor	*Monday*	*Tuesday*	*Wednesday*	*Thursday*	*Friday*
Art Lab S. Bohrer (No time frame; open enrollment)		12:00 noon- 2:00 p.m.		12:00 noon- 2:00 p.m.	
Dealing with Depression A. Ivanoff (6 weeks; continued from 4/10)		1:00-3:00 p.m. Lounge			
Women in Transition E. Goldsmith (4 weeks; starts 5/7)	1:00-3:00 p.m. Conference Room		1:00-3:00 p.m. Conference Room		
Social Growth thru Drama S. Tollefson (First 4 weeks)		1:30-2:30 p.m. Mural Room		1:30-2:30 p.m. Mural Room	
Changing Self-Image S. Tollefson, K. Romstad (4 weeks)		2:30-4:00 p.m. Lounge		2:30-4:00 p.m. Lounge	
Affiliation Skills Training *(Overcoming Shyness)* M. K. Bakker (6 weeks; starts 5/22)		2:30-3:30 p.m. Mural Room		2:30-3:30 p.m. Mural Room	

Pre-Orgasmic Women's Group E. Goldsmith (6 weeks; starts 5/7)	5:30-7:00 p.m. Enrollment by appointment with Elsa Goldsmith	5:30-7:00 p.m.
Sexual Response Group (with partners) for Women E. Goldsmith (6 weeks)	" " " " " "	
Sexual Response Group (without partners) for Women E. Goldsmith (6 weeks)	" " " " " "	
Personal Issues Surrounding Sexuality E. Goldsmith (6 weeks)	" " " " " "	
Sexual Enhancement for Men S. Breit, E. Goldsmith (6 weeks)	Enrollment by appointment with S. Breit	
Sexual Dysfunction Clinic S. Tollefson	Enrollment by appointment with Sue Tollefson	

Figure 7.1 *(continued)*

Class/Instructor	*Monday*	*Tuesday*	*Wednesday*	*Thursday*	*Friday*
Back Pain Management	Enrollment by appointment with Dr. Dworkin (543-3796 – Cl 2)				
Food Habit Management Program	Enrollment coordinated by Julie Waltz				
Information on ADP Food Habit Management Classes J. Waltz (1 session)		1:00-2:30 p.m. Conference Room			
Food Habit Management – Heart Saver L. Anderson (6 weeks; continued from 4/5)				4:00-6:00 p.m. Conference Room	
Food Habit Management J. Waltz (3 months; continued from 4/3)		10:00 a.m.-12:00 noon Conference Room			
Food Habit Management J. Waltz (3 months; continued from 2/5)	10:00 a.m.-12:00 noon Conference Room				

Maintaining Weight Loss J. Waltz (4 weeks; continued from 4/3)		3:00-4:30 p.m. Conference Room			
Changing Food Habits J. Waltz (6 weeks; starts 4/30)	4:00-5:30 p.m. Conference Room				
EVENING CLASSES					
Food Habit Management – Diabetes L. Anderson (6 weeks; continued from 4/5)				7:30-9:30 p.m. Conference Room	
Food Habit Management K. Romstad (3 months; continued from 4/2)	7:00-9:00 p.m. Conference Room				
Food Habit Management K. Romstad (3 months; continued from 3/13)		7:30-9:30 p.m. Conference Room			

Figure 7.1 *(continued)*

Class/Instructor	*Monday*	*Tuesday*	*Wednesday*	*Thursday*	*Friday*
Fears and Phobias E. Golosman (6 weeks; starts 5/3)				7:00-9:00 p.m. Room 131	
Fears and Phobias – Advanced Class E. Golosman (4 weeks)	Enrollment by appointment with E. Golosman				
Explorations in Human Territoriality E. Goldsmith (4 weeks; starts 5/10)				7:30-9:00 p.m. Room 233	
Overcoming Shyness (Small Talk and Conversational Skills) J. Walz (4 weeks; starts 4/30)	7:30-9:30 p.m. Mural Room				
Aggressiveness Training J. Walz (4 weeks; starts 5/7)	5:30-7:00 p.m. Mural Room				
Assertiveness Training J. Walz (4 weeks; starts 5/9)			7:00-9:30 p.m. Mural Room		

Dealing with Depression A. Ivanoff (6 weeks; continued from 4/10)		5:30-7:30 p.m. Conference Room			
Depression and Dialysis A. Ivanoff, F. Tenenbaum (6 weeks)				5:30-7:30 p.m.	
Stress Reduction and Stress Management C. Bakker (5 weeks; starts 5/8)		7:30-9:00 p.m. Mural Room			
Self-management for Mood and Behavior Change (4 weeks)	Enrollment by appointment with A. Ivanoff				
Marital Enhancement Program	Enrollment by appointment with Maureen Daniek				
Couples Communication S. Tollefson, Staff (6 weeks; continued from 4/3)		5:30-7:30 p.m. Lounge			

Figure 7.1 *(continued)*

Class/Instructor	*Monday*	*Tuesday*	*Wednesday*	*Thursday*	*Friday*
Couples Communication: Advanced Workshop S. Tollefson, R. O'Hara (4 weeks; starts 5/3)				5:30-7:30 p.m. Lounge	
Couples Negotiation C. B. and M. K. Bakker (6 weeks; starts 5/24)				7:00-9:30 p.m. Mural Room	

It is the objective of an education program, such as the ADP, to facilitate and encourage change and to avoid all quality control over the direction of such change. All efforts are directed at the enhancement of the students' autonomy, and the staff must therefore be trained in and reinforced for behavior that supports this objective.

Since most staff members have received training in more traditional settings, attention must be given to the elimination of behaviors which are incongruous with the program's model. New staff members are therefore trained to eliminate *psychiatrisms* and *psychologisms*–such as diagnostic labels, explanations of behavior in terms of unconscious motivation, and so on. This staff training includes modeling and role playing, as well as extensive discussions of the program's objectives and the reasons for the clearly articulated rules of interaction. To maintain this behavior, the staff remains on the alert for an occasional departure, such as a psychodynamic interpretation of someone's behavior. This draws an immediate response, and is often made light of through a humorous remark.

The fact that a professional is committed to the teacher role and understands the educational model does not eliminate the reality that she has some very definite ideas about the way people "should live their lives." Not infrequently, inservice training sessions are used to deal with such issues to clarify that staff members are free to entertain whatever personal ideals they feel committed to but that the intrusion of such ideals into their work represents an infringement on the student's prerogatives and an abuse of the teacher-student relationship.

This respect for the student's prerogatives must not be confused with playing a passive role. In the educational process the student is actively taught and, where useful, feedback is provided immediately and clearly. If a student accepts a homework task and returns on the agreed-upon time without having completed this work, the instructor is free to postpone meeting with the student until the homework is done. If the student must carry out certain tasks to acquire the desired skill, the teacher would be remiss if he did not confront the student with the fact that such skills are obtained only with practice. However, interpreting the motivation for such failure to do the homework would be totally out of keeping with the educational approach.

Another liability of a psychiatrically trained staff, which must be attended to during inservice training, is the inclination to perceive their clients as fragile and vulnerable. Such a perception is particularly destructive to the educational process, for it tends to elicit a rescue response which results in an immediate shift to a therapist-patient relationship. To cope with this problem, we established a routine which has functioned extremely well and which has probably contributed to the fact that none of the students has committed suicide while being involved in the ADP. This routine can be briefly described as follows. When a staff member has concern about the suicide potential of a student, a consultation must occur, either in a regular staff meeting or in an ad hoc discussion with the program director. The decision-making process which then takes place is based on the recognition that every human being has the capability of self-destruction and that a student's expression of this possibility is not in itself a basis for active intervention unless there are specific reasons–such as severe, acute despondency, or a profound depressive phase of a manic depressive illness–to believe that suicide is imminent and can be prevented by a relatively brief hospitalization. If the discussion leads to this conclusion, a deliberate shift to the medical model is recommended; the student is asked to seek assistance in the psychiatric emergency service and, where appropriate, relatives or friends are contacted to make sure that the necessary steps are taken. At the ADP this has been an extremely rare event.

Far more often the conclusion is that active intervention is not warranted and that the teacher-student relationship can be maintained. The adviser then acknowledges that she recognizes the seriousness of the student's concern and, if appropriate, renews the heuristic process to pinpoint the specific problems that have led to the consideration of suicide. This will often result in a reformulation of the student's objectives and a recommitment to the change strategy. The suicidal threats are neither reinforced nor challenged by the adviser. The fact that the student feels bad enough to consider suicide as a solution has at least one advantage: it implies that the student has little to lose and much to gain and that he therefore may be interested in a very intensive program to force major changes in a short time. Pointing this out can restructure the suicidal threats. Of course a student who wishes to receive drug treatment for the depression is never discouraged from doing so. It is simply pointed out that such service is not available at the ADP, but can be obtained at the psychiatric outpatient clinic, and that medications would not necessarily interfere with the individual's current educational program.

In addition to such an obvious problem as the threat of suicide, more subtle expressions of low self-esteem may disrupt the educational process. The instructor's fear of further decreasing a student's already low level of self-esteem may result in insufficient feedback on maladaptive behavior, exaggerated praise for achievements, and reduced expectations for class participation. Since self-respect is most effectively enhanced by progress in gaining coping skills and the described carefulness results in slower progress, it is clearly counterproductive to handle the student with kid gloves. Furthermore, it rarely escapes a student that she is treated in this manner, which confirms what she already suspected: that she is a weak person who does not command respect. It has been one of the more striking experiences at the ADP that a student with many problems and low self-esteem can be given direct and even unflattering feedback in an educational context, as long as it takes place in the pursuit of the student's own goals.

A final aspect of staff behavior which is carefully attended to at the Adult Development Program is the avoidance of professional arrogance. This arrogance may reveal itself in lack of courtesy, in letting students wait for appointments, or in signals which indicate the professional's belief that he has greater insight into the nature of human existence or better ability to predict the future than the customer. When an opportunity offers itself during staff discussions, any such inclinations are reduced by drawing attention to the fact that one's predictive abilities are exceedingly limited and that mental health professionals do not necessarily distinguish themselves by leading exceptionally well-organized lives. Finally, it seems useful to keep in mind that it is more certain that the staff owes a reasonably good living to our customers, rather than the other way around.

Program mechanics and policies. A number of practical policies help to further maintain a context which supports the educational model:

1. The students have access to their records. They are encouraged to read the notes made in their charts. If they do not approve of any of the entries, the adviser will rewrite them, provided that the content remains factual.
2. No reports are supplied to outside agencies which require evaluation or labeling of the student's behavior. Occasionally this creates difficulties relative to third party payments. When a diagnosis must be provided in order to receive reimbursement, the student is involved in this process. Usually the student has by this time thoroughly assimilated the nonpatient role and is able to recognize the labeling as an administrative necessity only.
3. Professionals in training, as well as new staff members in the process of becoming familiar with the program, are expected to participate in the classes in the same manner as other students. This is a particularly potent method for establishing an educational atmosphere. It avoids the inappropriate and incorrect dichotomy which implies that there are people with problems–patients–and people without problems–professionals–and instead conveys the fact that every person can benefit from improving her coping skills. It also makes the new staff members try out themselves every technique they will be offering later to their students. Teaching effectiveness and enthusiasm are greatly enhanced by personally experiencing the utility of what is taught.

4. Often students at the ADP become highly skilled in the methods taught in the classes. Several of these students have served as assistant teachers, and occasionally the full responsibility for a class has been turned over to such a student. In fact, teaching is such an effective learning device that students are frequently given the homework assignment of teaching the class content to someone else.

Remaining Problems

The history of the delivery of mental health services and its close linkage with medicine creates a number of serious problems for a program such as the ADP. Obviously, an individual's failure to cope effectively with his situation can, and sometimes does, result from a physical disease process, rather than from inadequate or maladaptive learning experiences. These behavior-affective diseases are best treated in a medical setting. However, by far the majority of the problems which humans encounter in their lives, regardless of how disabling they may be, result from lack of knowledge or coping skills (Phillips, 1978). The educational approach, therefore, is the most appropriate system of delivering useful services. However, by legislative fiat and deeply entrenched tradition, a label of illness must often be obtained if a person wishes to receive the insurance benefits to which she is entitled. Such a label is incorrect, and in addition creates a counterproductive contextual element which is incongruous with the educational approach. However, the skill deficits can be as disabling as diseases, and they do require professional services. This is not the place for an extensive discussion of the policies governing health insurance and other social benefits. However, it is clear that the contingencies created by such social realities must be drastically altered if the delivery of effective services designed to enhance human functioning and well-being are to become more available and cost effective.

References

Argyle, M., Trower, P., & Bryant, B. Explorations in the treatment of personality disorders and neuroses by social skills training. *British Journal of Medical Psychology,* 1974, *47,* 63-72.

Austin, W. T., & Bates, F. L. Ethological indicators of dominance and territory in a human captive population. *Social Forces,* 1974, *52,* 447-455.

Authier, J., Gustafson, K., Guerney, B., Jr., & Kasdorf, J. A. The psychological practitioner as a teacher: A theoretical-historical and practical review. *The Counseling Psychologist,* 1975, *5,* 31-50.

Bakker, C. B., & Armstrong, H. E., Jr. The adult development program: An educational approach to the delivery of mental health services. *Hospital and Community Psychiatry*, 1976, *27*, 330-334.

Bakker, C. B., & Bakker-Rabdau, M. K. *No trespassing: Explorations in human territoriality.* San Francisco: Chandler & Sharp, 1973.

Bakker, C. B., Bakker-Rabdau, M. K., & Breit, S. Measurement of assertiveness and aggressiveness. *Journal of Personality Assessment,* 1978, *42*, 277-284.

Bakker-Rabdau, M. K. Assertiveness-aggressiveness training: A territorial model. In C. B. Bakker (Chair), *Human territoriality.* Symposium presented at the VI World Congress of Psychiatry, Honolulu, September 1977. In World Psychiatric Association, *Abstracts of the VI World Congress on Psychiatry,* Basel, Ciba-Geigy, 1977, No. 277, p. 54.

Bateson, G. *Steps to an ecology of mind.* New York: Ballantine, 1972.

Bennett, N. *Teaching styles and pupil progress.* Cambridge: Harvard University Press, 1976.

Benson, H., Shapiro, D., Tursky, B., & Schwartz, G. Decreased systolic blood pressure through operant conditioning techniques in patients with essential hypertension. In R. C. Katz & S. Zlutnick (Eds.), *Behavior therapy and health care: Principles and applications.* New York: Pergamon Press, 1975.

Ellis, A., & Harper, R. A. *A guide to rational living.* Hollywood, Calif.: Wilshire Book Co., 1971.

Heimberg, R. G., Montgomery, D., Madsen, C. H., & Heimberg, J. S. Assertion training: A review of the literature. *Behavior Therapy,* 1977, *8*, 953-971.

Hobbs, N. Helping disturbed children: Psychological and ecological strategies. *American Psychologist,* 1966, *21*, 1105-1115.

Ivey, A. E. *Microcounseling: Innovations in interview training.* Springfield, Ill.: Charles C. Thomas, 1971.

Kanfer, F. H., & Goldstein, A. P. (Eds.). *Helping people change: A textbook of methods.* New York: Pergamon Press, 1975.

Kelly, G. *The psychology of personal constructs: A theory of personality* (1st ed.). New York: Norton, 1955.

McClelland, D. C. *Power: The inner experience.* New York: Irvington Publishers, 1975.

McFall, R. M., & Lillesand, D. B. Behavioral rehearsal with modeling and coaching in assertion training. *Journal of Abnormal Psychology,* 1971, *77*, 313-323.

McFall, R. M., & Twentyman, C. T. Four experiments on the relative contributions of rehearsal, modeling, and coaching to assertion training. *Journal of Abnormal Psychology,* 1973, *81*, 199-218.

Meichenbaum, D. A. Self-instructional approach to stress management: A proposal for stress inoculation training. In C. D. Spielberger & I. G. Sarason (Eds.), *Stress and anxiety* (Vol. 3). New York: Halsted, 1976.

Meichenbaum, D. A. *Cognitive-behavior modification: An integrative approach.* New York: Plenum Press, 1977.

Miller, J. G. *Living systems.* New York: McGraw-Hill, 1978.

Phillips, E. L. *The social skills basis of psychopathology: Alternatives to abnormal psychology and psychiatry.* New York: Grune & Stratton, 1978.

Pomerleau, O., Bass, F., & Crown, V. The role of behavior modification in preventive medicine. *New England Journal of Medicine,* 1975, *292,* 1277-1282.

Schachter, S., & Singer, J. E. Cognitive, social, and physiological determinants of emotional state. *Psychological Review,* 1962, *69,* 379-399.

Verhulst, J., & Verstraeten, D. Territoriality and marital therapy. In C. B. Bakker (Chair), *Human territoriality.* Symposium presented at the VI World Congress of Psychiatry, Honolulu, September 1977. In World Psychiatric Association, *Abstracts of the VI World Congress on Psychiatry,* Basel, Ciba-Geigy, 1977, No. 278, p. 54.

Williams, R. B., Jr. Psychophysiological processes, the coronary-prone behavior patterns, and coronary heart disease. In T. M. Dembroski, S. M. Weiss, J. L. Shields, S. G. Haynes, & M. Feinleib (Eds.), *Coronary-prone behavior.* New York: Springer-Verlag, 1978.

Wilson, E. O. *Sociobiology: The new synthesis.* Cambridge: Harvard University Press, Belknap Press, 1975.

Part Three

Specific Populations and Problems

Part Three is comprised of four chapters which represent attempts to improve the technology and practical utility of behavioral group therapy by extending its use to new clinical populations and problems.

Chapter 8 by Thomas Coates and Cheryl Perry describes the use of Bandura's social learning/reciprocal influence model in developing cardiovascular risk reduction programs in four behavioral areas: nutrition, exercise, smoking, and weight control. Of particular importance are the extension of behavioral group procedures (particularly modeling, behavioral rehearsal, and reinforcement) into school-based primary prevention programs and the amount of attention given to making these programs interesting and enjoyable for participants and therapists alike.

In Chapter 9, Henne Queisser, Hubert E. Armstrong, Jr., Wayne Smith, and Gay Davis present a detailed outline of a psychoeducational skills training program for individuals with epilepsy. Although this chapter might just as well have been included in Part One, it was placed in Part Three because of its special emphasis on tailoring the skills training program to the needs of epileptic individuals, for whom both the nature of epilepsy and society's reaction to the disorder contribute to a high incidence of difficulties in the areas of interpersonal functioning, stress management, mood control, problem solving, and memory.

David Roth, in Chapter 10, reviews the therapeutic implications of the various environmental-behavioral and cognitive models of depression and notes the relative scarcity of pertinent group psychotherapy research in this area. He offers specific recommendations for the behavioral group treatment of depression, focusing not only on the primary group interventions to be employed but also on guidelines for selecting the most appropriate group members and strategies for shaping and reinforcing therapeutically efficacious interactions among group members.

Chapter 11 by Gary Williamson and Francisco Barrios describes the authors' experiences as behavioral consultants to a day hospital program for chronic psychiatric patients. Although this chapter is somewhat different from other chapters in this volume in that the interventions described were largely unsuccessful, it is included here because the difficulties and frustrations which the authors encountered may be instructive to readers who are interested in doing behavioral consultation. Of particular note are the chapter's discussion of some major pitfalls (e.g., design error, staff resistance, patient countercontrol, weak system support) which must be taken into account in planning intervention programs of this type, as well as detailed suggestions about how to maximize the chances that the consultation will be effective.

Chapter 8

Multifactor Risk Reduction with Children and Adolescents: Taking Care of the Heart in Behavioral Group Therapy

Thomas J. Coates
Cheryl Perry

Abstract

*We have used a social learning/reciprocal interaction model to develop cardiovascular risk reduction programs in four behavioral areas: nutrition, exercise, smoking, and weight. Emphasized are ways of designing programs that can be delivered to groups of students in school to lead to behavior change. In addition, we have focused on the "heart" in other ways: to make the programs enjoyable and interesting for participants and for therapists. Preliminary results along with suggestions for future work are described.**

*The work reported here represents the efforts of the Youth Education Group of the Stanford Heart Disease Prevention Program: Thomas J. Coates, Cheryl Perry, Joel Killen, and Lee Ann Slinkard. We also thank some of our former and current collaborators and colleagues: Robert W. Jeffery, Brian G. Danaher, Melbourne F. Hovell, Alfred McAlister, Carl E. Thoresen, and Nathan Maccoby. This research was supported in part by Grant No. NIH HL21906 - O1A1 from the National Heart, Lung, and Blood Institute to the Stanford Heart Disease Prevention Program and by a contract from the National Interagency Council on Smoking and Health to Cheryl Perry.

INDICATIONS OF PREVENTABILITY

Cardiovascular disease is the leading cause of death and premature death in the United States. In 1975, for example, 52.5% of all deaths in the United States were attributed to diseases of the heart and blood vessels; 994,513 individuals died because of heart disease (American Heart Association, 1976). As is the case with many chronic illnesses today, a number of environmental factors and behavior patterns are believed to be related to the incidence of cardiovascular disease. Cigarette smoking, elevated blood pressure, elevated levels of serum cholesterol (especially low-density lipoproteins–LDL cholesterol), and diabetes are regarded as the principal risk factors for premature cardiovascular disease (Blackburn, 1974; Kannel & McGee, 1979; Kuller, 1976; Stamler, 1973). More recent evidence has suggested that high levels of high-density lipoproteins (HDL cholesterol) decrease risk of cardiovascular disease (Miller, 1979). The Type-A, or Coronary Prone Behavior Pattern, and elevated levels of serum triglyceride are less certain but potentially important risk factors (Blackburn, 1974, 1978a, 1978b).

Specific behavior patterns appear to be associated with increases or decreases in these principal risk factors. Blood pressure, for example, is presumably influenced by overweight, lack of physical activity, and excessive dietary salt intake (Report of the Task Force on Blood Pressure Control, 1977). Possible relationships between transient or chronic stress, Type-A behavior, and blood pressure have also been suggested (Shapiro & Surwit, 1976). Increased LDL cholesterol is presumably related to excess cholesterol and saturated fat in the diet, lack of physical activity, overweight, and lack of fibrous foods in the diet (Keys, 1970). Increased HDL cholesterol (which is negatively related to cardiovascular disease) is presumably influenced by increases in physical activity, reductions in dietary saturated fat, increases in dietary fiber, and decreases in weight (Blackburn, 1974, 1978a, 1978b).[1]

If cardiovascular risk factors are related to specific behavior patterns, the possibility exists that cardiovascular disease is preventable if persons can be helped to: (a) refrain from smoking; (b) maintain ideal weight; (c) increase consumption of dietary fiber; (d) decrease consumption of dietary cholesterol, saturated fat, salt, and sugar; (e) increase aerobic physical activity; (f) reduce stress; and (g) modify specific coronary-prone behaviors (e.g., time urgency, hostility).

The Stanford Three-Community Study was a 2-year bilingual mass-media health education campaign conducted in two California communi-

---

[1] These conclusions are not uncontested, however (cf. Mann, 1974, 1977).

ties (Farquhar, Maccoby, Wood, Alexander, Breitrose, Brown, Haskell, McAlister, Meyer, Nash, & Stern, 1977; Maccoby, Farquhar, Wood, & Alexander, 1977; Stern, Farquhar, Maccoby, & Russell, 1976). A third California community served as a control. Personal counseling of high risk individuals was also used. The campaigns were designed to increase participants' knowledge of risk factors for cardiovascular diseases and to change risk producing behaviors. The media campaign alone resulted in knowledge and reported changes in behavior. Adding the intensive instruction program enhanced the efficacy of the program for those participants initially evaluated to be at high risk. Overall reductions in cardiovascular risk factors were demonstrated.

The Stanford Five-City Multifactor Risk Reduction Program is currently underway in five (two treatment and three control) California communities. The objectives and scope of the research conducted in the three-city program have been broadened. Community organization, education for health professionals, direct adult instruction, and programs in the schools are being added to a mass media campaign. Knowledge and behavior changes are expected to lead to changes in cardiovascular risk factors, which in turn are expected to lead to reductions in morbidity and mortality due to cardiovascular diseases.

Youth Education

The youth education component of the five-city project is essential to its overall success for three key reasons:

1. Early and widespread efforts for young persons may help promote behaviors, knowledge, and attitudes that can be maintained throughout adolescence and into adult life.
2. There is converging evidence from a number of sources to indicate that young persons manifest behavior patterns which, if continued, will place them at increased risk for cardiovascular disease (cf. Golubjatnikov, Paskey, & Inhorn, 1972; Lauer, Connor, Leaverton, Reiter, & Clarke, 1975; Williams & Wynder, 1976).
3. By working with children and adolescents, it may be possible to influence significant numbers of adults in the community. School personnel, by becoming involved in teaching these programs, may be influenced to change specific behavior patterns. Parents, also, might be influenced to encourage heart-healthy family habits when their children receive specific training in the schools (Coates, Slinkard, Jeffery, Hovell, & Kennigsberg, Note 1).

The goals of the Youth Education Program, presented in Table 8.1, are consonant with the major risk factors for cardiovascular disease. There

Table 8.1
Goals of the Youth Education Program

Smoking prevention:	To discourage the initiation of the smoking habit prior to the time it begins.
Smoking cessation:	To assist students who have acquired the smoking habit to quit.
Obesity reduction:	To assist overweight and obese children and adolescents in achieving a normal body weight and percent body fat.
Nutrition education:	To reduce intake of saturated fat, dietary cholesterol, salt, and refined sugar; to increase consumption of complex carbohydrates.
Physical exercise:	To encourage aerobic and regular physical exercise.
Stress management:	To teach children and adolescents methods of stress management.

is little question that interesting and innovative methods can be devised to teach students *about* cardiovascular disease and the dangers of smoking, overweight, poor nutrition, minimal physical activity, and excess stress. The difficulty—and the challenge—arises from the need to develop programs that (a) can be administered to groups of students in classrooms, (b) will lead to attitude and knowledge change, and (c) will lead to behavior change that is sustained over time and generalized to other settings such as the home.

A Model for Change

Principles of behavior change, combined with techniques derived from group therapy, have been used to devise programs that will help students change behaviors and maintain those changes over time and in different settings. We have adopted Bandura's social learning/reciprocal influence model as a useful theoretical foundation upon which to build these school-based primary prevention programs (Bandura, 1977a, 1977b, 1978).

Advantages of Group Programs

We hypothesized that programs based on this model are best conducted in groups. The group situation provides the opportunity to develop certain behaviors using the input of a variety of persons in the treatment group. Modeling and role playing by different group members present a variety of behaviors for other members of the group to observe, imitate, and improvise upon. Vicarious reinforcement occurring when a person observes role plays can motivate later performance. Behavior modeled in the group can become the norm and possibly will be reinforced outside of treatment sessions. In this way a structure will have been developed to encourage use of target action outside of group treatment sessions.

The Interactive System

The reciprocal influence model posits an interactive system with three principal components: (a) the physical and social environment; (b) the person's competencies, beliefs, attitudes, expectations, and emotions; and (c) the person's behavior. Figure 8.1 presents a schematic overview. The *environment* presents information (models, antecedents, and consequences) to the person about behaviors that are expected, normative, or likely to be rewarded or punished (e.g., a peer offers a cigarette). The *person* evaluates these environmental influences in relation to personal competencies (e.g., to resist peer pressure to smoke), perceptions, values, expectations, beliefs, emotions, and physical states. These filters determine the specific *behavior* the person will perform.

Finally, behaviors do not occur in isolation, but rather are clustered with other specific kinds of behavior. Participation in athletic activities, for example, increases the probability that a student will not smoke (Creswell, Hoffman, & Stone, 1970; Horn, Courts, Taylor, & Solomon, 1959). Similarly, smoking tobacco increases the probability of later alcohol and drug use (Kandel, 1975). Thus, specific behaviors performed by a person will increase or decrease the probability that other specific behaviors will occur.

Influence is multidirectional. The environment influences person and behavioral variables, while at the same time behaviors influence and modify the environment and also person variables. This multidirectional influence process sets the stage for further action. An obese adolescent male, for example, may have learned specific weight loss skills (e.g., declining food offers from concerned grandparents or friends). These skills may not be used unless environmental conditions (e.g., encouragement from parents, the promise of financial reward, frequent contact with the therapist) support their use. Once the adolescent actually performs a difficult action with support from the environment, he may begin to evaluate his abilities differently ("I am the kind of person who can control what I eat"). The circle has been completed. The environment served as a source of skills and also supported their use. Using these skills facilitated changes in the environment and also in person variables (what the adolescent believed he was capable of doing).

Figure 8.1. A Social Learning/Reciprocal Interaction Model

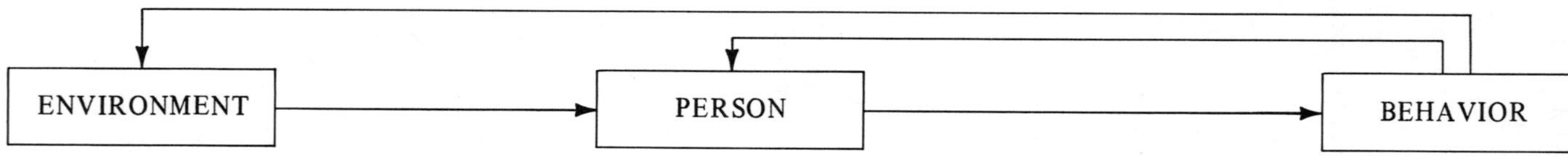

ENVIRONMENT

- Distal
- Proximal
  - –Models
  - –Antecedents
  - –Consequences

PERSON

- Competencies
  - –What can the person do?
- Perceptions
  - –Environment as perceived
  - –Behavior as perceived
  - –Self as perceived
- Values
  - –Placed on personal conduct, stimuli perceived, and consequences received
- Expectancies
  - –Stimulus–outcome and behavior–outcome expectancies
- Beliefs
  - –Knowledge, rules for personal conduct
- Emotions
- Physical reactions and dependencies

BEHAVIOR

- Performance
  - –What does the person do?

Behavioral sources of influence are especially important in a program designed to encourage behavior, knowledge, and attitude changes. Bandura (1978) has recently emphasized the concept of self-efficacy in his writings on processes governing behavioral production and change. Competence in dealing with myriad environmental influences requires that persons be skilled in improvising and adjusting actions in accordance with changing environmental circumstances. Self-efficacy is concerned with judgments about how well one can organize and execute courses of action required to deal with situations containing many ambiguous, unpredictable, and often stressful elements.

Bandura (1977b) proposed that self-efficacy has diverse and important effects on behavior and thoughts. Perceptions of self-efficacy (a) influence choices of activities and environmental settings, (b) determine how much effort persons will expend and how long they will persist when confronted with obstacles, and (c) influence their thought processes and emotional reactions when they are anticipating or completing transactions with the environment.

Persons can learn about environments, behaviors for coping with the environment, and personal efficacy through verbal transmission, persuasive communication, and interpretations of physiological status. Observing others and performing actively, however, are far superior for teaching new knowledge and behaviors, for motivating persons to perform these actions, and for learning about one's abilities to interact effectively with the social and physical environment. Both modeling and rehearsal combine several dimensions (e.g., verbal, visual, physiological), including the experiential, to teach and motivate persons to behave in specific ways or to create novel ways of interacting with the environment to achieve personal goals (Bandura, 1977a).

Learning and Performance

A second important feature of the social learning/reciprocal interaction model involves the distinction between learning and performance. The environment teaches specific behavioral skills (e.g., how to engage in specific cognitive skills such as perceptions of self-efficacy) through direct experience and observation of others. Media provide the opportunity to be exposed vicariously to a wide variety of models and experiences. Incentives can promote learning by encouraging a person to attend to and think about experiences.

Contrary to Socrates' dictum, however, knowledge is not a virtue. Performance does not occur simply because a person knows what to do. Again, the example of the overweight adolescent male enrolled in a weight loss program will be helpful here. Methods for refusing offers of food (a specific lesson in that program) can be taught via role models on videotape and through guided practice with other group members. Whether the person uses these skills in a specific situation, however, depends on current interactions between the environment, the person, and his performance. Specific physical cues might occur first (e.g., grandparents offer food at dinner). The person interprets these cues in terms of their meaning and significance to him (e.g., "If I accept now, my grandparents will continue to offer me food" or "If I refuse, my grandparents might be insulted"). Specific actions follow in response (he tells his grandparents "No, thanks, I'm full"). His parents support his refusal by telling his grandparents about how many pounds he has lost and how his social life has improved since he joined the weight loss program. These actions can also influence the environment (e.g., future food offers may be fewer in number), as well as the adolescent's beliefs about his abilities to take action (self-efficacy) which will lead to weight reduction. Such beliefs are validated when the environment (e.g., the bathroom scale and a significant potential friend) tells the adolescent that indeed he has been successful in losing weight.

Design of Group Treatment Programs

The reciprocal perspective considers that personal, environmental, and behavioral sources of influence should not be separated theoretically or therapeutically. Because they function interdependently, elements of the system can never function autonomously. A complete account of the state of the individual requires simultaneous consideration of all elements of the system. Most important, it is hypothesized that interventions using multiple sources of influence concurrently will be more powerful immediately and in the long term than programs which rely on one source exclusively.

The social learning/reciprocal interaction model has three implications for the design of treatment programs for children and adolescents in educational settings. The programs should:

1. Use behavioral rehearsal with corrective feedback as the major method for teaching skills. Behavioral rehearsal is especially important in giving persons the opportunity to practice target behaviors, to experience success in performing them, and to experience efficacy and enjoyment from performing them.
2. Structure the environment so that skills are used consistently and persistently outside of teaching sessions. The environment will be changed as persons perform target actions: this environmental change provides additional support for the clients to continue using those actions in the future. Environmental sources of influence retain enormous power in influencing behavior and thoughts. At the same time, successful performance of target behaviors in the natural environment should encourage changes in person

variables: beliefs, perceptions, values, and emotions. Especially important may be the enhancement of self-efficacy: the person's belief that she can perform desired actions (Bandura, 1977b).

3. Have a design that enhances the development of personal influence skills (competencies, knowledge, perceptions, beliefs, expectations, values, and emotions). According to the reciprocal influence model, a minimum amount of information is required to accomplish behavior change objectives. Programs that emphasize experiential methods of learning (e.g., modeling and role playing) are the most persuasive sources of information about the world and about oneself (Bandura, 1977a). Feedback from personal actions is the primary vehicle by which persons learn about the world and their ability to interact with it.

Table 8.2 presents some strategies derived from a social learning/reciprocal interaction model that are potentially useful for developing group treatment programs for children and adolescents in schools. Behavioral sources of influence promote action (and environmental and personal change) by prescribing target behaviors that students are likely to use, by encouraging public commitment to perform those actions, and through behavioral rehearsal. Environmental sources of influence provide the cues and incentives to perform desired actions. Personal sources of influence result from environmental and behavioral influences and are, in turn, essential for maintenance and generalization of treatment effects.

We have used this social learning perspective to develop four group treatment programs in the areas of nutrition and exercise for elementary students, smoking prevention and cessation for high school students, and weight loss and nutrition change for high school students.

THE HEART-HEALTH PROGRAM FOR ELEMENTARY STUDENTS[2]

Two objectives guided the design of the Heart Health Program: (a) to increase elementary students' consumption of complex carbohydrates and to decrease their consumption of foods containing excess saturated fat, cholesterol, sodium, and sugar and (b) to increase the students' activity levels at the noon recess. We expected knowledge and attitude changes for students participating in this program.

The program was developed to teach concepts and to promote significant behavior change through eight 45-minute class sessions over a 4-week

---

[2] This program was developed by Thomas J. Coates, Lee Ann Slinkard, and Robert W. Jeffery (see Coates et al., Note 1).

Table 8.2
Strategies Derived from a Social Learning/Reciprocal Interaction Model that are Potentially Useful in Building School-based Primary Prevention Programs

Behavioral Sources of Influence	
Target behaviors:	Behaviors varying in ease of performance, degree of intrinsic reinforcement, and degree to which they evoke reinforcement from environment.
Behavioral commitment:	A public (verbal or written) commitment to practice specific behaviors.
Behavioral rehearsal:	Opportunity to practice and master target behaviors with feedback and reinforcement.
Environmental Sources of Influence	
Models:	Persons similar to participants modeling target actions.
	Program endorsement by significant persons.
Media:	Television, newspapers, posters, school newspapers, other prompts, and reminders in the natural environment.
Environmental cues:	Social or physical cues in the environment indicating that a specific course of action is expected or will be rewarded (e.g., family members subscribe to and begin to practice target behaviors).
Feedback:	Displays of progress.
Incentives:	Material or financial reward; social reinforcement from peers and family; vicarious reinforcement given to significant others.
Personal Sources of Influence	
Knowledge:	Opportunity to learn about the direct health hazards and benefits associated with specific behaviors.
Behavioral competencies:	Personal ability to perform actions needed to perform target behaviors.
Efficacy expectations:	Person's belief that he can perform behavioral competencies.
Values:	Placed on personal conduct and available consequences.
Beliefs:	About target actions; about rules for action.
Emotions:	Feelings of self-growth, self-confidence associated with target behaviors.
Physical reactions and dependencies:	Reactions to and dependencies on chemical substances.

period. Six nutrition classes were followed by six exercise classes. The reciprocal interaction model guided the development of the overall curriculum and each specific class. Table 8.3 presents the principal components as they were applied in this program.

As an example, the third class session introduced the concept of saturated fat and its relationship to atherosclerosis. Students first discussed items normally eaten for lunch. This was followed by an explanation of fat, the differences between saturated and unsaturated fat, and usual sources of each (specification of target behaviors). Food items were then divided into three parts: those containing large amounts of saturated fat, those containing predominantly unsaturated fats, and those containing relatively no fats (models). Students next divided into groups of three, inspected each others' lunches, discussed the relative amounts of fat contained in each lunch, and decided how to prepare a heart-healthy lunch (rehearsal). Students were given a handout specifying food substitutions to decrease saturated fat content of their lunches (family involvement). Finally, each student chose a personal goal (behavior commitment) indicating a specific food substitution to decrease the saturated fat content in his lunch. These goals were written on daily goal sheets which were checked by the teachers but kept by the students.

Three elements of this program seemed most critical: behavioral commitment, feedback and incentives, and family involvement. Students often have difficulty translating general principles into specific behavior changes. The daily goal sheets, on which students made a written commitment to substitute specific heart-healthy foods for items normally in their lunches and to engage in heart-healthy playground activities, seemed essential in promoting behavior change. Students knew precisely what had to be done to meet program goals and to earn available social rewards.

The feedback system, which provided students with objective information regarding their progress, seemed essential in motivating improvement. Students typically expressed amazement at how far they had progressed and also at how far they had to go to eat and exercise in heart-healthy ways. An unplanned bonus was achieved when we presented data separately for boys and girls. The competition between the sexes was keen.

Finally, the token incentive system proved to be enormously popular. We used stickers, buttons, and a rubber stamp—all containing a heart (the logo of the Stanford Heart Disease Prevention Program). The instructors circulated intermittently on the playground, dispensing hearts to students whose lunches contained a specified number of heart-healthy food items or who were engaging in heart-healthy activities. Students who earned heart stickers plastered them all over their lunch pails and notebooks. Students also insisted on heart stamps on their books, papers, and hands. These were important both in motivating and in cueing performance of desired actions. These strategies were especially effective

Table 8.3
Components of the Heart-Health Program for Elementary Students

Behavioral Sources of Influence	
Target behaviors:	To eat low cholesterol, low fat, low salt, and low sugar foods *that they liked.*
	To play heart-healthy activities (e.g., handball, football, soccer) that they enjoyed at lunch.
Behavioral rehearsal:	Food taste tests, practice in asking parents to pack specific foods in lunches, ordering heart-healthy foods in restaurants, reading labels, playing and organizing heart-healthy activities.
Behavioral commitment:	Daily writing of a specific food goal and a specific activity goal for the next day.
Environmental Sources of Influence	
Models:	Instructors who were undergraduate students and who practiced heart-healthy eating and exercise; athletes who came to class to give testimonials; outstanding students who were used as models for others; modeling of each concept or skill.
Family:	Families informed of program and asked to sign consent forms; PTA organized and asked to assist; handouts with simple prescriptions for change sent home.
Feedback and incentives:	Outcome data plotted and displayed during the course of the treatment; token rewards (e.g., stickers and stamps) handed out to students with heart-healthy lunches and exercise.
Personal Sources of Influence	
	Rationale for program and for target behavior changes presented in context of positive health and prevention of cardiovascular disease; competencies and efficacy enhanced through training in skills needed to have heart-healthy foods in lunches and to play heart-healthy activities at lunch time.

when instructors circulated unpredictably on the playground during lunch periods, dispensing hearts and stamping hands and foreheads.

Because sending handouts home is a relatively weak method for involving families, we employed two more powerful strategies to encourage behavior change in the home. First, the token reward system reinforced at school behavior that had to occur at home. Having heart-healthy lunches, made by parents or students, required students to negotiate that specific food items at least be available. PTA members as a group were encouraged to participate by calling friends and informing them of the program.

Two 1-year studies were conducted sequentially in one elementary school district adjacent to Stanford University. In each year three fourth-grade classes in one school and three fifth-grade classes in a second school were involved. A time series multiple baseline design was employed: lessons in the fifth-grade groups were lagged behind lessons in the fourth-grade classes.

To evaluate the program, we used direct observation, paper-and-pencil measures, and phone interviews to parents. While the students were eating lunch, observers approached each student individually and wrote down the contents of his or her lunch. Following lunch, individual students were observed for a 1-minute period. At each 5-second interval the observer made a judgment regarding the student's activity level (sitting, standing, walking, climbing, running, standing still but moving upper trunk) according to preestablished operational definitions. Observer agreement, checked once weekly, was consistently high (mean percent agreement=.97, *SD* = .02). Following lunch, the observers inspected the trash and tabulated the items found. Paper-and-pencil questionnaires were administered before and after the classes to assess changes in knowledge, food and activity preferences, and home eating habits.

Foods in lunches were classified as heart healthy (target food items in the lessons) or non-heart healthy. Figure 8.2 presents the nutrition behavior changes for Study 1. At both schools the average daily number of heart-healthy foods in students' lunches increased during the program, remained high following the program, and maintained above baseline levels when follow-up data were collected. The postprogram data were collected at the end of the spring term, while the follow-up data were collected in the fall following summer vacation. Students also increased in knowledge and in reported performance for heart-healthy foods and activities after the program. These results also were maintained at follow-up. In phone interviews parents reported that they also were eating more heart-healthy foods at home both immediately following the program and at follow-up (see Figure 8.3) (cf. Coates, Slinkard, Jeffery, Hovell, Kennigsberg, Note 1). Equally important was the fact that both students and parents reacted favorably to the program.

Figure 8.2. Average Number of Heart-Healthy Items in Students' Lunches in Schools 1 and 2

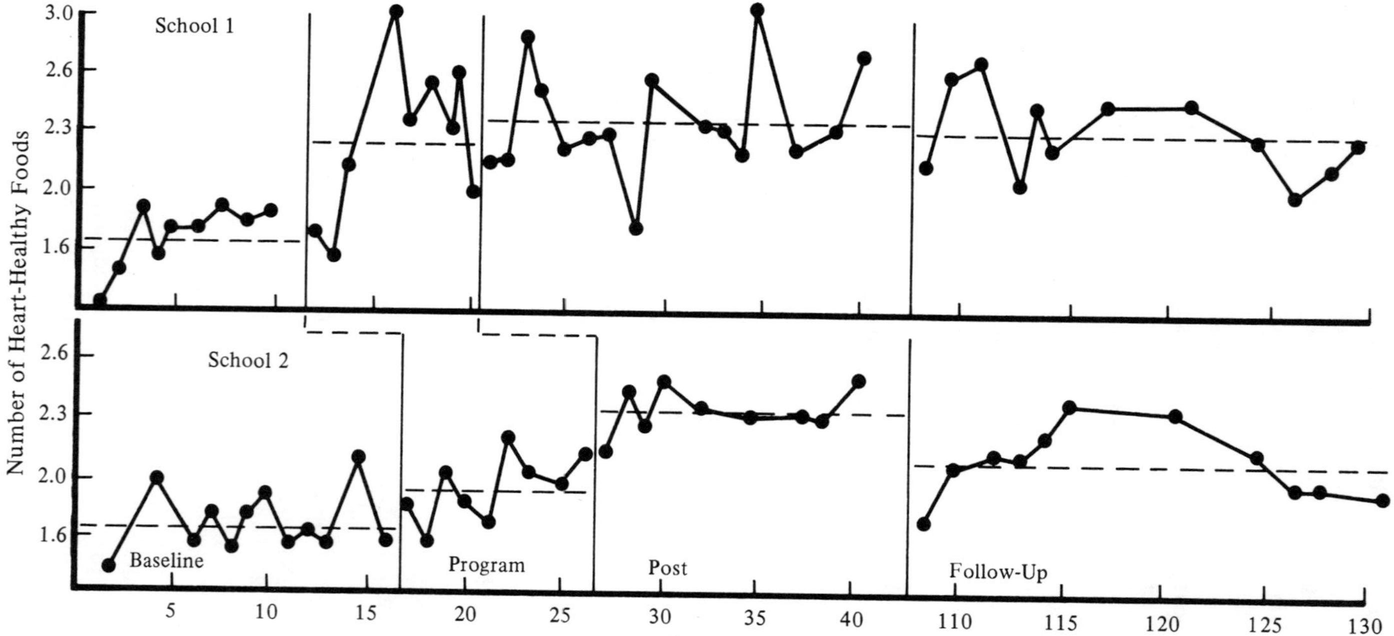

Note. Postprogram data were collected immediately following the end of class sessions, while the follow-up data were collected in the fall following summer vacation (Study 1).

Figure 8.3. Mean Percentage of Heart-Healthy Foods Reported to be Eaten by Family Members in a Single Day Before and Following the School-based Nutrition Program (Study 1)

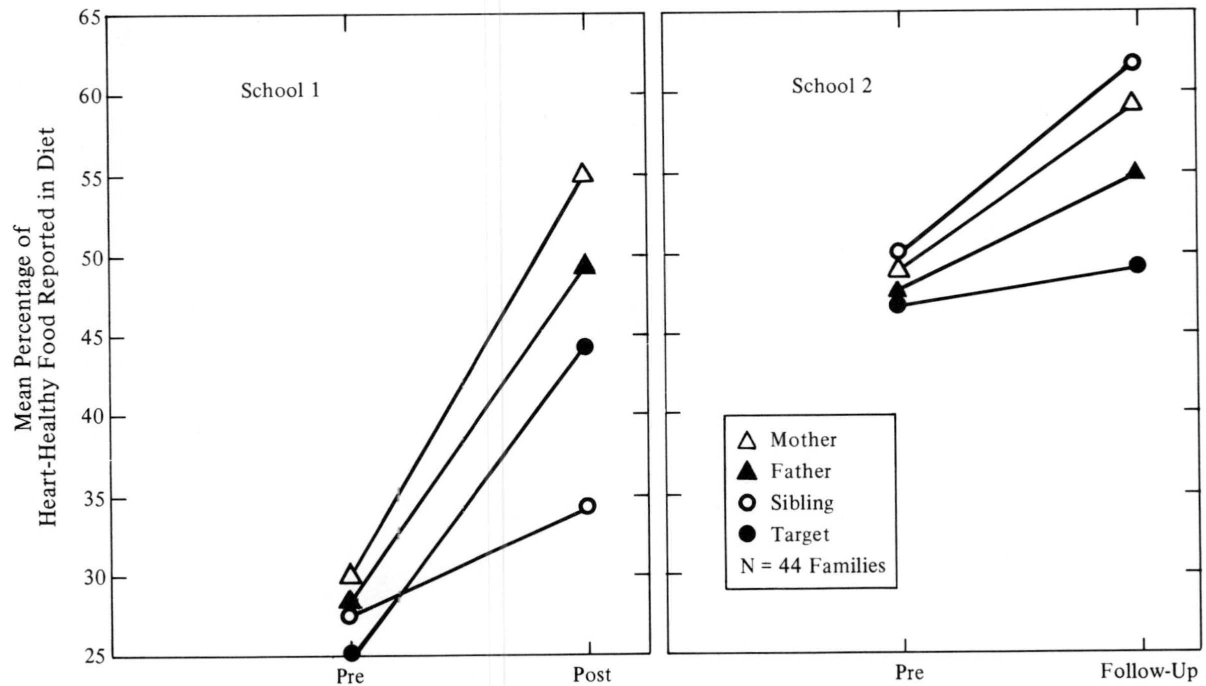

There was one major exception to these positive results: students' lunchtime activity levels did not increase during or after the program. In fact, both boys and girls were quite sedentary throughout. The average time spent sitting and standing ranged from 60% to 70% per day. During follow-up observations in the fall we found that boys' activity levels had increased significantly; they were now sitting and standing only about 40% of the time. The reason was quite simple (as it often is in retrospect): they were playing soccer and football in the fall, while in the spring they were playing baseball. Rather than teaching students specific heart-healthy exercises, we decided that the most effective strategy for helping students to be more active would be to encourage them to play games that involve more activity.

The first year's experience led us to revise the program in the following ways: (a) nutrition and activity lessons were integrated so that students would be encouraged to change both simultaneously; (b) the number of in-class lessons was reduced to two per week; and (c) the third class each week, conducted at lunchtime on the playground, was designed to encourage all students to join in more active games. The sequence of classes during the second year was as follows:

1. Introduction–basic physiology of the cardiovascular system, the heart as a muscle, risk factors, everyday and occasional foods, goal setting.
   Taste Test: fresh vegetables.
2. Exercise–self-assessment of personal activity, importance of exercise for heart health, psychological benefits of exercise, characteristics of heart-healthy exercise, and goal setting.
   Taste Test: protein food used for energy in exercise.
3. Cholesterol–review, cholesterol as a risk factor, LDL and HDL cholesterol, how to increase and decrease cholesterol (food and exercise), everyday and occasional foods, and goal setting.
   Taste Test: pastry made with egg white, vegetable oil, and wheat flour.
4. Saturated and Unsaturated Fat–fat as a nutrient, difference between saturated and unsaturated fat, how exercise affects fat levels, label reading, how to control fat intake, and goal setting.
   Taste Test: fresh fruit.
5. Sugar–simple and complex carbohydrates, sources of sugar, label reading, heart-healthy alternatives, relation of sugar to obesity, disadvantages of obesity, and goal setting.
   Taste Test: food with natural sugar (e.g., trail mix).
6. Salt–a problem when too much salt is consumed, blood pressure and its causes and problems, sources of salt, label reading, alternatives to salt, and goal setting.
   Taste Test: food flavored with spice instead of salt.

7. Cardiovascular System and Exercise—anatomy of heart, veins, and capillaries; what happens when a person exercises; fitness tests; heart-healthy exercise; and goal setting.
   Taste Test: carob chip whole wheat cookies.
8. Overview of Risk Factors—cholesterol, smoking, blood pressure; eating and exercise behaviors related to each; alternatives to common foods and sedentary patterns; diploma; and goal setting.

This program is currently being evaluated with three fourth-grade classes at one school and two fifth-grade classes at the second school. Short-term results in nutrition, presented in Figure 8.4, were promising. To evaluate the program, we used a time series design with time-lagged controls and an untreated control. Three fourth-grade classes at School 1 and two fifth-grade classes at School 2 were the first and second experimental units. The third experimental unit, a fifth-grade class at School 2, controlled for the reactive effects of observation and information alone. They received instruction in the program only after the experimental classes were finished. They also received only the information about target foods and exercises; goal setting and feedback were not used with this class. As shown in Figure 8.4, Groups 1 and 2 changed significantly, while the control class did not. Unfortunately, students again did not modify activity patterns. As with the previous program, the long-term outcomes remain the key to the entire program.

This program is significant in two important respects. First, it employs behavioral principles within the context of regular educational programs to promote behavior change among all students. Working with groups of students rather than individuals may have enhanced program effects. Undoubtedly, students modeled and reinforced changes in fellow students. Second, it demonstrated the potential efficacy of using in-school programs to promote family change. Both student and family change were essential in ensuring the maintained changes we observed at follow-up.

SMOKING PREVENTION FOR JUNIOR HIGH STUDENTS: PROJECT CLASP[3]

Project CLASP (Counseling Leadership About Smoking Pressures) was developed to prevent the onset of smoking among adolescents. The components of the program, as derived from the social learning model, are presented in Table 8.4. The program differed from the traditional smoking education programs in five ways. First, the program did not emphasize the negative effects of smoking on health. Students already know the evils of smoking; the elementary school health programs and mass media have

[3] This program was developed by Cheryl Perry, Alfred McAlister, Lee Ann Slinkard, and Joel Killen (see also McAlister et al., 1979).

Figure 8.4. The Ratio of Heart-Healthy to Other Foods in Students' Lunches (Study 2)

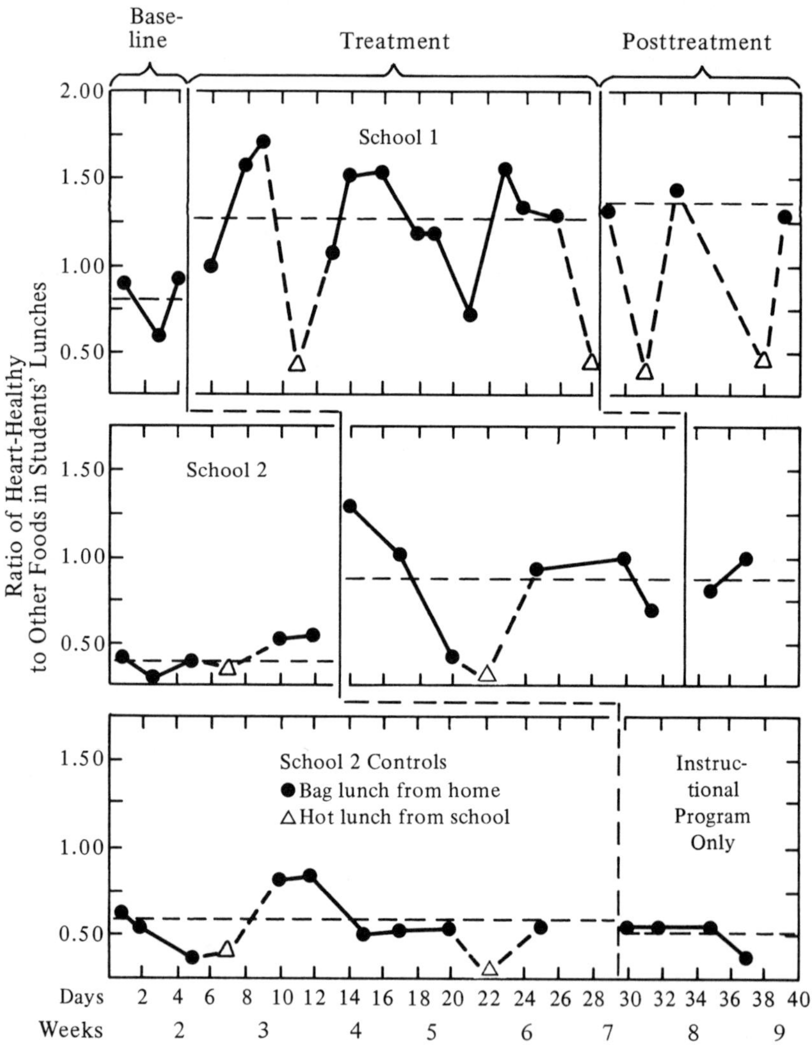

transmitted that knowledge quite adequately (Thompson, 1978). Second, the program exposed students to the social pressures to smoke tobacco and taught them skills for dealing with these pressures. Adapted from works by Evans (1976) and Evans, Hansen, and Mittelmark (1977), this orientation provided that students be "inoculated" with a small dose of the "disease" (social pressures) and an effective "antidote" (social skills). Third, the teams of high school

Table 8.4
Components of the Junior High and High School Smoking Prevention Programs

Behavioral Sources of Influence	
Target behaviors:	To identify pressures to smoke and ways to deal with them. To get feedback on own health and learn methods to help other people quit or not start smoking.
Behavioral rehearsal:	Devise and recite methods to resist advertising and peer pressure.
Behavioral commitment:	Indicate verbally to class how to resist pressures and how to help others quit or not start smoking.
Environmental Sources of Influence	
Models:	College and high school students' demonstration of biofeedback techniques, monitoring of session to explain immediate health effects.
Family:	Students' discussing ways to deal with parental smoking and parental attitudes.
Feedback:	Immediate feedback on students' carbon monoxide levels, pulse rate, skin temperature, lung capacity, and blood pressure, with nonsmokers and smokers compared.
Personal Sources of Influence	
	Students learning to identify pressures to smoke, learning the immediate effects of smoking, learning skills to quit smoking, and learning ways to help others to quit or not start.

students taught social skills for handling the pressures to smoke to junior high school participants. High school students are excellent models for their younger peers at an age when adults increasingly exert less influence over adolescents. Fourth, modeling, behavioral rehearsal, and role playing were the primary teaching methods employed. Lectures and worksheets were rarely used in the classroom. Fifth, the program treated the entire school population rather than any subgroup identified as being particularly at risk. It was hoped that working with the whole school would help create a nonsmoking school environment that would encourage the use of skills taught in the classroom.

Year 1: Planning the Program

During the first year a committee of teachers, personnel from Stanford University, and staff from the Lung Association designed the basic program. Junior and senior high school students consistently outnumbered adults during the planning stages. The primary outcome of the first year's efforts included scripts for lessons to be conducted in each day of the intervention. A 2-day pilot test of the project helped us to identify and solve some political and logistical problems involved in school-based interventions, class schedules, bussing, lunches, and on-site training and supervision.

Year 2: Implementation

The second year of the program (1977–1978) involved 15 seventh-grade classes (N = 289) at one junior high school. Students at two other junior high schools served as no-treatment controls. Two teams of tenth-grade students from a nearby high school conducted the program in all seventh-grade classes seven times during the school year. Seventh grade was selected because this marks the time when more and more students take up the smoking habit (United States Public Health Service, 1976, 1977). This seems to be associated with the transition from elementary to junior high school. The program was conducted in each seventh-grade class during one regular class period in each treatment day. Three consecutive sessions were held early in the school year before the end of September. Four follow-up sessions were scheduled at 1- to 2-month intervals throughout the school year.

Peer leader recruitment and training. The high school student peer leaders were recruited in several ways. An attractive and popular teacher spoke about the program in her (or his) classes, public address announcements were made daily, and posters were distributed on campus. Interested students filled out an application form indicating their personal interest in being peer leaders. A committee of students, the attractive teacher, and CLASP personnel made the final choice. Eight to 10 students per team were selected.

Formal training of the peer leaders took place in two parts. Two afternoon workshops prior to the September program constituted the first part. During these workshops the students on the committee and CLASP personnel introduced the project and modeled methods for implementing the first 3 days of the program. The peer leaders then chose one of six roles to model:

1. The main peer leader (manager and organizer for the team) who opened the class sessions.
2. The advertising specialist who specified ways that media and advertising influence the decision to smoke and methods to counter these influences.
3. The peer pressure specialist who thought of situations in which peer pressure occurs and methods students can use to handle this pressure.
4. The adult-model specialist who identified attractive and potent adult models who smoke and identified ways to counteract the influence of these models.
5. The cool-rebellion specialist who thought of situations in which young people rebel against authority or smoke as a way to be "cool" and planned possible ways to deal with these situations.
6. The other-drugs specialist who handled questions relating to the use of other drugs and pointed out the similarities between the pressures to smoke and to use other drugs.

A typical day's script had three parts: small group discussion, slide show, and discussion of slide show. The script for each class session outlined the points each peer leader should make. The peer leaders, however, were encouraged to translate the lesson plan into their own language in order to make the points to the class. On the second day of the 3-day program class sessions and methods for handling discipline problems were discussed, modeled, and practiced.

On-site feedback formed the second aspect of training. Stanford graduate students monitored the classroom sessions. Snack breaks and lunches were devoted to problem solving, planning, and reinforcement. These sessions were crucial to maintaining the enthusiasm and energy levels among the peer leaders.

Learning to resist peer pressure. During the first 3 days of the program the seventh-grade students were introduced to the pressures to start smoking tobacco that they will meet and were trained in methods for coping with them. The program began with 3 days of treatment in the fall, just after the students had arrived. This was essential, since without treatment, the number of smokers in the seventh-grade would have tripled by Christmas. This process of the program involved:

1. Testimonials by high school students who experimented with tobacco and who became regular smokers.
2. A warning to the seventh grade students that they could get pressured to smoke and get hooked.

3. A slide presentation on the pressures to smoke, after which peer leaders modeled ways to deal with these pressures.
4. Role-play sessions in which the seventh-grade students practiced coping with these pressures.
5. A commitment by each seventh-grade student to refrain from becoming regular cigarette smokers.

The seventh-grade students learned through this process to identify pressures in their environment that encourage tobacco use and to practice over and over the behaviors for coping with these pressures. The four follow-up sessions reinforced these concepts in more complex social situations by using movies, slides, contests, and small group discussions.

Evaluation

Students in the treatment and in two control schools were asked in a short questionnaire to report smoking rates in September, December, and June. In addition, carbon monoxide breath tests were administered to 50% of the students in September, 10% in December, and 100% in June. The correlation between self-report of smoking behavior and carbon monoxide readings was significant beyond the .001 level. Significant differences between treatment and control schools were found in June, 1978, and in December, 1978 (see Table 8.5).

Table 8.5
Results from Year 2 Test of Junior High School Smoking Program

Percent of students responding yes to the question: "Have you smoked during the past month?"

	Treatment	*Control 1*	*Control 2*
September, 1977	4.0	4.6	10.4
June, 1978	10.9	19.2	17.6
December, 1978	13.8	21.7	28.7

Percent of students responding yes to the question: "Have you smoked during the past week?"

	Treatment	*Control 1*	*Control 2*
September, 1977	2.0	0.9	5.2
June, 1978	5.3	10.3	11.8
December, 1978	8.1	10.4	19.1

Several factors contributed to the initial success of Project CLASP. First, the porject was built around a well-conceptualized theoretical model. Second, the curriculum, while formed from this theoretical framework, was written by adolescents and teachers using language and ideas they understood. Third, the project was given considerable attention by school personnel and community agencies. The school principal, the Lung Association, and local media consistently supported the program. Fourth, the participants in the project were given opportunities for input; this shifted the control of the program away from the project directors and toward the participants and peer leaders. Fifth, the participants, peer leaders, and committees were given both formal and informal feedback on a regular basis. Finally, staff meetings and treatment sessions were not the only times we got together. Parties, dinners, and data-coding sessions were held to extend involvement and establish interpersonal ties. These informal relationships were an important part of the formal agenda of Project CLASP.

SMOKING PREVENTION IN HIGH SCHOOLS[4]

It was anticipated that a smoking program among high school students might encourage them to become peer leaders in Project CLASP. Also, because the smoking rate continues to climb during the high school years (United States Public Health Service, 1976, 1977), this project was designed to prevent the onset of tobacco use among high school students.

The program had five major objectives. Students were to learn to: (a) identify the pressures to smoke in their environment, (b) describe methods and behaviors to deal with these pressures, (c) identify the immediate health hazards of smoking, (d) describe an effective approach to quit smoking, and (e) identify ways in which high school students can take an active role in reducing the incidence of smoking. Finally, the students were to make commitments to reduce the amount of smoking, quit smoking, or not start smoking. Health education teachers and Stanford graduate students served as facilitators for this predominantly student-centered curriculum. The basic components of this program are outlined in Table 8.4.

The program was conducted over 1 week in five separate class sessions. It began with small group discussions on the pressures to smoke tobacco. Each group was responsible for answering a specific question regarding social pressures, which in turn was reported back to the entire class. A

---

[4] This program was developed by Cheryl Perry, Brian Danaher, Joel Killen, and Lee Ann Slinkard.

slide show depicting the various pressures was shown to reinforce the ideas and solutions presented.

The next day was spent on cigarette advertising. Students viewed newspaper ads and then identified: (a) to whom the ad was addressed, (b) what the ad was *really* selling, and (c) ways to verbally counteract cigarette advertising. A class-wide discussion followed each ad. To summarize, the movie *Too Tough to Care* (Sid Davis Productions, 1975) was shown; this definitely added humor to the session since it tears apart cigarette advertising in a blatant and corny manner.

Biofeedback was the topic of the third day. Stanford graduate students administered tests of carbon monoxide, blood pressure, skin temperature, lung capacity, and pulse rate to all the students. Smokers and nonsmokers were compared on each of these measures. A chart displaying these comparisons was hung in the classroom as a visual reminder to the students.

In the final two days the teacher led a discussion on the immediate health effects of smoking and introduced the topic of cessation. Self-management strategies, including self-monitoring, goal setting, imagery rehearsal, and self-reinforcement, were discussed, modeled, and role played. The program ended with small group brainstorming sessions centered on three topics: (a) What can you, as a high school student, do to help younger people refrain from starting to smoke? (b) What can you do to help people your age to quit smoking? (c) What can you do to help people your parents' age to quit smoking?

Both the high school teachers and the Stanford students received training prior to program implementation. Project coordinators modeled the program for them during a 1-hour evening or afternoon session. On-site feedback was given to the teachers and college students during the first class sessions.

Evaluation

The program evaluation consisted of self-report, knowledge questionnaires, and carbon-monoxide breath tests of all students in the health classes in three treatment and two control schools. These measures were taken at the beginning and end of the school semester (September, 1978, and January, 1979). Significant differences between treatment and control schools were found on self-report measures, carbon monoxide breath tests, and the knowledge questionnaire, using χ^2 analysis on the post-treatment data. The results of the knowledge questionnaire are presented in Table 8.6. Of particular interest are the differences in knowledge between treatment and control schools and between smokers and nonsmokers. Table 8.7 presents responses to the questions about smoking, validated through the carbon monoxide tests. These data indicate that the program led to significant changes in knowledge, attitudes, and smoking behavior.

With the expensive and questionable costs of conducting cessation programs, projects aimed at prevention should be taken seriously. This project, using a minimal but behaviorally based treatment, appears to have lowered the rate of smoking among students in the treatment program. The program employed high status models (Stanford students) to conduct a portion of the treatment in an informal manner and structured the other components of the curriculum to be primarily student led. Each day was varied in content and in method and included: small group discussions, slide shows, movie, self-administered biofeedback tests, brainstorming sessions, and modeling and role playing, which served to keep interest focused on the subject matter. The shift to the student's needs, interests, and control differentiates this program from the typical approach which emphasizes health hazards of smoking.

WEIGHT LOSS FOR ADOLESCENTS[5]

The Adolescent Weight Loss Program was designed to help adolescents lose weight and to maintain significant weight loss by learning and practicing (a) healthy eating habits (including low caloric but especially heart-healthy foods) and (b) healthy activity patterns. We have sought to do more than help students change specific habits and lose weight. We also have attempted to teach them a set of analytic problem solving skills so that in the future they will be able to face different eating and exercise problems and manage them successfully.

Our research in adolescent weight loss has been progressing over the past 5 years (Coates & Thoresen, in press–b; Thoresen & Coates, 1978). Two important realizations have guided our efforts. Despite the fact that we labeled ourselves behaviorists, we recognized that we had consistently failed to take advantage of the powerful influence of models and behavioral rehearsal in designing our treatment programs (Jeffery & Coates, 1978). We also learned, once again, the importance of critical distinctions between *learning* weight loss skills, *practicing* those skills in the short term, and *using* those skills long enough to obtain and maintain clinically significant improvement. Encouraging young dieters to perform, over a short or long period of time, the difficult actions needed to lose clinically significant amounts of weight and to maintain those losses is the difficulty. Adolescents especially seem to experience difficulty in practicing eating and exercise habits that will permit them to lose weight even in the short run (Coates & Thoresen, in press–a).

---

[5] This program was developed by Thomas J. Coates, Robert W. Jeffery, Lee Ann Slinkard, Joel Killen, and Brian G. Danaher (see Note 2).

Table 8.6
Results on the Knowledge Test from the High School Smoking Prevention Program

	Comparisons between Schools			*Treatment and Control Schools Combined*		
	Treatment	*Control*		*Nonsmokers* N=610	*Smokers* N=122	
1. What are the reasons people your age smoke?	2.39	1.96		2.20	2.08	n.s.
2. What happens to your blood pressure if you smoke?	.89	.62	$\chi^2=91.1$ $p<.001$	.79	.57	$\chi^2=23.6$ $p<.001$
3. What happens to the carbon monoxide in your blood?	.87	.60	$\chi^2=91.0$ $p<.001$	.76	.63	$\chi^2=7.47$ $p<.005$
4. What happens to your pulse rate?	.81	.52	$\chi^2=88.3$ $p<.001$	.68	.52	$\chi^2=10.71$ $p<.001$
5. What happens to your skin temperature?	.65	.12	$\chi^2=268.0$ $p<.001$	.38	.33	n.s.
6. What happens to your lung capacity?	.88	.69	$\chi^2=49.8$ $p<.001$	.81	.63	$\chi^2=17.9$ $p<.001$

7. Is it difficult for people your age to quit?	.50	.52		.50	.48	n.s.
8. What is the best way to quit?	.81	.52	$\chi^2=26.6$ $p<.001$	.67	.57	$\chi^2=4.79$ $p<.05$
9. What is your general opinion of smoking?	3.44	3.26		3.55	2.51	$\chi^2=10.93$ $p<.005$
10. What can a high school student do to reduce smoking?	.88	.52	$\chi^2=66$ $p<.001$	.74	.54	$\chi^2=8.49$ $p<.005$
Total Score =	12.12	9.32	$p<.01$	8.86	11.11	$p<.01$

Note. A complete description of the scoring procedures is available from the first author upon request.

Table 8.7
Rates of Reported Smoking in Treatment and Control Schools following the High School Smoking Program

Percent responding yes to the question: "Have you smoked in the past month?"			
	Treatment	*Control*	χ^2
September	29.2	26.3	
January	23.6	30.4	4.98, $p<.05$
Percent responding yes to the question: "Have you smoked in the past week?"			
	Treatment	*Control*	χ^2
September	19.5	21.6	
January	16.3	21.9	4.47, $p<.05$
Percent of students with carbon monoxide breath tests greater than 10 parts per million.			
	Treatment	*Control*	χ^2
Postprogram	5.2	13.3	17.9, $p<.001$

Teaching Skills and Maximizing Performance

During the past 2 years we have devoted our efforts to developing procedures to teach weight loss skills effectively and to maximize student performance of those skills. Table 8.8 presents an integrated overview of the components in the weight loss program based on the social learning model. We developed a series of videotapes to model important weight loss skills. Concepts and skills such as stimulus control, aerobic exercise, food portion, and how to interact with parents were presented more concretely and dramatically by example rather than by lecture. Behavioral rehearsal, rather than conversation, became the primary vehicle by which students learned and mastered these important skills.

Developing strategies to help students use the weight loss skills required more ingenuity. Our analysis of the literature suggested that focusing on weight loss (rather than habit change alone) and producing weight loss early in a program could be important (Coates & Thoresen, in press–a; Jeffery, Thompson, & Wing, Note 3). Wing and Jeffery (in press) found that large monetary contracts for weight reduction or habit change enhanced short-term losses significantly. Jeffery and Wing (1979)

Table 8.8
Components of the Adolescent Weight Loss Program

Behavioral Sources of Influence	
Target behaviors:	Problem solving focus: To analyze individual eating patterns and then to devise and try out various solutions.
Behavioral rehearsal:	Focus on one basic skill per class; emphasis on opportunities to practice important skills with feedback.
Behavioral commitment:	Students' (and parents') completion of a written agreement each week, specifying behavior change goals to be accomplished during that week.
Environmental Sources of Influence	
Models:	Videotape presentations of skills critical to controlling eating and exercise; testimonials presented by former students and parents who were successful; successful students used as models for other students.
Family:	Parents attending separate sessions to learn what they must do to support children's weight loss; parents given handouts of materials presented in class; contingency system implemented to reward parents for making specified behavior changes at home.
Feedback:	Students coming in each morning or afternoon to weigh in and to compare weights with previous days and weeks; students encouraged to graph weight and post it in a public place.
Incentives:	Students depositing sums of money at beginning of program, which is refunded throughout program for progress (weight loss or reported habit change).
Personal Sources of Influence	
	Knowledge of weight loss skills, cardiovascular benefits of appropriate eating and exercise; learning skills for refusing food offers, negotiating with family and friends, ordering in restaurants, and identifying and solving personal eating and exercise problems.

reported that frequent treatment contact promoted more rapid habit and weight changes than less frequent contact. We reasoned that contingency contracts for weight loss might provide students with the focus and motivation needed to experience success and then help them to continue losing weight.

Year 1: Frequency of Contact and Monetary Rewards

We have conducted two studies to date to explore these promising leads. In the first study we used a 2 X 2 factorial design to examine the relative efficacy of two reward contingencies (for weight loss vs. habit change) and two levels of therapeutic contact (one vs. five times per week). Basic personal influence skills were taught in ten 1-hour lessons; five additional meetings were devoted to problem solving and practice in weight loss skills.

Thirty-six adolescents, 13 to 17 years of age, ranging from 9% to 100% above average weight for height, age, and sex, were randomly assigned to one of the four treatment groups. All subjects were required to deposit the equivalent of 15 weeks of their allowance or estimated earnings from part-time employment. Subjects in groups stressing weight loss received deposit refunds for achieving weight loss of at least 1 pound per week. In the habit change groups subjects received refunds for keeping caloric intake below an individually established goal level. Subjects in the daily contact groups visited the laboratory at Stanford each morning or afternoon to receive refunds for meeting weight loss or habit change goals. Weekly contact subjects visited once a week for refunds, if earned. Subjects could earn back their entire deposit with perfect adherence to goals. Students in the Daily-Weight Loss Group lost an average of 15.1 pounds (SD = 13.2) during the program, while subjects in the other three groups lost 50% or less weight on the average (Daily-Habit Change: M = 8.4, SD = 14.6; Weekly-Weight Loss: M = 4.3, SD = 8.4; Weekly-Habit Change: M = 5.7, SD = 19.8). Students returned 6 months later for follow-up assessments. Because of changes in height and age, we used comparisons of changes in percent overweight to evaluate the short- and long-term efficacy of the program. Data indicating percent overweight at the beginning of the program, 15 weeks later at the end of treatment, and at the 6-month follow-up are presented in Figure 8.5. A 2 (frequency) X 2 (reward) X 3 (pre-post-follow-up) repeated measures analysis of variance revealed a significant trials effect only ($F(2, 58) = 3.07, p < .05$). Paired comparisons, analyzed by t-tests, revealed that the Daily-Weight Loss Group was the only group to show significant changes from pre- to posttreatment and from pretreatment to follow-up.

Figure 8.5. Percent above Ideal Weight Before and Following the Program and at 6-Month Follow-Up for Students in Each of the Treatment Groups (Study 1)

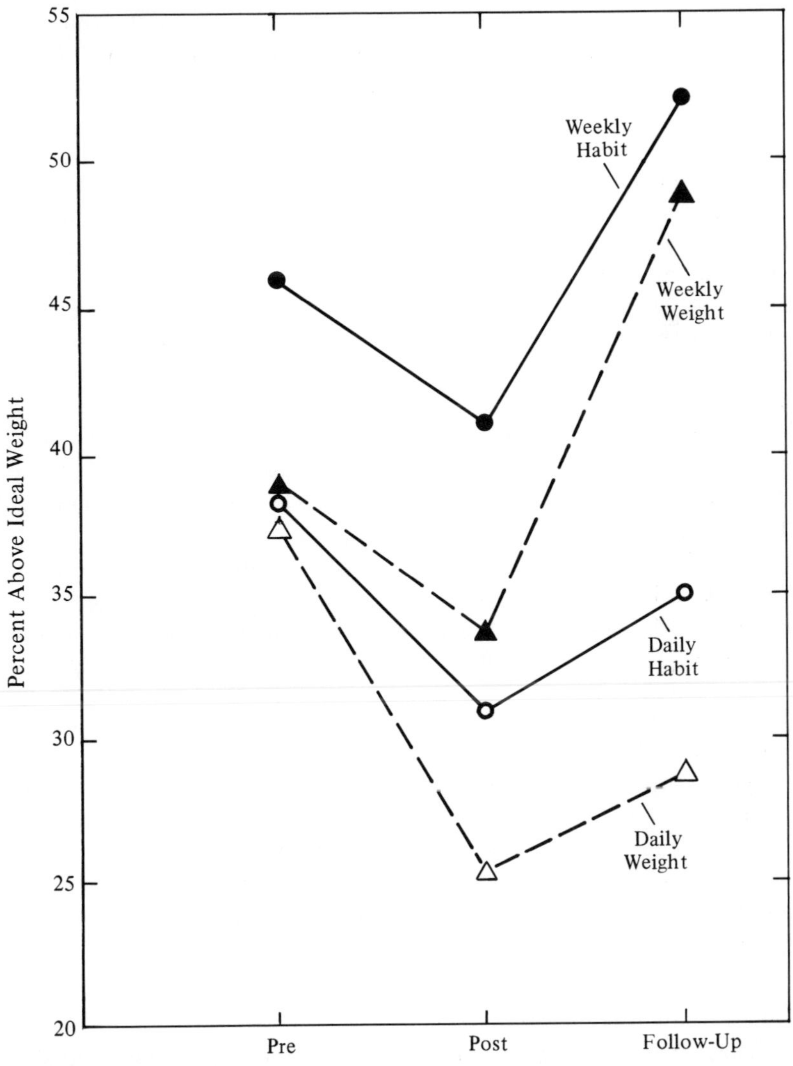

Year 2: Parent Training

The next phase was designed to determine if training for parents would enhance environmental support and ultimately weight loss and habit change. Students were divided into two groups: those whose parents attended classes and those whose parents did not attend classes. All students attended classes once weekly and weighed in 5 days per week. They deposited $100 and could earn back $1 per day on each day that their weight was 1 pound below their weight 1 week ago on the same day.

Parents attended a weekly class, separate from their sons and daughters, to learn skills necessary for enhancing and supporting their children's weight losses. Classes were separate because we wanted each group to develop its own source of identity and because parents and students each had separate sets of skills to learn and practice. The parents and students were brought together only when we determined that a joint class would enhance behavior change and weight loss. This determination was made by deciding if the skills to be targeted required parents and children to learn and practice them together. Examples included learning how to negotiate goals and rewards (Session 7) and how to reinforce each other when ordering food in restaurants (Session 13). These outcomes suggested that a program combining training in personal influence skills with environmental support to encourage use of those skills might be the most effective with the largest proportion of students. The sequence of classes for parents and adolescents is presented in Table 8.9. Classes for both parents and students emphasized heart-healthy nutrition and exercise, as well as behavioral weight loss skills. Parents' roles in providing a heart-healthy home environment were stressed. In classes held separately and together we conducted training in problem solving and mutual reinforcement. We wanted adolescents and parents to learn how to communicate about problems relative to program goals, to devise and to implement solutions, and to mutually reinforce each other's progress and efforts.

The short-term results parallel those found in the Daily-Weight Loss Group in Year 1. Both groups responded to the powerful short-term treatments of frequent contact and reinforcement for weight loss. We expect, however, that differences will be noted at follow-up; students whose parents participated will have maintained weight losses, hopefully because of changes in their home environments.

TAKING CARE OF THE "HEART"

These four programs represent a beginning. From our results, it appears that: (a) behavioral principles can be developed into practical programs, (b) such programs can be administered within regular school channels and structures, (c) all students can participate, and (d) programs can lead to health behavior, knowledge, and attitude change that is maintained. The programs represent an important extension of behavioral group therapy. Obviously, their implementation and ultimate efficacy

Table 8.9
Class Schedule for Adolescent Weight Loss Program

Week	*Students*	*Parents*
1	Introduction–calories and weight	Introduction–calories and weight
2	Reinforcing yourself and your parents for change	Reinforcing yourself and your children for change
3	Exercise	Exercise
4	What to eat–calories and heart health	What to eat–calories and heart health
5	Problem solving	Problem solving
6	Problem solving	Problem solving
7	Problem solving workshop–parents and students together	
8	Fat and thin thoughts	Positive and negative thoughts about your children
9	Using imagery for problem solving and self-control	Using imagery for problem solving and self-control
10	Pot luck barbecue	
11	Problem solving workshop–parents and students together	
12	Exercise workshop	Reinforcing exercise
13	Eating in restaurants	
14	No class	
15	Criteria for success and self-evaluation	Criteria for success
16	No class	
17	Individual interviews and problem solving	
18	No class	
19	Final class–disco party and buffet	
20	Final weigh-in	

rely heavily on group processes including mutual modeling, mutual reinforcement, and behavioral rehearsal. The programs presented here demonstrate that it is possible to use these same principles and processes to achieve change in large numbers of children and adolescents in educational settings.

The methodology, however, represents only one part of the program. There is another dimension—the "heart"—mentioned only in passing and certainly more difficult to replicate. To reproduce the programs without consideration for the "heart"—enthusiasm, energy, and emotional commitment—may not lead to the same results. These more elusive factors may in fact be the life blood which makes programs work and which gives them impact and power.

We did not develop the program in isolation from the ultimate consumers or the persons who had practical experience with the students. The entire first year of Program CLASP was spent working out the details of the program with students, teachers, and parents. Theoretical principles were translated into strategies and scripts that were specifically designed to be relevant to student recipients.

We have tried to make the programs fun and enjoyable for the participants. If students dislike the presence of researchers from outside their school, they are not likely to respond strongly to even the most cleverly designed program. Even a neutral response may not be sufficient to motivate change. By contrast, enthusiasm and positive anticipation help students stay attentive during sessions, maintain a sense of identity with the program, regard the teachers highly, and take seriously the information presented. Enthusiasm and positive anticipation may even be necessary antecedents to short- and long-term behavior change.

Some components of the programs are designed solely to make them fun and enjoyable. In the Heart Health Program, for example, the teachers wore T-shirts with the heart logo. We passed out food frequently in class and allowed the students to practice heart-healthy activities outside. Games, small group activities, and charts and posters were used to provide variety and a break in normal class routines. Program CLASP departed from traditional lessons in its liberal use of nontraditional teaching techniques: peer leaders, media presentations, models, and role plays. Even a dated movie served a purpose. Students could laugh at the movie and even learn a few things while doing it. In the weight loss program we organized a barbecue and also dinner at a local restaurant.

Implementing any of these programs requires the recruitment and coordination of large numbers of persons. The programs also take time; persons must be prepared to work over a long period. Obviously, these persons are key. Their enthusiasm, commitment, and energy can mean the difference between success and failure in any specific program.

Two activities helped us to maintain energy and commitment. First, we encouraged the idea that participating in the programs is a privilege not open to everyone for the asking. We received a large number of applications for peer leaders, but there were relatively few positions available. Those chosen had to work, but they also enjoyed the advantage of certain rewards such as release from school, trips to present the program to other groups, and dinners and lunches with program leaders. Undergraduate and graduate students worked on the other three programs. Maximum responsibility such as teaching classes required maximum time. Persons wanting to work with us had to first spend time assisting with data collection, editing, and transcription. Working on program development sometimes came next, while teaching was reserved for those who demonstrated that they could accept the responsibilities and carry them out effectively.

Finally, we tried to encourage involvement and build interpersonal ties among program leaders. We believe that such efforts build spirit and cooperative behavior among a team that will influence the way the program is carried out. The enthusiasm of the leaders cannot help but influence the students. Program leaders participated in the development of the program and were encouraged to add new elements to the program, as well as to express opinions and criticisms. Parties, lunches, and dinners were scheduled frequently to promote informal and candid interactions in settings other than research meetings.

Health and illness are, in many ways, behavior problems. Young people and the whole society need effective programs. We hope that our modest efforts can contribute to their realization.

Reference Notes

1. Coates, T. J., Slinkard, L. A., Jeffery, R. W., Hovell, M., & Kennigsberg, M. The heart health program: Improving nutrition and activity among elementary students. Unpublished manuscript, Johns Hopkins Medical School, 1979.
2. Coates, T. J., Jeffery, R. W., Slinkard, L. A., Killen, J. D., & Danaher, B. G. Monetary incentives and frequency of contact in weight reduction in adolescents. Unpublished manuscript, Johns Hopkins Medical School, 1979.
3. Jeffery, R. W., Thompson, P., & Wing, R. Effects on weight reduction of strong monetary contract for calorie restriction or weight loss. Unpublished manuscript, Stanford University, 1978.

References

American Heart Association. *Heart facts.* Dallas, Tex.: Aita, 1976.

Bandura, A. Self-efficacy: Toward a unifying theory of behavior change. *Psychological Review,* 1977, *84,* 191–215. (a)

Bandura, A. *Social learning theory.* Englewood Cliffs, N.J.: Prentice-Hall, 1977. (b)

Bandura, A. The self-system in reciprocal determinism. *American Psychologist,* 1978, *33,* 344–358.

Blackburn, H. Progress in the epidemiology and prevention of coronary heart disease. In P. Yu & J. F. Goodwin (Eds.), *Progress in cardiology.* New York: Lea & Febiger, 1974.

Blackburn, H. Physical activity and cardiovascular health: The epidemiological evidence. In I. Landry & L. Irban (Eds.), *Physical activity and human well-being.* Miami: Symposia Specialists, 1978. (a)

Blackburn, H. Prevention of coronary heart disease: The need, potential, strategy, impediments, and implications. In G. Schettler (Ed.), *Changes in the medical panorama.* Stuttgart, Germany: George Thieme, 1978. (b)

Coates, T. J., & Thoresen, C. E. Obesity in children and adolescents: The problem belongs to everyone. In B. Lahey & A. Kazdin (Eds.), *Advances in clinical child psychology.* New York: Plenum, in press. (a)

Coates, T. J., & Thoresen, C. E. Treating obesity in children and adolescents: Is there any hope? In J. M. Ferguson & C. B. Taylor (Eds.), *A comprehensive handbook of behavioral medicine.* Jamaica, N.Y.: Spectrum, in press. (b)

Creswell, W. H., Hoffman, W. J., & Stone, D. B. *Youth smoking behavior: Characteristics and their educational implications.* Urbana, Ill.: University of Illinois Anti-Smoking Education Study, 1970.

Davis, S. (Producer). *Too Tough to Care.* Santa Monica, Calif.: Sid Davis Productions, 1975. (Film)

Evans, R. I. Smoking in children: Developing a social-psychological strategy of deterrence. *Journal of Preventive Medicine,* 1976, *5,* 122–127.

Evans, R. I., Hansen, W. B., & Mittelmark, M. B. Increasing the validity of self-reports in a smoking in children investigation. *Journal of Applied Psychology,* 1977, *62,* 521–523.

Farquhar, J. W., Maccoby, N., Wood, P. F., Alexander, J. K., Breitrose, H., Brown, B. W., Haskell, W. L., McAlister, A. L., Meyer, A. T., Nash, J. D., & Stern, M. P. Community education for cardiovascular health. *Lancet,* 1977, *1,* 1192–1195.

Golubjatnikov, R., Paskey, T., & Inhorn, S. L. Serum cholesterol levels of Mexican and Wisconsin school children. *American Journal of Epidemiology,* 1972, *96,* 36–39.

Horn, D., Courts, F. A., Taylor, R. M., & Solomon, E. Cigarette smoking among high school students. *American Journal of Public Health,* 1959, *49,* 1497–1511.

Jeffery, R. W., & Coates, T. J. Why aren't they losing weight? *Behavior Therapy,* 1978, *9,* 856–860.

Jeffery, R. W., & Wing, R. Frequency of therapist contact in the treatment of obesity. *Behavior Therapy,* 1979, *10,* 186–192.

Kandel, D. Stages of involvement in adolescent drug use. *Science,* 1975, *190,* 912–914.

Kannel, W. B., & McGee, D. C. Diabetes and cardiovascular risk factors: The Framingham study. *Circulation,* 1979, *59,* 8–12.

Keys, A. Coronary heart disease in seven countries. *Circulation,* 1970, *42,* 1-211. (Supplement)

Kuller, L. H. Epidemiology of cardiovascular diseases: Current perspectives. *American Journal of Epidemiology,* 1976, *104,* 425–456.

Lauer, R. M., Connor, W. E., Leaverton, S. E., Reiter, M. A., & Clarke, W. R. Coronary heart disease risk factors in school children: The Muscatine study. *Journal of Pediatrics,* 1975, *86,* 697–706.

Maccoby, N., Farquhar, J. W., Wood, P. D., & Alexander, J. Reducing the risk of cardiovascular disease: Effects of a community-based campaign on knowledge and behavior. *Journal of Community Health,* 1977, *3,* 100-114.

Mann, G. V. The influence of obesity on health. *New England Journal of Medicine,* 1974, *291,* 178–185.

Mann, G. V. Diet and obesity. *New England Journal of Medicine,* 1977, *296,* 812.

McAlister, A., Perry, C., & Maccoby, N. Adolescent smoking: Onset and prevention. *Pediatrics,* 1979, *63,* 650–658.

Miller, N. E. The evidence for antiatherogenicity of high density lipoprotein in man. *Lipids,* 1979, *13,* 914–919.

Report of the Task Force on Blood Pressure Control. *Pediatrics,* 1977, *59,* 797–820.

Shapiro, D., & Surwit, R. S. Learned control of physiological function and disease. In H. Leitenberg (Ed.), *Handbook of behavior modification and behavior therapy.* Englewood Cliffs, N.J.: Prentice-Hall, 1976.

Stamler, J. Epidemiology of coronary heart disease. *Medical Clinics of North America,* 1973, *57,* 5–16.

Stern, M. P., Farquhar, J. W., Maccoby, N., & Russell, S. H. Results of a two year health education campaign on dietary behavior: The Stanford three-community study. *Circulation,* 1976, *54,* 826–833.

Thompson, E. L. Smoking education programs, 1960–1976. *American Journal of Public Health,* 1978, *68,* 250–257.

Thoresen, C. E., & Coates, T. J. What does it mean to be a behavior therapist? *The Counseling Psychologist,* 1978, *7,* 1–19.

U.S. Public Health Service. *Teenage smoking: National patterns of cigarette smoking, ages 12 through 18 in 1972 and 1974.* U.S. Department of Health, Education, and Welfare, Public Health Services, National Institute of Health. DHEW No. (NIH) 76-931, 1976.

U.S. Public Health Service. *Cigarette smoking among teenagers and young women.* U.S. Department of Health, Education, and Welfare, Public Health Services, National Institute of Health. DHEW No. (NIH) 77-1203, 1977.

Williams, C. L., & Wynder, E. L. A blind spot in preventive medicine. *Journal of the American Medical Association,* 1976, *236,* 2196–2197.

Wing, R., & Jeffery, R. W. Outpatient treatment for obesity: A comparison of methodology and clinical results. *International Journal of Obesity,* in press.

Chapter 9

Psychoeducational Skills Training for Individuals with Epilepsy

Henne R. Queisser
Hubert E. Armstrong, Jr.
Wayne R. Smith
Gay R. Davis

Abstract

*A psychoeducational skills training program implemented at the University of Washington Epilepsy Center provides individuals with needed skills for greater life competence. Individuals with epilepsy have a high incidence of psychosocial dysfunction which has not been responsive to traditional intrapsychic interventions. The relevance of the psychoeducational skills training approach, the development of the curriculum, the content of courses, and initial evaluation efforts are described.**

A high rate of psychosocial dysfunction has been noted among individuals with epilepsy. A study by Rodin, Shapiro, and Lennox (1977) indicated that over 50% of their sample of epileptic patients displayed social or psychological problems. The psychosocial problems experienced by individuals with epilepsy have been attributed to a variety of factors

---

*This project was supported by NIH Contract Number N01-NS-6-2341, National Institute of Neurological and Communicative Disorders and Stroke, PHS/DHEW, U.S.A.

including psychiatric dysfunction (Bear & Fedio, 1977; Flor-Henry, 1976), an "epileptic personality" (Suter, 1956), cognitive impairment (Rennick, 1975; Rodin, Shapiro, & Lennox, 1977), and reactions to social stigma (Epilepsy Foundation of America, 1975; Goldin & Margolin, 1975). Interventions generally have been limited to such traditional methods as psychodynamic therapy, psychopharmacology, peer counseling, and advocacy.

As an alternative to traditional formulations and interventions, the University of Washington Epilepsy Center offers a psychoeducational skills training program in which many of the psychosocial problems evident in our population are conceptualized as skill deficits. The program's aim is to promote personal and interpersonal competence in dealing with problems in living. In this chapter we will discuss the relevance of the psychoeducational model to the problems presented by epilepsy populations, describe the development of the Skills Project, and describe some of the techniques and interventions taught in our classes.

THE PSYCHOEDUCATIONAL MODEL AND EPILEPSY

The psychoeducational model is a hybrid which integrates social learning theory with educational theory and technology. A basic assumption of the model is that most problematic behavior is learned or reflects the absence of desirable behavioral skills. Most maladaptive behaviors are viewed as the result of inadequate or dysfunctional learning, rather than the result of disease processes, deep-seated conflicts, or enduring personality traits. The implication is that dysfunctional behaviors may be replaced by more adaptive behaviors if the individual is exposed to appropriate learning experiences. Even those maladaptive behaviors which originate from biological dysfunction often may be "overcome or partially compensated through appropriate training in more skillful response alternatives" (Goldsmith & McFall, 1975, p. 51).

In the psychoeducational approach appropriate learning experiences are provided through a skills training paradigm which relies heavily on cognitive-behavioral and educational methodology (Armstrong & Bakker, 1971; Bakker & Armstrong, 1976). An individual's behavioral repertoire is assessed and specific deficits in interpersonal functioning and self-control skills are identified. Needed information and opportunities to practice new behaviors are then provided in an educational atmosphere.

Skills training has been used effectively to ameliorate various psychosocial deficits in a wide variety of populations ranging from relatively well-adjusted college students to chronic, institutionalized psychiatric patients (Chaney, O'Leary, & Marlatt, 1978; Goldsmith & McFall, 1975; Hersen & Bellack, 1976; Twentyman & McFall, 1975; Richey, Note 1). To our knowledge, the delivery of skills training activities in a carefully designed psychoeducational context has been limited to psychiatric populations, though Armstrong and Booth (1979) described its application to an obese clientele. There have been no studies reporting psychoeducational skills training with an epileptic population. This absence is surprising because the skills training model appears particularly applicable to the psychosocial problems associated with epilepsy.

Both the nature of epilepsy and society's reactions to the disorder contribute to a high incidence of skills deficits in the areas of interpersonal functioning, stress management, mood control, problem solving, and memory. For example, individuals with early onset of seizures may experience social stigmatization and rejection during childhood and adolescence. The result is decreased opportunity for the learning of social skills basic to successful functioning as adults. These social skill deficits frequently result in incapacitating social anxiety and dysfunctional interpersonal behavior, such as uncontrolled anger or extreme timidity. Overprotection by families further hinders the acquisition of problem solving and independent living skills, resulting in prolonged dependence on parents well into adulthood. Social isolation and withdrawal are common for people with epilepsy who lack interpersonal and independent living skills. Such individuals have few opportunities for rewarding experiences and frequently suffer from depression, loneliness, and low self-esteem (Libet & Lewinsohn, 1973).

Because frequent and severe seizures and long-term use of anticonvulsant medications are associated with cognitive deterioration (Tarter, 1972; Travers, Note 2), memory difficulties are a common complaint among epilepsy patients. Memory impairments not only create practical and vocational problems, but also contribute to high levels of anxiety, depression, and social isolation.

The unpredictable nature of seizures and the loss of control they represent contribute to dysfunctional moods and behaviors. Chronic anxiety and social withdrawal may be maintained by fear of having seizures in public. Feelings of helplessness, depression, or anger may result from the individual's perception of lack of control over important aspects of behavior. The individual fails to engage in prudent risk taking which is necessary, especially in the absence of appropriate modeling, if one is to learn to competently respond to a variety of problems in living (Seligman, 1975).

Many of the problematic behaviors seen in an epileptic population represent learned avoidance responses and a failure to gain access to opportunities for learning adaptive self-management and problem solving skills. Even those dysfunctional behaviors which have an organic basis, such as cognitive impairment, may be partially compensated for by training in alternative, adaptive behaviors.

THE FIELD SETTING

The University of Washington Epilepsy Center in Seattle, Washington, is one of the five comprehensive epilepsy programs funded by the National Institute of Neurological and Communicative Disorders and Stroke (NINCDS). The Center offers a wide spectrum of medical, rehabilitation, and other psychosocial services. It serves approximately 1,200 outpatients and has a 6-bed inpatient unit. The facility is located at Harborview Medical Center, directly across the street from Harborview Community Mental Health Center, where the Life Skills Program (LSP) is housed.[1] The LSP's close proximity has facilitated active collaboration between the staffs of both programs; indeed, the LSP served as a model for the development of the Epilepsy Center's Skills Project.

The mean age of the Center's adult population is 29.85 years. The mean education is 11.54 years. Most patients are Caucasian; only 4.5% represent other racial groups. About 34% of the patients are married, 44% are single, and the remaining 22% are divorced, widowed, or separated. Approximately 15% of the Center's population have additional developmental disabilities such as mental retardation or cerebral palsy.

As a tertiary referral facility, the Epilepsy Center treats a patient population which is likely to have complex and often interrelated medical and psychosocial problems. The majority of adult Center patients, about 64%, experienced the onset of their seizures before the age of 20, with mean age of onset being 16.7 years. Mean duration of the seizure disorder is 12.75 years. Analysis of a sample of 50 new patients indicates that 64% had one or more seizures per month on the average over the 6 months preceding their initial clinic visit. Thus, the Center's patients have generally severe and long-standing seizure disorders.

The Center's population exhibits a relatively high rate of psychosocial dysfunction. Over 48% are unemployed. Twenty-four percent have a history of chronic psychiatric problems needing inpatient care, and an additional 41% have experienced intermittent emotional difficulties requiring outpatient treatment. Almost 40% have a history of impulse control problems such as substance abuse, criminal activities, violent outbursts, or suicide attempts. Twenty-one percent are not functioning independently for their age and are living with parents or in institutions.

DEVELOPMENT OF THE SKILLS TRAINING PROJECT

In response to the needs expressed by both patients and Center staff, the Skills Project was initiated in July, 1977. The project offers patients

---

[1] See Chapter 6 in this volume.

various classes organized around relevant psychosocial target areas. We highlight the educational format of the Skills Training Project by calling it a School for Behavior Change, by stressing that participants are taught techniques for changing their own behavior, and by communicating our confidence that participants' goals can be reached through the techniques taught in class. This educational/skills training perspective allows us to avoid the negative implications many of our patients associate with traditional psychiatric interventions. To them, the "treatment" of "mental illness" is unacceptable and a misnomer. We emphasize that behavioral problems are a reflection of a failure to acquire desirable life skills, rather than symptoms of an emotional disorder.

By means of a structured interview with 36 of the Center patients who initially indicated interest in the skills training program, we identified problem areas our patients most frequently endorsed, and we subsequently developed the following classes: Problem Solving, Remembering More, Relaxation and Stress Management, Dealing with Depression, Assertiveness Training, Decreasing Depression/Increasing Assertiveness, Shyness and Conversation Skills, and Communication Skills.

During the first 1½ years of the Skills Project's operation these classes have been presented to more than 80 individuals for a total of approximately 1,100 participant-hours. Classes typically meet 1½ hours per week for 6 to 8 weeks. Participants in the classes are self-referred as a result of announcements placed in the outpatient clinic and inpatient unit, or they are referred by clinic social workers, vocational rehabilitation counselors, neuropsychologists, or physicians. A screening interview is conducted to assess the student's mental status, verbal and written skills, current behavioral problems, goals, and desire to participate. Requirements for enrollment in classes include that the patient: (a) desires to change some aspect of his behavior, (b) is able to participate in class without being disruptive, and (c) is not too cognitively or psychiatrically impaired to benefit from participation. The latter category includes those individuals who are unable to read and write, who have IQs below 80 and/or have severe memory impairments (neuropsychological test results are available for most Center patients), or who are actively psychotic. These individuals are referred to more appropriate programs in the community. Only about 5% of potential students screened have not met enrollment criteria. A comparison between class participants and nonparticipating adult clinic patients indicates that patients who particpate in the skills classes tend to be younger (mean age is 27 years), have slightly more education, are more likely to have never been married, have a higher frequency of current unemployment, and have a higher incidence of past psychiatric treatment.

For each class there is a multilevel evaluation procedure. This consists of attempts to document each patient's level of participation in the class: whether or not the didactic presentations are understood, whether or not competence in the target skills is displayed in class, the extent to which generalization occurs in situations outside class, and the degree to which the behavior changes affect the quality of the student's life and seizure frequency.

The decision to include seizure frequency as an outcome variable stems from observations in the literature that emotional tension may trigger seizures (Standage, 1972) and that behavioral techniques, including enhancement of certain self-control skills, may significantly improve seizure control (Cabral & Scott, 1976; Daniels, 1975; Mostofsky & Balashak, 1977; Parrino, 1971; Wells, Turner, Bellack, & Hersen, 1978; Zlutnick, Mayville, & Moffot, 1975). Thus, we reasoned that reduction of tension, whether accomplished directly through stress management training or indirectly as a result of improved personal and interpersonal functioning, could decrease seizure frequency.

CURRICULUM

Behavioral Rehearsal Classes

Rehearsal of new social behaviors is emphasized in three of our classes: Communication Skills, Shyness and Conversation Skills, and Assertiveness Training. Basic principles of behavior therapy are implemented in these classes: problematic or deficit behaviors are pinpointed; alternative responses are discussed and specified; and with the aid of modeling, coaching, and feedback from class instructors and other participants, individuals practice new responses, usually in a role-play format. Videotape feedback is employed. Care is given to begin rehearsal with very specific and limited behaviors which promise easily attainable success and to gradually work toward more complex tasks. Each successfully completed step is positively reinforced by praise from instructors and class participants. Self-praise and other positive cognitions are actively encouraged. Participants are urged to try these new behaviors outside of class and to report back at future sessions.

Communications Skills class. In the Communications Skills class, participants first identify significant people in their lives with whom they would like to be more comfortable and communicate better. Students are taught principles of active listening, effective communication, and negotiation skills. Role playing in an exploratory context then allows participants to learn new ways of approaching these individuals. Learning to communicate better with parents and other family members is a commonly expressed goal of students in this class. Self-report data collected at the end of the course consistently indicate improved relationships with target individuals.

Shyness and Conversation Skills class. The Shyness and Conversation Skills class allows participants to pinpoint deficits in their own conversation skills, such as introducing new topics, ending conversations, asking questions, and paraphrasing (Gambrill & Richey, 1976). Behavioral rehearsal with the aid of videotape replay is employed to increase competence in these skills and desensitize participants to the interpersonal risks involved in engaging in conversations and making new acquaintances. Relaxation techniques and the use of positive self-statements are taught as methods to decrease the anxiety experienced in social situations. Lectures and discussions provide information on activities and places to visit to maximize social encounters. Informal analysis of videotapes appeared to show an increase in conversation skills. More formal evaluations of the effectiveness of this class are currently under way.

Assertiveness Training class. The Assertiveness Training class is aimed at developing appropriate responses to a variety of troublesome social situations in which the student judges that her response is less than satisfactory. The class follows the format of typical assertion training groups reported in the literature (Galassi, Galassi, & Litz, 1974; Lomont, Gilner, Spector, & Skinner, 1969). Participants first respond to a series of vignettes requiring an assertive response and are asked to rate the degree of difficulty they would expect to experience in each situation. An inventory of target situations is thereby generated for each student. Situations requiring students to inform others of their seizures and situations which involve asking for help, particularly related to being incapacitated by a seizure, are most frequently endorsed as problematic by our class participants.

Lectures and structured exercises are used to train the students to reliably discriminate between assertive, nonassertive, and aggressive responses. In one such exercise students are given a series of 15 verbally presented situations (Lange & Jakubowski, 1976) that require some degree of assertiveness, along with a sample response; they are asked to label the response as either assertive, aggressive, or unassertive. After simply reviewing and discussing the responses, seven of our first eight students improved their ability to accurately discriminate the types of responses in another set of sample situations. Students are next trained to recognize the faulty or irrational beliefs which implicitly underlie unassertive and aggressive behavior and then are taught to actively challenge these beliefs (Lazarus & Fay, 1975). After students have grasped what assertiveness is and have learned to identify some of their own unassertive beliefs, the basic skills for behaving assertively are explained and demonstrated. These skills include recognizing annoyance and irritation, identifying and keeping to a single issue, using phrases such as "I would like" and "I think . . . ," using phrases which simultaneously communicate courteousness and seriousness, and employing bargaining skills. The remainder of the sessions are devoted to behavioral rehearsal of situations that students have identified as troublesome for them.

Measures of student attendance and verbal participation in the Assertiveness class suggest that this is one of the most popular classes with patients of the Center. Participants most frequently indicate that behavioral rehearsal is the most helpful aspect of this class, as well as the Communications Skills and Shyness and Conversation Skills classes.

Depression Classes

Dealing with Depression class. The Dealing with Depression class teaches the relationship between behavior and moods, as well as techniques for establishing new habits that are likely to enhance participants' positive moods. Interventions for this class are derived from several theories of depression: Beck's cognitive model (Beck, Rush, Emery, & Shaw, 1979); Seligman's theory of learned helplessness (Seligman, 1975); and Lewinsohn's pleasant events model (Lewinsohn & Graf, 1973), which postulated that a low rate of response-contingent positive reinforcement is responsible for depressive behaviors. Didactic material on these theories adapted from Hays & York (1978) is presented to class members. Cognitive interventions include thought stopping (Rimm & Masters, 1974); rational-emotive therapy techniques, which involve analysis and challenging of faulty beliefs (Ellis & Harper, 1975); and the replacement of negative thoughts with positive self-evaluative cognitions (Mahoney, 1971; Todd, 1972). Opportunities for positive reinforcement are increased by encouraging participants to schedule pleasant events and by devoting a portion of each class session to assisting students in choosing realistic pleasant activities for the following week. Completion of scheduled events is supported by praise from class members and instructors.

The Depression Adjective Checklist (DACL) (Lubin, 1965) and the Beck Depression Inventory (BDI) (Beck, Ward, Mendelsohn, Mock, & Erbaugh, 1961) are used to evaluate the course's impact. For our first seven participants, the DACL has indicated no significant changes between preclass and follow-up administrations. However, the BDI indicates significant reductions in depression at the 8-week follow-up as compared to preclass levels ($t\ (6) = 2.95, p < .05$, two-tailed).

Decreasing Depression/Increasing Assertiveness class. Because some studies have indicated a relationship between unassertiveness and the maintenance of depression (Wolpe & Lazarus, 1968), we have offered a combined assertiveness/depression course. In the Decreasing Depression/ Increasing Assertiveness course the basic interventions of the Dealing with Depression class are supplemented by didactic presentations on assertive techniques and inclass practice of assertive responses. The Rathus Assertion Schedule (Rathus, 1973), the Depression Adjective Checklist, and the Beck Depression Inventory are used to evaluate the outcome of this class.

Both depression measures indicated that the eight participants were experiencing relatively low levels of depression at the onset of the first class, and this level did not change significantly at 8-week follow-up administration. The results of the Rathus Assertion Schedule indicated some increased assertiveness at the 8-week follow-up, but this was not statistically significant.

Relaxation and Stress Management Class

In the Relaxation and Stress Management class relaxation, desensitization, and cognitive self-control techniques are taught to participants to foster effective tension control in stressful situations. Individuals are trained to recognize and monitor their tension levels and precipitants of stress. Progressive relaxation (Bernstein & Borkovec, 1973; Wolpe, 1958), deep breathing exercises, and pleasant imagery are practiced in class. The relationship between thoughts and the experience of stress is discussed. Active challenging of faulty beliefs which heighten or precipitate anxiety is presented. Participants are encouraged to use coping statements to replace negative cognitions such as "I can't handle this" with more positive thoughts such as "I've dealt with situations like this in the past and I'm sure I can handle this now" (Meichenbaum, 1977). Although a few participants have experienced seizures during the progressive relaxation exercises—a finding which is consistent with recent research in this area (Lubar, Note 3)—most class members have felt that relaxation and deep breathing exercises are the most helpful elements of this class. The use of positive self-statements is rated the next most useful. As yet, our preliminary data have indicated that the class has had no impact on seizure frequency.

Remembering More Class

A concern frequently voiced by the Center's adult epileptic patients is their increased forgetfulness. Therefore, it is not surprising that the Remembering More class has been most frequently requested by Center patients. The goals of the Remembering More class are to decrease the incidence of forgetfulness and to improve basic organizational skills. This class is organized around the three steps in remembering presented by Lorayne & Lucas (1974): (a) Get It (receiving the information), (b) Link It (increasing associations with familiar material), and (c) Hook It (retrieving the information).

Emphasis is placed on covert verbal repetition and self-directed questions and answers to help direct and maintain attention to important aspects of the task at hand. The routine use of calendars, efficient note taking, and planning and scheduling of daily activities is prescribed. Homework includes such assignments as taking notes on evening television news and remembering the names of two recent acquaintances without writing them down. The techniques themselves are easily grasped by our participants who, when prompted, can quickly learn to adapt the techniques to the situation at hand. Students in the class are administered a practical memory test consisting of items testing recognition and recall. Data for the first seven students indicate significant improvements from preclass to postclass administrations of alternate forms of the test ($\alpha = .05$, Wilcoxon, two-tailed).

Preliminary experience suggests that patients cognitively acquire memory-enhancing skills and display some competence in performing the skills in analog situations in class. The greatest difficulty in training individuals to employ memory-enhancing techniques is teaching them to recognize the discriminative cues indicating *when* to use a particular encoding or retrieval strategy.

Problem Solving Class

The Problem Solving class has had a slightly different format than many of the other classes. It has been an on-going group which meets twice a week for 1 hour. Target behaviors for the class include increased frequency of effective decision making and decision-related information gathering, as well as reduced frequency of impulsive or passive avoidance responses.

A majority of the class sessions utilize the five steps to problem solving outlined by Goldfried and Davison (1976): (a) General Orientation, which consists primarily of inhibiting immediate responses and preparing to engage in problem solving; (b) Problem Definition; (c) Generation of Alternatives or Brainstorming; (d) Decision Making; and (e) Verification, or testing the utility of the decision. A problem solving work sheet containing these five steps is used by each student. When a class member brings up a problem he wishes to solve, the problem solving process is initiated, with special emphasis on input from group members on Steps *b* and *c*, that is, asking questions to elicit a complete behavioral description of the problem and brainstorming alternative courses of action. It usually comes as a surprise to a new class member that the class can generate such a large range of alternative courses of action, and with little encouragement, students quickly get involved in offering creative solutions.

Students have brought up a wide range of concerns in the Problem Solving class, including problems specific to epilepsy, such as responding to a relative or friend who is inappropriately protective, increasing independence, dealing with an acquaintance who suddenly withdraws after witnessing a seizure, responding when someone inappropriately calls an ambulance after a seizure, or deciding whether or not to take medications. However, a majority of the problems are ones everyone commonly encounters—how to increase the number of friends; how to reduce television watching; how to communicate with a parent; and whether to go back to school, change jobs, or move.

Preliminary evidence suggests that participants become more skillful in the problem solving process by participating in class 3 weeks or more and that these skills generalize to day-to-day problems encountered outside of class.

Group Process Variables

In all the skills training classes specific techniques are utilized to increase commitment and optimism, decrease resistance, facilitate the development of learning skills, and achieve intragroup cohesions and support among the participants. Most importantly, the group process conveys the message that students are important to the learning process and are viewed as self-directed adults capable of managing their own behavior.

This process begins during the initial screening interview when the student is asked to identify specific goals and classes she thinks may be helpful in achieving these goals. When appropriate classes are identified as helpful in the pursuit of these goals, students are provided a description of the class and expectations for students who participate. This includes an explanation of the didactic material to be covered, group activities, and the expectation that each student will provide evaluation data.

During the first meeting of a class students are provided an overview of the class content and processes. They discuss (in dyads and then with the group at large) their expectations for the class. Expectations which have been identified are openly discussed, questions are answered, and participants are told whether or not a specific event can be incorporated into the class or is clearly outside its scope. Every effort is made to include relevant information and activities that are requested by the students or to provide an alternative or referral if this is not possible. After the class content and format are agreed upon, a verbal (sometimes written) contract is arranged; it states what the facilitators and students will each bring to the learning organization. This may vary from being on time or completing homework assignments to agreement on the part of facilitators to provide written summaries of class content. Establishing congruence between students' expectations and class content results in the students' perceiving the goals of the learning experience to be their own goals and begins the process of acceptance on the part of the learners for shared responsibility of planning and operating the class. We believe this process has been largely responsible for a low attrition rate, active class participation, student satisfaction, excellent data collection, and minimal resistance.

Small group activities (dyads and triads) are utilized throughout the course. These mini-learning and support organizations increase the active participation of students by helping them utilize their own experiences as resources for learning. This provides an opportunity for students to increase their learning skills (e.g., identifying experiences and generalizing the learning to new situations) and to share the on-going responsibility for the process of mutual inquiry and skill building, which makes the course content more meaningful and integrated.

The classes reflect our view that it is important to utilize those special variables that are most conducive to growth and development in adult learning. These variables (Knowles, 1977) include: (a) the learners' desire to change; (b) a learning environment which attends to physical comfort and the establishment of mutual trust, respect, helpfulness, and acceptance; (c) the learners' perception that the goals of the learning experience are congruent with their goals; (d) the learners' sharing of the responsibility for the planning and process of the learning experience, which results in commitment to it; (e) the learners' active participation in the learning process; (f) utilization of the learners' past experience in the learning process; and (g) the learners' perceived progress toward their goals.

Students often look to a class as an opportunity to meet and talk with others who have epilepsy. For some it is the first time they have had the opportunity to meet other individuals with seizures and talk about epilepsy-related experiences. Although many of these interactions occur informally before and after class sessions, an effort is made to include discussions on dealing with epilepsy-related problems in each course.

For most participants the reassurance that comes from knowing others have epilepsy and are dealing with similar feelings, moods, and life events is an important purpose of the classes. Those who are functioning well provide models for those who are doing less well. Socialization before and after classes, small group activities in class, and a 6-week follow-up get-together provide many opportunities for students to share concerns, provide mutual support, and learn from one another. Peer pressure discourages the use of epilepsy as an excuse for dysfunctional behavior, encourages students' involvement in the classes, and promotes an optimism for gaining greater control over one's life. The latter is a critical issue for those who have a disorder which is often uncontrollable and unpredictable and leads many to view themselves as stigmatized and helpless.

Conclusion

Psychoeducational skills training has been effectively implemented in various settings with a variety of populations to strengthen psychological and social competence. Psychoeducational intervention for psychosocial problems of individuals with epilepsy shows promise as an important component in the rehabilitation of individuals with this disorder. The efficacy of this model lies in its capacity to promote adaptive behavioral repertoires for individuals who have either maladaptive or deficient interpersonal and self-management skills without the stigmatization which traditionally has accompanied efforts to remedy psychosocial dysfunction. In addition, it provides a mechanism for the implementation of group process variables which are a major component in the change process. Among these are the dynamics necessary for building an effective learning organization and group support system, gaining commitment through group contracting and manipulation of expectations, creating an optimism for change, and placing a portion of the responsibility for learning in the hands of the adult learner.

The Skills Training Project at the University of Washington Epilepsy Center offers a psychoeducational curriculum which has resulted in increased competence in assertiveness, memory, stress and mood management, conversation skills, and problem solving skills among epilepsy patients. We are excited by the promise of this program; by its adaptability to a wide range of settings, client populations, and client needs; and by its contribution to more comprehensive rehabilitation for individuals with epilepsy.

Reference Notes

1. Richey, C. A. *Utilizing self-reinforcement in group assertion training for women: A follow-up report.* Paper presented at the 9th Annual Convention of the Association for the Advancement of Behavior Therapy, San Francisco, December 1975.
2. Travers, R. D. *Factors affecting variations in response to anticonvulsants in a group of epileptic patients.* Unpublished Doctor of Medicine thesis, Yale University, School of Medicine, 1971.
3. Lubar, J. F. *Clinical applications of EEG feedback, including treatment of epilepsy.* Paper presented at Biomedical Services, Inc., Clinical Biofeedback Seminar, San Francisco, September 1978.

References

Armstrong, H. E., Jr., & Bakker, C. B. The adult development program: An educational approach to behavior change. In J. H. Straughan (Ed.), Proceedings of the Fall Conference, Association for the Advancement of Behavior Therapy-Northwest. *Behavior Therapy,* 1971, *2,* 431-432. (Summary)

Armstrong, H. E., Jr., & Booth, C. An educational program for obesity management: Preliminary report. *Psychotherapy: Theory, Research and Practice,* 1979, *16,* 280-285.

Bakker, C. B., & Armstrong, H. E., Jr. Implementation of an educational approach to the delivery of mental health services. *Hospital and Community Psychiatry,* 1976, *27,* 330-334.

Bear, D. M., & Fedio, P. Quantitative analysis of interictal behavior in temporal lobe epilepsy. *Archives of Neurology,* 1977, *34,* 454-467.

Beck, A. T., Rush, J., Emery, G., & Shaw, B. F. *Cognitive therapy of depression.* New York: Guilford, 1979.

Beck, A. T., Ward, C. H., Mendelsohn, M., Mock, J., & Erbaugh, J. An inventory for measuring depression. *Archives of General Psychiatry,* 1961, *4,* 561-571.

Bernstein, D. A., & Borkovec, T. D. *Progressive relaxation training: A manual for the helping professions.* Champaign, Ill.: Research Press, 1973.

Cabral, R. J., & Scott, D. F. Effects of two desensitization techniques, biofeedback and relaxation on intractable epilepsy: Follow-up study. *Neurology, Neurosurgery, & Psychiatry,* 1976, *39,* 504-507.

Chaney, E. T., O'Leary, M. R., & Marlatt, G. A. Skill training with alcoholics. *Journal of Consulting and Clinical Psychology,* 1978, *46,* 1092-1104.

Daniels, L. K. The treatment of grand mal epilepsy by overt and operant conditioning techniques: A case study. *Psychosomatics,* 1975, *16,* 65-67.

Ellis, A., & Harper, R. A. *A new guide to rational living.* Englewood Cliffs, N.J.: Prentice-Hall, 1975.

Epilepsy Foundation of America. *Basic statistics on the epilepsies.* Philadelphia: F. A. Davis, 1975.

Flor-Henry, P. Lateralized temporal-limbic dysfunction and psychopathology. *Annals of the New York Academy of Science,* 1976, *280,* 777-797.

Galassi, J. P., Galassi, M. D., & Litz, M. C. Assertive training in groups using video feedback. *Journal of Counseling Psychology,* 1974, *21,* 390-394.

Gambrill, E. D., & Richey, C. A. *It's up to you: Developing assertive social skills.* Millbrae, Calif.: Les Femmes, 1976.

Goldfried, M. R., & Davison, G. C. *Clinical behavior therapy.* New York: Holt, Rinehart & Winston, 1976.

Goldin, G. J., & Margolin, R. J. The psychosocial aspects of epilepsy. In G. Wright (Ed.), *Epilepsy rehabilitation.* Boston: Little, Brown, 1975.

Goldsmith, J. B., & McFall, R. M. Development and evaluation of an interpersonal skills-training program for psychiatric patients. *Journal of Abnormal Psychology,* 1975, *84,* 51-58.

Hays, V. L., & York, H. *A psychological program for dealing with depression.* Seattle: University of Washington Press, 1978.

Hersen, M., & Bellack, A. S. Social skills training for chronic psychiatric patients: Rationale, research findings, and future directions. *Comprehensive Psychiatry,* 1976, *17,* 559-580.

Knowles, M. S. *The modern practice of adult education.* New York: Association Press, 1977.

Lange, A. J., & Jakubowski, P. *Responsible assertive behavior: Cognitive/behavioral procedures for trainers.* Champaign, Ill.: Research Press, 1976.

Lazarus, A., & Fay, A. *I can if I want to.* New York: William Morrow, 1975.

Lewinsohn, P. M., & Graf, M. Pleasant activities and depression. *Journal of Consulting and Clinical Psychology,* 1973, *41,* 261-268.

Libet, J. M., & Lewinsohn, P. M. The concept of social skill with special reference to the behavior of depressed persons. *Journal of Consulting and Clinical Psychology,* 1973, *40,* 304-312.

Lomont, J. F., Gilner, F. H., Spector, N. J., & Skinner, K. K. Group assertion training and group insight therapies. *Psychological Reports,* 1969, *25,* 463-470.

Lorayne, H., & Lucas, J. *The memory book.* New York: Ballantine, 1974.

Lubin, B. Adjective checklists for the measurement of depression. *Archives of General Psychiatry,* 1965, *12,* 57-62.

Mahoney, M. J. The self-management of covert behavior: A case study. *Behavior Therapy,* 1971, *2,* 575-578.

Meichenbaum, D. H. *Cognitive-behavior modification: An integrative approach.* New York: Plenum Press, 1977.

Mostofsky, D., & Balashak, B. I. Psychobiological control of seizures. *Psychological Bulletin,* 1977, *84,* 723-750.

Parrino, J. J. Reduction of seizures by desensitization. *Journal of Behavior Therapy and Experimental Psychiatry,* 1971, *2,* 215-218.

Rathus, S. A. A 30-item schedule for assessing assertive behavior. *Behavior Therapy,* 1973, *4,* 398-406.

Rennick, P. M. Psychosocial evaluation of individuals with epilepsy. In G. Wright (Ed.), *Epilepsy rehabilitation.* Boston: Little, Brown, 1975.

Rimm, D. C., & Masters, J. C. *Behavior therapy: Techniques and empirical findings.* New York: Wiley, 1974.

Rodin, E. A., Shapiro, M. A., & Lennox, K. Epilepsy and life performance. *Rehabilitation Literature,* 1977, *38,* 34-39.

Seligman, M. E. P. *Helplessness: On depression, development and death.* San Francisco: Freeman, 1975.

Standage, K. F. Treatment of epilepsy by the reciprocal inhibition of anxiety. *Guy's Hospital Report,* 1972, *121,* 217-221.

Suter, C. Psychological factors in epilepsy. *Neuropsychiatry,* 1956, *3,* 153-171.

Tarter, R. E. Intellectual and adaptive functioning in epilepsy: A review of 50 years of research. *Diseases of the Nervous System,* 1972, *33,* 763-770.

Todd, F. J. Coverant control of self-evaluative responses in the treatment of depression: A new use for an old principle. *Behavior Therapy,* 1972, *3,* 91-94.

Twentyman, C. T., & McFall, R. M. Behavioral training of social skills in shy males. *Journal of Consulting and Clinical Psychology,* 1975, *43,* 384-395.

Wells, K. C., Turner, S. M., Bellack, A. S., & Hersen, M. Effects of cue-controlled relaxation on psychomotor seizures: An experimental analysis. *Behaviour Research and Therapy,* 1978, *16,* 51-53.

Wolpe, J. *Psychotherapy by reciprocal inhibition.* Stanford: Stanford Press, 1958.

Wolpe, J., & Lazarus, A. A. *Behavior therapy techniques.* New York: Pergamon Press, 1968.

Zlutnick, S., Mayville, W. J., & Moffot, S. Modification of seizure disorders: The interruption of behavioral chains. *Journal of Applied Behavior Analysis,* 1975, *8,* 1-12.

Chapter 10

Suggestions for Behavioral Group Therapy of Depression

David Roth

Abstract

Behaviorally oriented theoreticians and researchers have examined the clinical phenomenon of depression for well over a decade. Various models have been offered to account for the range of depressive symptomatology, experimentation has been conducted with the aim of testing the fundamental postulates of different conceptual systems, and a multitude of treatment interventions have been developed. The focus of this chapter is twofold. First, an attempt will be made to acquaint the reader with the major theoretical frameworks in this area. A differentiation will be made between those models which view environmental and behavioral factors as central to depression and those systems which regard specific cognitive dysfunctions as being causal determinants of depression.

*Second, the various therapeutic implications of the environmental-behavioral and cognitive models will be reviewed. Particular attention will be paid to the use of group treatments for depression. Issues which will be examined include the relative dearth of group-oriented interventions and the similarity between behavior therapy conducted individually and that done within a group context. Speculations regarding the state of the art and recommendations concerning the modification of existing group treatments will be offered.**

---

*I would like to offer my deepest appreciation to the following members of the Department of Psychiatry at Michigan State University for their many valuable ideas and thoughtful comments: Bertram Stoffelmayr, Alan Cooper, and Eleanor Fox. Additionally, I would like to thank Sandor Kornblith and Lynn Rehm for their encouragement and their helpful suggestions.

Depression has been a recognized entity for well over 2,000 years (Arieti, 1974). In the fourth century B.C. Hippocrates described a psychiatric disturbance which he termed "melancholis." The primary symptomatic expression of this disorder fell within the affective dimension. An excess of "black bile" was accorded an etiological status; relaxation, sexual and alcoholic abstinence, and a balanced intake of food constituted the recommended treatment. Scientific thought has progressed, experimental findings have accumulated, and vigorous debate has taken place since Hippocrates offered his seminal observations. While contemporary theoretical formulations and treatment strategies for depression proliferate, a number of facts cannot be overlooked.

Depression has begun to rival schizophrenia as the nation's most pervasive clinical disorder. Recent surveys by the National Institute of Mental Health estimate that approximately 10% of the general population will experience at least one debilitating depressive episode within their lifetimes. Furthermore, about 80% of the annually reported 22,000 suicides are associated with depression (cited in Friedman & Katz, 1974). Hence, the development of effective and cost-efficient treatment programs for depression could eliminate much personal suffering and potential self-harm.

Only within the past decade have behaviorally oriented investigators begun to develop and validate theoretical models of depression. The field has witnessed the birth of two comprehensive frameworks. On the one hand are a growing number of researchers who contend that depression is causally related to a loss or relative lack of reinforcement (Ferster, 1974; Lazarus, 1968; Lewinsohn, 1974a; Liberman & Raskin, 1971). Treatment focuses on the acquisition of skills which are apt to result in social reinforcement and the more frequent engagement in rewarding activities. Alternatively, it has been posited that depression results from a faulty and self-defeating style of interpreting reality and anticipating the future (Beck, 1967, 1974; Rehm, 1977; Seligman, 1974, 1975). In this model, modification of dysfunctional cognitive processes is viewed as the treatment of choice. The former perspective has been referred to as a *behavioral* or *environmental approach*, whereas the latter orientation has been designated a *cognitive* or *cognitive-behavior modification* model.

This chapter is written with the intent of examining the treatment of depression through the use of behaviorally oriented group psychotherapy. As noted earlier, psychotherapeutic treatments have been derived from both the behavioral and cognitive frameworks. In a comprehensive review of the treatment literature Rehm and Kornblith (1979) cited 46 depression studies. Only six (or 13%) of the published reports involve the use of a group format. To understand the current state of the art, one must closely inspect the various individual models of depression, carefully consider the behavioral tradition, and analyze methodological issues.

The remainder of this chapter will be divided into three main sections. The first two sections will provide the reader with a rudimentary grounding in the models of depression and their treatment implications, with particular attention being placed upon group psychotherapy programs. The final section will offer detailed recommendations for the behavioral group treatment of depression.

BEHAVIORAL THEORIES

Behavioral accounts of depression have experienced a quantum leap in terms of their sophistication, explanatory power, and conceptual breadth. Skinner (1953), probably the first behaviorist to address depression, vaguely equated it with loneliness and the general weakening of various aspects of an individual's behavioral repertoire. The depressive constellation was viewed as arising from the interruption of established response sequences which had culminated in social reinforcement. For example, certain grooming behaviors which had previously resulted in social acceptance at work would be expected to diminish in frequency if one were to lose a job. Skinner's formulation merits attention primarily in its attempt to underscore the relationship between loss of social reinforcement (i.e., extinction) and depression.

Fifteen years following Skinner's original work, Lazarus (1968) reintroduced and expanded upon the notion that depression was intimately related to the loss of reinforcement. Fundamentally, depression was construed as the result of an inadequate or insufficient amount of reinforcement. Disrupted social relationships were conceptualized as being only one of many losses which a given individual may experience. Others include loss of status, money, recognition, health, youth, job, and security. Furthermore, Lazarus stated that not only is the actual loss of reinforcers capable of engendering a depressive reaction, but so is the anticipated loss of gratifications. The employee who believes that she will be fired may become as depressed as the person who does in fact lose a job. Here is the beginning of a cognitive approach to the understanding of depression.

Reinforcement Deficits

Lewinsohn and his colleagues have developed what is perhaps the most comprehensive, clearly and precisely articulated behavioral model of depression. These behaviorists have incorporated both classical and operant conditioning processes within their theoretical framework. Lewinsohn, Weinstein, and Shaw (1969) and Lewinsohn (1974a) postulated that a low rate of response-contingent positive reinforcement acts as an eliciting stimulus for a variety of depressive behaviors, such as feelings of dysphoria, fatigue, and somatic complaints (e.g., insomnia, loss of appetite, constipation). Members of the social community respond to the depressive's cries for help with sympathy, support, and encouragement. Unfortunately, these same behaviors are viewed by Lewinsohn as further reinforcing the occurrence of the aforementioned symptomatology. Moreover, those giving help are likely to react to the continued emission of depressive behaviors with frustration and a diminished frequency of social contact with the depressed individual. This further loss of social reinforcement is hypothesized to account for other facets of the depressive constellation such as decreased verbal and motoric activity. That is to say, the depressed individual has essentially been placed upon an extended extinction schedule.

The acquisition of relatively low rates of reinforcement is thought to be a multiply determined phenomenon. Lewinsohn (1974a) offered a detailed description of various factors which are presumed to account for the total amount of positive reinforcement which an individual receives. In particular, obtained reinforcement is believed to be a function of the number of events which are potentially reinforcing, the number of available reinforcers, and the ability of an individual to emit behaviors which are likely to be rewarded. Potential reinforcers vary according to age, sex, learning history, and so forth. Available reinforcement is related to events such as the death of a spouse, a move to a new city, Sunday closing of stores, and attentiveness of family members. Assertiveness, occupational performance, and communicational ability are types of instrumental skills which are likely to result in social and material reinforcement. A systematically conducted series of basic research and treatment-oriented studies offers substantial support to the proposed model (Lewinsohn, Biglan, & Zeiss, 1976; Lewinsohn & Graf, 1973; Lewinsohn & Libet, 1972; Lewinsohn & Shaffer, 1971; Lewinsohn, Weinstein, & Alper, 1970; Lewinsohn, Weinstein, & Shaw, 1969). Since a review of the basic experimental literature is beyond the scope of this chapter, the interested reader is referred to Lewinsohn (1975). The remainder of this section will focus upon treatment strategies which have evolved from the reinforcement theory of depression.

Reinforcement-based treatments of depression share an important common denominator. The goal of the therapy program is to increase the amount of reinforcement which is received by the depressed individual. The manner in which this is achieved varies according to the specific presenting symptom pattern, the patient's environmental and familial situation, and the therapist's clinical orientation.

One group of behaviorists has advocated the contingent use of reinforcement. Treatment regimes are designed to eliminate rewards for depressive behaviors and provide rewards for designated nondepressive behaviors. Reisinger (1972) reported on just such a program. Using a single subject design, Reisinger was able to demonstrate a diminution of crying behaviors and an increase in smiling behaviors. Initially, the patient received token reinforcement contingent upon the emission of smiles and lost tokens (i.e., a response cost procedure) for crying episodes. Prior to discharge, the reinforcement contingencies were successfully shifted to ones which could be maintained or supported by the natural environment. Social reinforcement was employed in the place of tokens, and social inattention (i.e., extinction) supplanted the response cost procedure.

The utilization of therapy time to differentially reinforce specific depressive and nondepressive behaviors has also been explored. Employing a DRH procedure (i.e., differential reinforcement of high rates of behavior), Robinson and Lewinsohn (1973) were able to modify formal aspects of the speech pattern of a depressed patient. They employed a single subject withdrawal design in which extra therapy time was awarded contingent upon rates of speech in excess of the baseline speech rate. The experimental contingencies resulted in an acceleration of speech, whereas removal of these consequences led to a substantially slower rate of speech.

Several investigators have studied the effect of the differential reinforcement paradigm within the family system. Lewinsohn and his colleagues (Lewinsohn & Atwood, 1969; Lewinsohn & Shaw, 1969; Martin, Weinstein, & Lewinsohn, Note 1) have made extensive use of home observation in the treatment of depression. Visits are intended to provide diagnostic information regarding the functional relationship between the depressed individual and her family, as well as to enlist the support of significant family members in the treatment process. Visits are usually scheduled for dinner time, when most of the family is present. Intrafamilial interactions are analyzed in consecutive 30-second time periods through the use of a Behavioral Rating Scale (Lewinsohn, Alper, Johansson, Libet, Rosenberry, Shaffer, Sterin, Stewart, & Weinstein, Note 2). Individual-to-individual transactions are seen as having both a source (initiator) and an object (recipient of communication). Both the actions emitted by the source and the reactions of the object are coded. Actions include themes such as somatic complaints, criticism, praise, and requests for help. Reactions may be either positive (e.g., approval, laughter, interest, etc.) or negative (e.g., disapproval, interruption, disagreement, etc.). Following a baseline or data collection phase, the clinical team provides therapeutic recommendations to the family, and a treatment contract is negotiated.

Lewinsohn and Shaw (1969) described a case study which nicely illustrates the development and evaluation of a response specific, differential reinforcement program. Home observations revealed that much of the patient's time was spent discussing somatic complaints. Also, it was determined that the patient's husband was displaying relatively few positive responses to nondepressive talk. Treatment recommendations included a shift from somatic worries to discussions concerning daily events, a decrease in positive responses to bodily complaints (i.e., extinction), and a concomitant increase in verbal and nonverbal reinforcement for healthy conversations. By the end of treatment substantial improvement was witnessed in both the patient's depression and the couple's relationship.

Liberman and Raskin (1971) have also treated specific depressive behaviors through modification of the reinforcement contingencies provided by family members. From a home visit the therapist noted that sympathy, concern, and helpfulness were directed toward the patient for crying, pacing, withdrawal, and the like. Treatment involved ignoring the latter behaviors and providing abundant and immediate attention to such coping behaviors as cooking, cleaning, and caring for the children. Careful monitoring of the family's behavior pattern indicated that the therapeutic recommendations were followed. Decreases in depressive behavior and increases in healthy behavior occurred.

A somewhat different behavioral approach to the treatment of depression involves having clients engage more frequently in reinforcing or pleasant activities. Researchers and clinicians who ascribe to this treatment model argue that there is a relationship between the number of pleasant activities which individuals engage in and their mood. The research literature provides ample empirical support for this position (Lewinsohn & Graf, 1973; Lewinsohn & Libet, 1972; Rehm, 1977).

For more than 10 years Lewinsohn and his colleagues have employed activity schedules in order to compile individualized lists of reinforcing events. Typically, a patient is asked to complete the 320-item Pleasant Events Schedule (MacPhillamy & Lewinsohn, Note 3) during the intake interview. Items are rated in terms of frequency of occurrence and subjective pleasantness. The cross-product between frequency and pleasantness is thought to yield a measure of total obtained reinforcement. The 160 most rewarding activities are selected from the parent list, and patients are required to monitor their occurrence on a daily basis. Also, mood recordings are made at the end of each day with the aid of the Depression Adjective Check List (Lubin, 1965). This procedure continues for a period of 30 days. At the end of this time a correlational analysis is conducted between activity and mood. Treatment goals consist of increasing the frequency of items found to be significantly associated with a positive affective state.

More recently Lewinsohn, Zeiss, and Munoz (Note 4) have modified the basic pleasant events treatment module to facilitate the client's implementation of reinforcing activities. Notably, depressed individuals receive relaxation training with a focus upon employing this skill in naturalistic settings. This technique is designed to enhance the enjoyability of targeted activities. Also, individuals are required to read selected chapters from *How to Get Control of Your Time and Your Life* (Lakein, 1973).

These readings are intended to assist the client in planning activities and budgeting his time.

Variations of the rudimentary activity program have been reported by several authors. Depressed individuals often have difficulty participating more frequently in pleasant events, and it is not uncommon to hear a depressed patient complaining that she does not have the energy to see a favorite film or attend a potentially gratifying concert. Therapeutic interventions have therefore been constructed with the aim of reinforcing the engagement in pleasant events. Lewinsohn (1974b) reinforced patients' participation in certain pleasant events with therapy time. The procedure was found to be quite effective in increasing reinforced behaviors relative to nonconsequated activities. Analogously, Lazarus (1974) allowed patients to shadow him throughout the day. Therapist concern and interest were thereby employed to increase daily involvement in pleasant events.

Social Skills Deficits

Depression has been characterized as being related to a passive orientation to interpersonal transactions (Ferster, 1973). Feelings of helplessness, hopelessness, and despair may legitimately be a consequence of the depressive's unassertive posture toward his social world. On various self-report measures depressed individuals have been found to endorse items indicating a significantly less assertive style than was found for matched groups of nondepressed individuals. For instance, Sanchez and Lewinsohn (Note 5) reported that depressed individuals taking the Wolpe-Lazarus Assertion Inventory (Wolpe & Lazarus, 1966) indicated that they are less assertive than nondepressed individuals taking the inventory. Youngren and Lewinsohn (Note 6) found that depressed patients felt less comfortable behaving assertively than nondepressed patients. Standardized clinical interviews support the latter findings. Weissman and Paykel (1974) interviewed groups of depressed and nondepressed women over a 20-month period. The depressed population demonstrated significant social impairment in their various roles as mother, worker, community member, and wife throughout the entire length of the study (i.e., acute depressive episode, recovery period, and posttreatment follow-up).

An accurate self-portrayal based upon clinical interviews and self-report measures requires nondistorted self-awareness. Depressives are known to underestimate their abilities (Friedman, 1964). It is therefore quite noteworthy that researchers have been able to pinpoint discrete behavioral social skills deficits in depressed populations. Depressed individuals relative to nondepressed individuals have been shown to evidence: (a) slower rates of speech (Hinchcliffe, Lancashire, & Roberts, 1971); (b) fewer communicative hand and arm gestures (Ekman & Friesen, 1974); and (c) a longer latency between another's communication and their response, interactions with fewer social acquaintances, less overall activity, and fewer reinforcing responses (Libet & Lewinsohn, 1973; Libet, Lewinsohn, & Javorek, Note 7).

Casting the above into a reinforcement model, one can say that passive, unassertive, and dysfunctional communication styles are likely to result in the obtainment of few rewards. The social environment is apt to avoid continued interactions with the socially deficient depressive (Coyne, 1976, 1977; Hammen & Peters, 1977, 1978). Isolation and consequent loss of opportunities to engage in pleasant and gratifying activities can be expected. Furthermore, the depressed individual has little chance to observe and practice reinforcement-engendering social skills. Once such a cycle has developed, it tends to take on a self-perpetuating course.

The implementation of social-skills-oriented treatment programs for depressed individuals has resulted from a recognition of this relationship. It is important to note that one of the primary goals of such an intervention strategy is to enable an individual to emit those behaviors which are likely to be rewarded and maintained (e.g., by attention, conversation, collective activities, etc.) within the social community. A diminution of depression is thereby expected to follow. Thus, the acquisition of a socially skillful repertoire is seen as one means by which a reinforcement deficit can be redressed.

Lewinsohn, Weinstein, and Alper (1970) conducted one of the first group psychotherapy treatment outcome studies in this area. Nine depressed individuals met weekly for a 3-month period in what was termed a self-study group. The principal therapeutic strategy was to provide patients with information about their behaviors and the effect which these behaviors have upon other group members. The aforementioned Behavioral Rating Scale (Lewinsohn et al., Note 2) was employed to quantify various aspects of each member's interactional style. Following this assessment phase, individualized goals were negotiated for each patient. The therapist and peers then proceeded to reinforce goal-relevant behaviors and to extinguish behaviors which were incompatible with the agreed-upon goals.

For instance, observational records indicated that one of the most active initiators was engaged in conversation by other group members a very limited amount of the time. Group participants informed this individual that they found his comments to be quite critical and therefore minimized their interactions with him. Coders provided a concrete graphic record which supported this feedback. The behavioral goal which this individual elected to pursue involved a decrease in put-downs and an increase in positive, supportive comments. The group successfully encouraged and reinforced his efforts in this direction. Similar programs were designed for members demonstrating a restricted interactional style, limited communicational initiations, etc. A decrease in depression was reported for the group as a whole.

Lewinsohn (1974b) described a number of quite interesting procedural modifications of the basic self-study group. First, in Self-Study Group II, members were required to engage in sensitivity training group exercises to foster communicational interchanges. Moreover, beginning with Session 9, patients were required to wear a single earphone which enabled therapists behind a one-way mirror to communicate with them and thereby offer immediate, goal-related communication training. Unfortunately, Lewinsohn failed to provide sufficient information to allow evaluation of the efficacy of this clinical strategy.

Rehm, Fuchs, Roth, Kornblith, and Romano (in press) conducted a study comparing the differential effectiveness of a self-control and an assertiveness training program for depression. Participants in the social skills condition met for 6 weekly, 1½-hour group therapy sessions. The fundamental focus of this module involved teaching group members to behave more assertively in each of the following four areas: requesting that another person change her behavior, refusing an unreasonable request, expressing disapproval, and expressing affection. Emphasis was placed both upon appropriate verbal content and nonverbal behaviors such as loudness, affect, eye contact, gesturing, and latency of response. Therapeutic tactics included role playing, modeling, and behavioral rehearsal of standard, as well as subjectively, problematic situations. Group members engaged in role-play exercises with one another, at times in pairs and at other times in front of the group. Subjects were required to provide specific feedback about all assertive responses. Members of this group evidenced a decrease in depression and a substantial increase in the quality of their assertiveness.

Hilford (1975) utilized a multidimensional treatment of depression on an inpatient psychiatric unit. The therapy group consisted of four or fewer patients and met 1 hour each day for a 3-week period. Certain aspects of this program involved cognitive control procedures (i.e., "worry less," "think positively," etc.) and will not be reported upon here. One of the major thrusts of this clinical regime concerned patients' relationships with each other. Patients were informed verbally and through written instructions that depression could be alleviated by the development of more social relationships and through gaining a perspective upon the effect which one's behavior has upon other people. All in all, subjects in the treatment groups experienced significantly greater changes in self-report of depression and in ward behavior than members of no-treatment and contact-treatment control groups.

Social skills training for depressed patients has also been conducted by Shaw (1977). Subjects met in a group, two times a week, over a period of 1 month. Shaw employed a complex behavioral treatment package based upon the collected works of Lewinsohn and his colleagues. In addition to receiving social reinforcement and communication training, group members constructed verbal contracts and attempted to increase their activity level. Formal and informal aspects of the group process were not reported upon in this publication. As a whole, this group experienced a slight decrease in both self-report and interview ratings of depression.

Close inspection of the aforementioned social skills training programs reveals a number of interesting similarities and differences. First, it appears that certain therapy modules assume an idiographic stand with respect to treatment goals. Lewinsohn (1974b), Lewinsohn, Weinstein, and Shaw (1969), Hilford (1975), and Shaw (1977) hoped to instill the set that a relationship exists between one's behavior and the amount of attention, support, criticism, and so forth which one receives from his social environment. Different individuals were seen as evidencing different skill deficits. Treatment goals were therefore tailored for each depressed group member. On the other hand, Rehm et al. (in press) employed a specific treatment package. All participants in the therapy group attempted to develop more effective assertion skills. Although this allowed for more intensive training, those individuals with diverse interactional difficulties would probably fare better in a broad-based therapy group. Specific pretreatment diagnostic testing would probably foster a better match between a patient's needs and the type of group experience she would receive. Differences in posture are also evident among the above cited groups in the area of treatment techniques. Each author referred to various procedures which are currently employed in individually oriented psychotherapy (e.g., role playing, modeling, etc.). However, little more than a passing reference was made to the manner in which group process was employed. Although Rehm et al. (in press), Lewinsohn (1974b), and Lewinsohn, Weinstein, and Alper (1970) described the active therapeutic use of group transactions, one is left wondering whether group process was an artifact of the situation, a helpful but unnecessary treatment component, or an important clinical strategy in its own right.

COGNITIVE THERAPIES

Depression has traditionally been conceptualized as an affective disorder. Feelings of dysphoria, hopelessness, pessimism, grief, and so forth have been viewed as central to the depressive syndrome. Quite recently a number of clinicians and theoreticians have challenged this age-old belief (Beck, 1967, 1974; Rehm, 1977; Seligman, 1975). It is their contention that the affective, behavioral, and physiological features of depression are secondary to a dysfunctional mode of interpreting reality. Fundamental to depression is cognitive distortion, a disorder of thinking.

Cognitive-Behavioral Model

One of the first and perhaps the most influential advocates of a cognitive-behavioral model of depression is Aaron Beck. For Beck (1967, 1974) the core of depression is the cognitive triad. Operationally, the cognitive triad refers to an organized belief system consisting of negative evaluations about oneself, one's future, and one's environment. The depressed individual regards himself as a loser, as inadequate, and as the cause of most unpleasant experiences; views the world as being filled with burdens, frustrations, and complications; and anticipates a future which is likely to be as bad as or worse than the present.

Beck (1967, 1972, 1976) maintained that the distorted cognitions of the depressed individual are derivatives of various faulty cognitive mechanisms. Four such mechanisms are *arbitrary inference, selective abstraction, overgeneralization,* and *magnification.* By *arbitrary inference,* Beck means that a conclusion is drawn about reality which is not supported by objective information. For example, a client once informed me that by arriving late for a therapy session (my car had broken down on the highway), I was communicating that I no longer cared for him. *Selective abstraction* refers to a process whereby an individual focuses upon a single, negative aspect of a complex, multifaceted situation. The student who only looks over the mistakes made on a test is evidencing this cognitive distortion. *Overgeneralization* occurs when a person makes a broad global statement about his attributes on the basis of scant information. For instance, the patient who reports "I am an idiot" when he can't recall a telephone number is exemplifying the process of overgeneralization. Finally, *magnification* entails a grossly distorted evaluation of the significance of certain events and/or actions. A prime example of this process comes from a patient who told me that she was certain that her fiancé hated her because he had failed to answer one of her letters within 24 hours. The treatment implications of this model will be discussed later.

Learned Helplessness Model

Seligman (1974, 1975) has developed an alternative model of depression. The central construct within this system is that of learned helplessness. It is argued that the symptomological manifestations of reactive depression result from the belief that the rewards and punishments which an individual receives from her environment are not under her personal control. Thus, the depressive who is complimented by her boss is apt to believe that this praise would have been given regardless of her performance on the job. She does not see the praise as something she has earned; she does not feel in control.

To gain a better understanding of this formulation, one needs to examine the early research activities of Seligman and his colleagues. An experiment by Seligman and Maier (1967) exemplifies the basic methodological strategy which has been employed in many of the studies examining the phenomenon of learned helplessness. In this investigation dogs were exposed to either controllable, uncontrollable, or no shock. Each animal in the uncontrollable shock group was individually yoked with an animal in the controllable shock group, and thereby dogs in both conditions received the same number of shocks. The only factors differentiating animals in the two conditions was the controllability or uncontrollability of shock termination. This procedure has come to be known as the triadic design. In the second phase of this experiment all dogs were placed in a shuttle box, and signaled escape-avoidance conditioning was instituted. Relative to dogs in the other two conditions, animals who received inescapable shock pretraining evidenced a failure to escape or avoid the shock, as well as an impairment in their learning of the relationship between jumping across the shuttle box hurdle and safety from shock. This pattern of findings has been replicated with other populations of dogs (Overmier & Seligman, 1967), rats (Seligman, Rosellini, & Kozak, 1975), cats (Seward & Humphrey, 1967), and humans (Hiroto, 1974; Klein & Seligman, 1976; Miller & Seligman, 1973, 1975, 1976; Miller, Seligman, & Kurlander, 1975). On the basis of accumulated data, Seligman summarized the basic symptomological manifestations of learned helplessness, reporting that it is characterized by behavioral passivity, difficulty in learning the relationship between action and consequence, reduced aggression, loss of appetites and sexual interests, depleted supplies of brain neural transmitters, and a limited time course.

Seligman (1974, 1975) has proposed that four criteria must be met before one can extend analogue research to the realm of clinical phenomenon. Notably, parallels must exist between symptomatology, etiology, cure, and prevention. Given this perspective, Seligman (1975) and Seligman, Klein, and Miller (1976) described a group of symptoms and signs which were felt to be fundamental to reactive depression and which are strikingly similar to those embodied within the learned helplessness framework. Included within this list are passivity, negative cognitive set, a time course, anorexia, sexual deficits, and reduced levels of norepinephrine. Applying the triadic design methodology to groups of college students who had been designated as depressed and nondepressed (on the basis of Beck Depression Inventory scores), Klein and Seligman (1976) found quite interesting symptomatologic parallels between learned helplessness and depression. Namely, nondepressed subjects receiving inescapable noise pretraining and depressed subjects who did not receive prior exposure to uncontrollable aversive contingencies demonstrated similar deficits on a shuttle box task. Both groups evidenced significantly greater performance deficits than a group of nondepressed subjects who did not receive prior exposure to noise.

The overt and covert manifestations of learned helplessness are viewed by Seligman (1974, 1975) as being causally related to experiences in which the individual lacks control over significant facets of his life. Utilization of the triadic design enables one to demonstrate that it is not simply the exposure to aversive contingencies which results in depressive-like symptomatology; rather, it is the uncontrollability of these same experiences which is of crucial importance. Seligman has amassed a substantial body of literature in support of this contention (Maier, 1970; Overmier, 1968; Overmier & Seligman, 1967; Seligman & Beagley, 1975; Seligman & Maier, 1967; Seligman, Maier, & Geer, 1968). Extending the analogue model of learned helplessness to clinical reactive depression, Seligman (1975) asserted that central to depression are beliefs that one is unable to control the rewarding and aversive aspects of one's life. Lewinsohn (1974a) has argued that if the death of a loved one, a physical injury, growing older, etc., represents a significant loss in reinforcement, an individual will become depressed. Seligman, however, remarks that it is not the loss of reinforcement per se; instead, it is the consequent belief that one is no longer going to be able to control the occurrence of satisfactions which is likely to cause depression. Seligman (1975) further bolstered his argument by drawing heavily upon the writings of such researchers as Bibring (1953), Lichtenberg (1957), and Melges and Bowlby (1969), each of whom has stated that beliefs of personal helplessness are at the core of depression.

For Seligman the appropriate treatment and prevention of both depression and learned helplessness have as a goal the development of "competency" cognitions. Seligman, Maier, and Geer (1968) have shown that dogs receiving inescapable shock pretraining were able to overcome helplessness-induced performance deficits by being forced to engage in successful escape and avoidance responses within the shuttle box apparatus. Analogously, Klein and Seligman (1976) have demonstrated that success at simple discrimination problems has a therapeutic effect on depressed subjects' performances and produces a concomitant change in beliefs of control. Immunization against learned helplessness has also been experimentally demonstrated. Seligman and Maier (1967) found that performance deficits did not arise when dogs were exposed to inescapable shock training after having been given practice sessions within the shuttle box situation. That is to say, learned helplessness can be prevented by providing animals with experiences in which they can control their environment. Seligman, Klein, and Miller (1976) speculated that similar processes tend to immunize individuals against depression. It is their contention that a history of mastery experiences provides one with a generalized set that she is largely in control of significant aspects of her life. The person who views herself as being in charge of life would be more apt to entertain beliefs that discontrols were situational (e.g., "I am unhappy on this job because of my dictatorial boss") rather than personal and would be expected to bounce back, instead of giving up, when confronted with a meaningful loss.

At best, the above represents only a cursory overview of the thought and research of Seligman and his colleagues. Within the past year numerous criticisms have been directed at the learned helplessness theory of depression (Costello, 1978; Depue & Monroe, 1978). Also, Seligman has reformulated significant aspects of his conceptual framework (Abramson, Seligman, & Teasdale, 1978). Since an examination of these issues is beyond the scope of this chapter, the interested reader is referred to the primary sources for an illuminating and thought-provoking discussion of the relevant issues.

Self-Control Model

Recently Rehm and his colleagues have developed an alternative model of depression. This formulation represents a modified version of Kanfer's (1970a; 1970b; 1971) theory of self-control. Essentially, it is proposed that depression results from the action of one or more dysfunctional cognitive mechanisms. Rehm (1977) enumerated six dysfunctional self-control processes. It is hypothesized that depressed individuals frequently attend to negative events; attend to immediate (as opposed to delayed) consequences of behavior; set overly stringent evaluative criteria; attribute positive outcomes to external factors and negative outcomes to internal factors; emit few self-rewards; and emit many self-punishments. Depressed individuals do not evidence all of these self-defeating cognitive patterns, but a diversity of self-control deficits is felt to be the rule rather than the exception.

An examination of the relationship between the symptomatology of depression and dysfunctional self-control processes should help identify the critical elements of this model. The depressed person's negative view of the world can be regarded as a manifestation of selective attention to negative events. Pessimism may be viewed as resulting from a restricted attention to the future consequences of one's behavior. Feelings of inadequacy or incompetence can be expected to develop out of overly high and excessively stringent self-evaluative standards. Guilt, one of the hallmarks of depression, is a product of frequent internal ascriptions of responsibility for negative events. Finally, slowed down and lessened rates of behavior are construed as being a function of the infrequent emission of self-rewards (e.g., praise) and the frequent occurrence of self-punishment (e.g., criticism). The interested reader is referred to Rehm (1977) for a more comprehensive exposition of this system.

Treatment Programs

The goal of cognitive therapy is to modify the depressed person's distorted mode of thinking. As might be expected from the diversity of theoretical models in the area, there are no universally recommended methods for achieving these objectives. Treatment programs depend upon such factors as the severity of a patient's depression, idiosyncratic patterns of ideation, and the therapist's conceptual framework. While traditional behavioral treatments focus directly upon observable response patterns, there are surprisingly few cognitive therapy regimes which focus exclusively upon faulty styles of thinking. More often than not, the cognitive-behavior therapist makes use of both performance based on cognitively oriented tactics and the cognitions themselves.

The various cognitive treatments of depression can be compared on the basis of their therapeutic scope. At one extreme are treatments which attempt to increase and/or decrease the frequency of a delimited set of cognitions (Johnson, 1971; Mahoney, 1971). Importantly, the patient usually does not learn a broadly applicable therapeutic strategy for changing related symptomatology. A number of case studies have employed this type of intervention. For example, Johnson (1971) reported a case study in which he treated a suicidally depressed college student by applying the Premack Principle. The patient was required to generate a list of positive self-statements (low-frequency behavior) and review these items immediately prior to urination (high-frequency behavior). While no objective data were reported, the patient indicated that he felt less depressed following treatment. A similar procedure was employed by Mahoney (1971). Positive self-statements were placed on index cards and inserted into a package of cigarettes. One card was left blank. Before smoking a cigarette the patient was required to read the statement on the top card. The blank card cued the patient to generate a new, positive self-statement. Substantial increases in spontaneously occurring positive thoughts were reported throughout the length of treatment and at the 4-month follow-up. Variants of this technique have made use of temporal cueing (Vasta, 1976) and have been included within complex treatment packages (Taylor & Marshall, 1977; Schmickley, Note 8).

The aforementioned studies represent an interesting extension of operant procedures to covert phenomena. However, it has been my own experience that while patients will practice or rehearse positive statements, they often indicate that these descriptions do not fit with what they are feeling. I have been told more than once, "If you want me to say these sentences to myself, I will; but, I really don't feel that they *fit* my self-image." Another group of patients with whom I have employed this strategy have reported that they were at a loss regarding how to proceed with this technique on their own. A procedure which I have found to be effective in addressing the dual problems of self-statement/self-image fit and generalization involves training depressed individuals to generate empirically valid positive self-statements. One means of attaining this goal involves having patients actively contrast current levels of performance with past patterns of behavior. Improvements or progress are fashioned into positive self-statements. Various sorts of cueing procedures are additionally employed. For example, a depressed student whom I had counseled was repetitively dropping out of various courses and extracurricular activity programs. Upon close inspection, it was discovered that this individual would severely criticize himself for the mistakes or informational deficits which are to be expected of a novice student or neophyte hobbyist. Being unable to accurately serve a tennis ball his first time on the court or having difficulty explaining the nature of schizophrenia in his third psychology class would characteristically result in self-denigration, frustration, and eventual avoidance of the discomforting activity. Treatment was initiated by requiring this individual to record the amount of time it took to answer accurately certain types of calculus problems. Increased speed of performance was defined as a measure of progress and was to result in a positive self-statement, such as "I'm beginning to really understand this material" or "I'm getting better at math." This procedure was analogously employed in other academic and nonacademic pursuits. In addition, this individual was required to spend two 5-minute periods each day reviewing his earlier accomplishments. He chose the middle of the day and before sleep, since these were times when he would typically engage in dysphoria-inducing ruminations. Upon the completion of treatment, the patient reported that his self-concept had vastly improved; moreover, at graduation he indicated that he had not dropped out of any other classes. Essentially, what I have described is one of many procedures which utilize a highly generalizable set of rules in order to help the depressed person to develop a positive self-image.

A second group of behavioral-cognitive clinicians and researchers have attempted to teach the depressed individual comprehensive strategies which can be used to identify and modify maladaptive cognitive patterns. This perspective is quite consistent with the recent trend in cognitive behavior therapy toward the implementation of self-control (Mahoney & Thoresen, 1974) and coping skills programs (Meichenbaum & Cameron, 1974). As applied in the treatment of depression, an individual learns: (a) that a relationship exists between thought, affect, and behavior;

(b) how to recognize broad classes of faulty ideation; (c) how to critically evaluate negative thoughts; and (d) how to generate a realistic, positive self-concept.

One of the earliest group psychotherapy outcome studies to employ aspects of this model was conducted by Shipley and Fazio (1973). These authors viewed depression as a loss of reinforcement engendered by faulty interpretations of one's physical or social environment and/or inappropriate responses to the environment. Group members learned how to analyze depression-inducing situations, develop alternative solutions for the resulting problems, and utilize positive or negative feedback to evaluate the adequacy of their actions. Significantly greater increases in Minnesota Multiphasic Personality Inventory depression scale scores were evidenced by subjects receiving this treatment than by individuals in a waiting list control group or individuals receiving supportive psychotherapy. It is important to note that the authors failed to mention the extent to which group members participated in each other's treatment programs. Reading between the lines, it appears that little group interaction was encouraged as "the *therapist* recommends a mutually agreed upon set of appropriate actions to be attempted by the client before the next session" (p. 373, italics added for emphasis).

Beck and his colleagues have made substantial contributions to the field of cognitive behavior therapy. A multitude of innovative, cognitively oriented treatments for depression have been developed and submitted to empirical testing by this research group. Beck is one of the foremost advocates of teaching depressed individuals broad-band cognitive skills. On a global level the patient entering this form of cognitive therapy learns how to both *identify* and *modify* depression-engendering cognitions. The dual processes are clearly interdependent; but for the sake of expositional clarity, they will be discussed separately.

From the onset of therapy the depressed person is trained to recognize that a relationship exists between her thinking and her depressive affect and behavior. By examining relevant experiences, the patient is shaped into adopting the therapeutic set that feelings of dysphoria and sadness, diminished interests, suicide attempts, and the like are a result of her maladaptive ideation. The therapist helps the depressed individual to identify three types of dysfunctional cognitions. First, the depressed person learns to recognize the occurrence of idiosyncratic or automatic thoughts, feelings, and images. For instance, if a co-worker passes in the hall without offering a greeting, the depressed individual may think, "He's angry with me for something I've done wrong." Activity monitoring has been shown to be a highly effective means of pinpointing relevant idiosyncratic thoughts (Rush, Khatami, & Beck, 1975). Second, after the patient is able to recognize various automatic thoughts, the therapist proceeds to a higher-order identification of distorted styles or patterns of thinking. Here, the depressed individual becomes aware of such mechanisms as arbitrary inference, selective abstraction, overgeneralization, and maximization. The clinician may point out, "From our discussions today it seems quite clear that you are habitually nonattentive to compliments, praise, and success. In fact, you seem to only search for personally significant failures." Finally, the depressed individual, through guided self-discovery, gains an awareness of basic underlying beliefs, attitudes, and assumptions. This is the first exposure for many individuals to their fundamental beliefs about their ineptness, the insurmountable burdens found in the environment, and their pessimistic outlook on the future. It is important to underscore that an enduringly successful treatment regime is a direct function of a patient's ability to track and to remain cognizant of the general class of depression-inducing ideation.

Depressed persons receiving Beck's form of cognitive therapy also receive training in the modification of dysfunctional thinking. The communication of a therapeutic attitude and the utilization of specific cognitive-behavioral techniques are employed in this undertaking. In order to actively participate in the therapy process, one must impose an objective perspective upon his thinking. The therapist attempts to achieve this goal by helping an individual distance himself from his cognitions. This can be done by fostering a differentiation between the external environment and one's interpretation of this experience. Also, the therapist frequently comments upon the reality of thoughts and images.

Armed with the beliefs that thought and feeling are intimately related and that ideation is often inaccurate, the patient is prepared to make use of additional treatment strategies. Essentially, the depressed individual is trained to employ various cognitively oriented procedures with which negative thoughts can be more comfortably and frequently monitored. Patients are formally schooled to use empirical and logical analysis to confirm or disconfirm negative cognitions. For instance, on a homework assignment a patient may be required to use the "triple column" technique. The procedure entails the recording of depression-engendering situations and thoughts in Columns 1 and 2. Column 3 is reserved for an analysis of the reasonableness of the negative thought. The "alternative" technique may be concurrently applied with the aim of developing adaptive, realistic interpretations for life situations. Thoughts of competence and enjoyment may be fostered through the implementation of the "mastery and pleasure" technique. Using this procedure, the patient is asked to record each day's activities and indicate which were pleasurable and which involved a mastery of the situation in which they took place. Interventions are based upon the patient's severity of depression and idiosyncratic cognitive style. The interested reader is referred to Beck, Rush, Shaw, and Emery (Note 9) for a detailed description of available treatment tactics. What must be pointed out, however, is that the long-lasting success of this treatment program results from an internalization of formal homework assignments and a consequent replacement of dysfunctional ideation with more functional thought processes.

Two clinical outcome studies have compared Beck's cognitive behavior therapy with an antidepressant medication program (Rush, Beck, Kovacs, & Hollon, 1977; Kovacs & Rush, Note 10). In both studies subjects met individually with a therapist. Participants in the cognitive therapy condition received a maximum of twenty 1-hour psychotherapy sessions. Depressed individuals within the pharmacotherapy regime met with a therapist for approximately 15 to 20 minutes each week to review their responses to the medication. Kovacs and Rush (Note 10) cited marked alleviation of depressive symptomatology on self-report measures and clinical interviews for subjects in both treatment conditions. Rush et al. (1977) reported that greater gains were evidenced by depressed persons receiving cognitively oriented psychotherapy at both termination and follow-up. In both investigations it was found that a disproportionately greater percentage of individuals receiving medication (Tofranil) terminated treatment prematurely. Taken as a whole, these results seem to indicate that cognitive therapy is at least as curative as Tofranil in the treatment of depression.

Shaw (1977) examined the relative efficacy of a cognitive and a behavioral treatment for depression. The cognitive regime was fashioned after Beck's therapeutic approach, and the behavioral strategy was based upon the work of Lewinsohn. Participants met two times each week over a 4-week period with therapy being conducted within a group context. Unfortunately, Shaw failed to indicate the nature of the group process, and therefore it is unclear whether aspects of group interaction were programmed to function as a therapeutic vehicle or whether the group may have been modeled along the lines of a therapeutic classroom, with participants addressing all questions and concerns to the group leader or teacher. All that is mentioned by Shaw is that Meichenbaum (1972) "successfully used a group format and thus it was judged to be an appropriate modality" (p. 546). Apart from this methodological oversight, results indicated that subjects in the cognitive treatment demonstrated significantly greater improvements on conventional depression measures than individuals receiving behavioral treatment.

Rehm and his colleagues also have developed a rather comprehensive treatment package for depression which is derived from their theoretical formulation of this disorder. Two studies have been conducted with the aim of establishing the efficacy of the overall "self-control" treatment strategy (Fuchs & Rehm, 1977; Rehm et al., in press). Depressed individuals attended 6 weekly 1½-hour, highly structured treatment sessions. Therapy was conducted within a group context. In successive 2-week blocks, participants learned to both recognize and modify cognitive dysfunctions involving selective attention, self-evaluation, and self-reinforcement. Each block had didactic and experiential features. Participants were first presented general information concerning the nature of a specific cognitive deficit and were then asked to explore the ways in which their own behavior was consonant with the construct under discussion. Subsequently, experientially oriented homework exercises were assigned. For example, after examining their tendency to attend to negative events, patients were required to maintain a daily log of their positive activities; or, following a discussion of stringent self-evaluative standards, group members were asked to set a long-term goal for themselves and to break this goal down into realistically achievable subgoals. Group members were encouraged to help each other in the design and implementation of various homework projects, as well as to help one another outside the group. The second treatment session of a 2-week block was spent reviewing the previous week's work, as well as discussing highlights of the homework assignment. Homework exercises were prepared, with the intent of helping depressed individuals identify frequently used cognitive mechanisms and modify these same dysfunctional patterns.

Fuchs and Rehm (1977) assigned depressed, outpatient women to one of three conditions. Individuals received self-control training, nondirective psychotherapy, or were placed on a waiting list. Subjects receiving cognitive therapy evidenced significantly greater improvement on self-report measures of depression than individuals in the other two conditions. These findings were largely replicated in a later study by Rehm et al. (in press) in which the self-control protocol was contrasted with a social skills training program.

In summary, various treatments for depression have been designed with the aim of altering faulty cognitive styles. Psychotherapeutic strategies have differed in a number of ways. First, various clinicians have focused their attention upon the modification of a delimited set of cognitions. Alternatively, investigators have opted to provide the depressed individual with highly generalizable skills which can be used to identify and remedy broad classes of thoughts. Second, treatment formats have been quite varied. Certain therapeutic interventions have been highly individualized, whereas others have favored exposing depressed individuals to the same structured array of tactics. Finally, few cognitive treatments of depression have employed a group context. Even when this modality has been utilized, however, published accounts have rarely included details of the nature of the group interaction. The reader is left wondering whether individually developed strategies were applied to group members in a group setting or whether group interaction was promoted as a vehicle for reducing depressive symptomatology.

Overview and Recommendations

Theoretical formulations concerning the nature of depression have proliferated during the past decade. Models which account for depressive symptomatology in terms of environmental factors (Ferster, 1973; Lazarus, 1968; Lewinsohn, 1974b) and dysfunctional cognitive styles (Beck, 1967; Rehm, 1977; Seligman, 1975) have been developed. Contemporary conceptual frameworks have resulted in the design and implementation of quite disparate treatment programs. Although a survey of the relevant literature indicates the existence of a sizable body of behaviorally oriented treatment studies (Rehm & Kornblith, 1979), it is noteworthy that only 13% of the published clinical studies involved the use of a group modality. These findings are quite perplexing and are suggestive of a number of critical questions. Why is there a dearth of behaviorally oriented group psychotherapy research? Is there a need to systematically evaluate the efficacy of behavioral group psychotherapy for depressed persons? Finally, upon which specific parameters should researchers in this area focus their attention? The remainder of this chapter will be spent examining each of these questions.

The scarcity of pertinent group psychotherapy research can partially be understood by addressing the methodological concerns of behavioral investigators. In conducting a psychotherapy study, the researcher hopes to be able to conclude that a particular treatment intervention has resulted in the alleviation of depressive symptomatology. Rigorous attempts are made to control for such nonspecific factors as the passage of time, the therapist's interpersonal style, unplanned environmental changes, and so forth. Unfortunately, the action of other group members can also be construed as one of those factors which may preclude the assignment of causal significance to a specific clinical strategy. By meeting with patients in a dyadic relationship (i.e., the therapist and a single patient), one can eliminate this source of variance. Analogously, many behavioral researchers have been schooled in the use of single-subject methodology. Briefly, advocates of this orientation contend that valuable information is lost by employing between-group experimental designs. Proponents of this position favor the use of an individual as her own control. The trend to utilize single-subject methodologies in order to evaluate treatment regimes is reflected in the publications of numerous researchers (Liberman & Raskin, 1971; Mahoney, 1971; Reisinger, 1972; Schmickley, Note 8).

Hence, the methodological biases of certain behavioral scientists and the fear of experimental impurity engendered by the group approach have contributed to the paucity of group psychotherapy studies of depression. This is an unfortunate state of affairs, since appropriate experimental controls are currently available for between-group designs. Also, the researcher can quite easily adapt single-subject methodologies to the group situation by simply enlarging her unit of investigation.

The decision to employ one treatment modality rather than another is also influenced by one's perception of the functional determinants of a given clinical disorder. Proponents of the viewpoint that depression is a consequence of assertion or social skills deficits are apt to favor an interpersonally oriented therapeutic modality. Group and family therapy clearly satisfy this requisite. Alternatively, researchers who understand depression from a cognitive perspective are likely to focus upon an individual's intrapsychic functioning. Individual therapy is often seen as being most suitable for just such an enterprise. An examination of Rehm and Kornblith's (1979) comprehensive review of the literature substantiates this hypothesis. Taken together, there are 10 family and group psychotherapy studies derived from a behavioral model and only 4 studies based upon a cognitive framework. The limited use of interpersonal modalities in the treatment of cognitive disorders is rather confusing. It is largely agreed that the manner in which one interprets reality is a function of various learning experiences. The individual who is told that he is forever making mistakes is likely to develop a generalized set of accepting responsibility for negative events. Alternatively, social learning factors can be marshaled to help individuals unlearn a cognitive dysfunction. It follows from this perspective that group interactions can be employed as one means of modifying faulty thinking.

Given the powerful impact which methodological concerns and theoretical biases seem to have had, the question arises as to whether a systematic investigation of behavioral group psychotherapy for depression is merited. It is my contention that a well-planned series of studies is justified on two levels. Depression is considered to be one of the most frequently experienced and potentially destructive of the psychiatric disorders. Since upwards of 10% of the general population can be expected to become severely depressed within their lifetimes, one can legitimize an examination of the cost-effectiveness of group psychotherapy. More importantly, however, the time has come for behaviorists to focus their attention upon the group itself. This is particularly true with regard to depression since this clinical disorder is not solely an intrapsychic phenomenon. Its clinical manifestations are intimately related to social transactions and therefore aptly suited to be treated within a group setting. Specifically, depressed individuals often believe that others in their social environment share a multitude of negative feelings about them. They feel they are disliked, thought to be inferior, looked at in a critical fashion, held in contempt, and the like. Moreover, many depressed individuals evidence quite dramatic social skills deficits. Conversations tend to be awkward and centered around various noxious topics (e.g., somatic complaints). Given the penchant of the depressed individual to isolate himself, it may be rather difficult to realistically correct these misconceptions and acquire socially adaptive skills. Following from this perspective, the group can be viewed as serving a variety of purposes. First, the depressed person is provided with the opportunity to explore many self-defeating beliefs, to learn that these beliefs are shared, and to learn that he is not alone. Also, it is in the group context, and not with the individual therapist, that the depressive is apt to engage in the type of deficit interactions which would characterize his relationships with peers. Therapeutic change is much a part of the group process. Peers can offer one another genuine feedback concerning both their assets and liabilities. Individuals may observe and model adaptive social skills. Not only does the depressed person have the opportunity to learn from others, but he also has the opportunity to help others to solve their problems. Such experiences are quite likely to enhance one's own self-esteem. Clinically relevant encouragement, support, and realistic feedback from group members are likely to motivate the depressed individual to engage in more adaptive and fewer maladaptive behaviors and thoughts. Importantly, changes which occur within this setting can be expected to transfer to the real world. Members are more apt to accept feedback from one another than from a paid professional. Once they have practiced asking a group member out on a date, it is easier for them to enact similar behaviors with a co-worker. Overall, the group context can foster interactions within a social microcosm, offer beneficial therapeutic experiences, and maximize treatment generalization.

Behavioral engineers should not only apply their skills to the development of discrete treatment packages, but also must begin to concern themselves with both interactional and structural aspects of the group context. Strategies must be developed to utilize group process as a therapeutic tool. Of equal importance, criteria which go beyond the diagnosis of depression must be established for group membership. What is being called for is a comprehensive conceptualization of behaviorally oriented group psychotherapy for depression. The ingredients of such a model should include a clearly articulated clinical intervention, behavioral strategies to promote therapeutically beneficial group interactions, and guidelines concerning group composition.

Group Process

Various authors have focused their attention upon the traditional manner in which behaviorists conduct group psychotherapy (Liberman, 1970; Lieberman, 1975; Rose, 1977; Shaffer & Galinsky, 1974; Yalom, 1970). Consensual agreement appears to have been reached in two key areas. First, the group leader tends to occupy an actively central role throughout the life of the group. It is the therapist who structures each session; who initiates, directs, and maintains communication; and who provides clinically relevant advice and therapeutically beneficial procedures for each of the group members. A corollary of this basic arrangement is that the interaction among group members is kept to a minimum. Rarely does one find published reports of behavioral groups in which the substance of group interaction is accorded at least equal therapeutic importance to the activities of the leader. It is as if group members were passive recipients of information emanating from the group therapist. Lieberman (1975) has aptly summarized this state of the art: "Those who have attempted to use a behavioral orientation in groups assume implicitly that the leader is in precise control of the situation—that he has the power to change, to desensitize, to model, and so forth, all the behavior that is desired. This implicit assumption limits the utility of behavioral approaches in the group context" (p. 443).

An alternative model would seek to diminish the individual therapist orientation while accentuating interpersonal interchanges (i.e., group process). How then does one begin to fashion a behavioral strategy which fosters therapeutic interactions among depressed individuals? Several researchers have analyzed the transactional patterns of groups which have been conducted by leaders of substantially different persuasions (Corsini & Rosenberg, 1955; Flowers, 1979; Lieberman, 1975; Yalom, 1970). On the basis of these investigations, various factors which are believed to promote clinically relevant change have been abstracted. Yalom (1970) enumerated 10 such "curative" factors. Included within this list are cohesion, altruism, universality, imparting of information, corrective recapitulation of family relationships, imitation, interpersonal learning, instillation of hope, social skills acquisition, and catharsis. Very similar factor clusters are offered by the other investigators.

One might consider utilizing the aforementioned parameters as a conceptual framework from which process-enhancing tactics are developed. The concrete operationalization of each factor would take on a different form depending upon the group's composition and goal (e.g., increased reinforcing activities, modification of dysfunctional cognitions, etc.). For example, the use of a buddy system could foster the development of cohesiveness. Buddies within a group which was aimed at increasing activity levels would be expected to help each other plan and/or carry out reinforcing activities. Group outings could also be planned. Analogously, buddies within a cognitively oriented group would be expected to acquire a working understanding of each other's belief systems, point out dysfunctional cognitions, and help correct faulty thinking. Libet and Lewinsohn (1973) have indicated that depressed individuals have a constricted range of interpersonal interactions. This behavioral style can be remediated by requiring group members to change buddies on a monthly basis. Hence, by designing relevant interactional strategies, the therapist is augmenting the treatment gains which the depressed individual would have made within the confines of the patient-therapist relationship.

Moreover, certain factors would be programmed to become active at specific points in the group's development. For example, early in treatment, individuals may be pessimistic, socially anxious, and reluctant to trust other members. At this point the behavioral engineer might implement strategies to instill hope, help members recognize their commonalities, and increase cohesiveness. For illustrative purposes, consider the following possibilities. The behavioral clinician could foster the instillation of hope or the establishment of positive expectancies by having successful members of a previous group detail and evaluate their experiences. For an assertion group the old members could comment upon the changes in their communicational styles. Individuals entering a cognitively oriented group could hear old members describing their newly acquired ability to differentiate between rational and depression-engendering thoughts. All new participants would be informed that meaningful change does not come easily but is quite rewarding when it is achieved.

Finally, individuals are likely to benefit from different aspects of the group interaction. Initially, a package could be designed which embodies the bulk of the curative factors. At a later time systematic dismantling strategies could be employed to determine the clinically effective elements of this shotgun approach.

To varying degrees, several of Yalom's fundamental factors are already incorporated within behavioral group treatments for depression. Members entering a behavioral group are usually offered a common treatment rationale. They may be told that depression results from faulty thinking or a limited social skills repertoire. Members are encouraged to incorporate this perspective by reflecting upon pertinent aspects of their own experience. Interpersonal feedback is a central feature of social skills acquisition groups (Hilford, 1975; Lewinsohn, 1974b; Lewinsohn et al., 1970; Rehm et al., in press). Participants are asked to help each other gain awareness of the consequences of their behavior. Imitation and interpersonal learning play a crucial role in these same groups. Depressed individuals are often required to role play problematic situations in front of the assembled group. Those who engage in the role-play exercise gain valuable practice in the enactment of troublesome behaviors, while other group members are able to learn through observation.

What is missing is an organized effort to employ process-oriented strategies within all behavioral group treatments of depression. Just as one habitually utilizes intervention packages, so should one make use of interaction packages. Such a package would detail relevant group exercises, interaction-facilitating group structures, and behaviors which the therapists could emit to encourage therapeutically efficacious social interchanges. Clearly, more research is needed in this area.

Group Composition

Depressed individuals who have received behaviorally oriented group psychotherapy have typically been chosen on the basis of their response to standardized clinical interviews (e.g., Hamilton Rating Scale for Depression, Hamilton, 1960) and/or self-report inventories (e.g., Beck Depression Inventory, Beck, Ward, Mendelson, Mock, & Erbaugh, 1961; Minnesota Multiphasic Personality Inventory–Depression Scale, Hathaway & McKinley, 1967). While diagnostic packages vary from study to study, inclusion criteria focus upon the *overall severity* of one's depressive symptomatology. Participants are generally required to be at least moderately depressed but not actively suicidal or in need of hospitalization. Having met these fundamental acceptance criteria, a subject will then be placed into a therapy group in which a predetermined intervention will be employed. Thus, a moderately depressed individual may receive social skills training (Hilford, 1975; Lewinsohn et al., 1970; Rehm et al., in press; Shaw, 1977), learn problem solving techniques (Shipley & Fazio, 1973), and/or examine and modify faulty styles of thinking (Fuchs & Rehm, 1977; Rehm et al., in press; Shaw, 1977). Therapy assignments are made without determining whether a given individual actually has a limited social skills repertoire, is generally unable to solve problems effectively, or interprets events in a depression-engendering manner.

Ideally, patients should be triaged into therapy groups which are designed to correct the specific cognitive or behavioral dysfunction that has resulted in a depressive episode. This goal can be realized by making use of pretreatment assessment packages which include two classes of measures. As noted earlier, a determination of the extent to which one is depressed should be made. The interested reader is referred to Rehm (1976) for an excellent review of available and recommended instruments. Having met a minimum cutoff level, the depressed individual would then be exposed to a second battery of diagnostic measures. Observational, interview, and self-report procedures would be utilized. Assertion deficits could be measured by the Wolpe-Lazarus Assertion Inventory (Wolpe & Lazarus, 1966), as well as by an audiotaped or videotaped situation test (Eisler, Hersen, & Miller, 1973; Rehm et al., in press). Activity level could be assessed by using a daily monitoring record and/or the Pleasant Activities Schedule (MacPhillamy & Lewinsohn, Note 3). The Self-Control Questionnaire (Fuchs & Rehm, 1977) could be employed to determine the presence of certain cognitive patterns. Other depression-inducing styles of thinking could be measured with experimentally derived inventories (Hammen & Krantz, 1976). Thus, a match could be fostered between a depressed person's idiosyncratic problem and the treatment intervention which she will receive.

Unfortunately, this endeavor may prove to be more difficult than it appears to be. The measurement of cognitive and behavioral dysfunctions while an individual is depressed may provide the clinician with spurious findings. The assessment of functioning during a depressive episode may result in the detection of state-related disabilities. For example, upon losing a spouse, a generally assertive person may become depressed, feel less self-confident, and become temporarily less assertive. This individual may be triaged into an assertiveness training group instead of a more appropriate grief or activities group. A potential solution to this problem would involve rigorous attempts by the researcher to determine both the causal factors which led to the depression and each individual's predepressive level of functioning. Such information could be obtained from a patient's family, friends, or work associates. Also, the directions for standardized measurement instruments could be reworded to pertain to one's nondepressed style of thinking and behaving. In this manner the therapist might be better able to determine those particular factors which underlie a given depressive episode.

A second problem must be solved before patients with specific dysfunctions can be effectively triaged into relevant psychotherapy groups, namely, that many depressed individuals are apt to evidence multiple symptoms. Into which therapy groups are these individuals to be placed? Research into a number of viable solutions is clearly indicated. First, perhaps the multiply symptomatic patient can be assigned to a relatively powerful treatment group. For instance, a socially isolated, unassertive, and cognitively impaired patient may benefit from an activity enhancement group. The performance of many pleasant events and the subsequent reflection upon one's active social participation may result in heightened assertiveness, self-confidence, and self-esteem. It is important to note that the efficacy of such a procedure would surely challenge the basic assumption of a patient-group matching approach. An alternative procedure would involve forming a heterogeneous group of multisymptomatic individuals. Various treatment strategies could then be systematically applied with the group as a whole. The advantage of this type of technique lies in the fact that individual members could function as task leaders when areas in which their own strengths lie are being addressed. Group members might learn more readily from a peer, and the task leader is likely to experience an increased sense of self-worth.

The latter point leads to an examination of the phenomenon of group composition from a different perspective. That is, can the judicious assignment of members to a group which is employing a given treatment strategy improve upon the efficacy of that same treatment intervention? What is needed is an empirical investigation of the effect which various membership constellations have upon treatment outcome, as well as an examination of the interaction between group membership and intervention strategy. By systematically selecting a particular group of depressed persons, one might be able to foster enhanced protherapeutic motivation, more intense learning experiences, and more meaningful exposure to relevant clinical issues. For example, consider the variable of a patient's age. Depressed persons entering an activities enhancement or assertion training group might benefit by sharing this experience with others who are at a similar developmental stage in life. Group cohesion, empathy, and motivation may be enhanced by the sharing of common life situations. Elderly depressed persons may prefer to focus upon the meaning of diminished instrumental skills, whereas college-aged persons may instead center their discussions around problems with dating, getting a first job, and so forth. Analogously, consider the variable of sex. A therapist may elect to assign an equal number of men and women to a cognitive modification group. Group participants can then examine various sex role beliefs with individuals who share those same attitudes, as well as with individuals whom they feel expect stereotypic behavior from them. Cognitive distortions can be addressed and modified. Depressed men may need to hear feedback from a woman in order to correct faulty and depression-engendering beliefs concerning their perceived need to always act strong, independent, and forceful. To take one final example, a problem solving group may benefit from a highly diverse cross-section of depressed individuals. Members should thus be able to offer advice based upon their own experiences. Practice in solving problems is fostered, altruistic feelings are encouraged, and group cohesion is strengthened. This may be preferable to a more restricted population in which members have quite similar seemingly unsolvable problems and can generate only a limited number of viable solutions.

By varying the membership within a group, the clinician can also help patients *maintain* already realized symptomatic relief. Certain group members may be asked to participate in two consecutively run groups. While both groups would utilize the same treatment intervention, an individual's role would be modified upon moving from one group to the next. In the second or follow-up group an old member may be defined as a cotherapist. He would be expected to assume many of the functions which the formal group leader would normally execute. He may also be assigned to work intensively with a number of new group members. By employing such a model, the old group member is required to overlearn important skills and should experience a heightened sense of importance in the role of teacher-adviser. Likewise, new group members should develop positive therapeutic expectancies, greater motivation to change, more opportunities to learn from interpersonal encounters, and so on.

Summary

In summary, behaviorists have been reluctant to adapt treatment methods in the group context. A multidimensional investigation of various parameters of behavioral group psychotherapy for depression is needed. In particular, behaviorists must be willing to develop comprehensive treatment packages which describe not only the primary treatment intervention, but which also detail strategies to enhance therapeutically efficacious intramember interactions and guidelines concerning the selection of group members.

Reference Notes

1. Martin, M. L., Weinstein, M. S., & Lewinsohn, P. M. *The use of home observation as part of the treatment of depression: The case of Mrs. B.* Unpublished manuscript, University of Oregon, 1968.
2. Lewinsohn, P. M., Alper, T., Johansson, S., Libet, J., Rosenberry, C., Shaffer, M., Sterin, C., Stewart, R., & Weinstein, M. *Manual of instruction for the behavioral rating used for the observation of interpersonal behavior.* Unpublished manuscript, University of Oregon, 1968. (Revised, 1971.)
3. MacPhillamy, D. J., & Lewinsohn, P. M. *Pleasant events schedule.* Unpublished manuscript, University of Oregon, 1971.
4. Lewinsohn, P. M., Zeiss, A. M., & Munoz, R. F. *Nonspecific improvement effects in depression using interpersonal skills training, pleasant activity schedules, or cognitive training.* Unpublished manuscript, University of Oregon, 1978.
5. Sanchez, V., & Lewinsohn, P. M. *Assertion in depressed versus nondepressed samples.* Unpublished manuscript, University of Oregon, 1976.
6. Youngren, M. A., & Lewinsohn, P. M. *The functional relationship between depression and problematic interpersonal behavior.* Unpublished manuscript, University of Oregon, 1978.
7. Libet, J., Lewinsohn, P. M., & Javorek, F. *The construct of social skill: An empirical study of several measures on temporal stability, internal structure, validity, and situational generalizability.* Unpublished manuscript, University of Oregon, 1973.
8. Schmickley, V. G. *A self-managed program for overcoming debilitating depression.* Paper presented at the American Personnel and Guidance Association Convention, Chicago, April 1976.
9. Beck, A., Rush, A. J., Shaw, B. F., & Emery, G. *Cognitive therapy of depression* (Treatment manual). Philadelphia: University of Pennsylvania, 1978.
10. Kovacs, M., & Rush, J. *Cognitive-behavioral psychotherapy versus antidepressant medication in the treatment of depression.* Paper presented at the meeting of the Eastern Psychological Association, New York, April 1976.

References

Abramson, L. Y., Seligman, M. E. P., & Teasdale, J. D. Learned helplessness in humans: Critique and reformulation. *Journal of Abnormal Psychology*, 1978, *87*, 49-74.

Arieti, S. (Ed.). *American handbook of psychiatry.* New York: Basic Books, 1974.

Beck, A. T. *Depression: Clinical, experimental, and theoretical aspects.* New York: Harper & Row, 1967.

Beck, A. T. *Depression: Causes and treatment.* Philadelphia: University of Pennsylvania Press, 1972.

Beck, A. T. The development of depression: A cognitive model. In R. J. Friedman & M. M. Katz (Eds.), *The psychology of depression: Contemporary theory and research.* New York: Wiley, 1974.

Beck, A. T. *Cognitive therapy and the emotional disorders.* New York: International Universities Press, 1976.

Beck, A. T., Ward, C. H., Mendelson, M., Mock, J., & Erbaugh, J. An inventory for measuring depression. *Archives of General Psychiatry,* 1961, *4,* 561-571.

Bibring, E. The mechanism of depression. In A. Greenacre (Ed.), *Affective disorders.* New York: International Universities Press, 1953.

Corsini, R., & Rosenberg, B. Mechanisms of group psychotherapy: Processes and dynamics. *Journal of Abnormal and Social Psychology,* 1955, *51,* 406-411.

Costello, C. G. A critical review of Seligman's laboratory experiments on learned helplessness and depression in humans. *Journal of Abnormal Psychology,* 1978, *87,* 21-31.

Coyne, J. C. Depression and the response of others. *Journal of Abnormal Psychology,* 1976, *85,* 186-193.

Coyne, J. C. Toward an interactional description of depression. *Psychiatry,* 1977, *39,* 14-27.

Depue, R. A., & Monroe, S. M. Learned helplessness in the perspective of the depressive disorders: Conceptual and definitional issues. *Journal of Abnormal Psychology,* 1978, *87,* 3-20.

Eisler, R. M., Hersen, M., & Miller, P. M. Effects of modeling on components of assertive behavior. *Journal of Behavior Therapy and Experimental Psychiatry,* 1973, *4,* 1-6.

Ekman, P., & Friesen, W. V. Nonverbal behavior and psychopathology. In R. J. Friedman & M. M. Katz (Eds.), *The psychology of depression: Contemporary theory and research.* New York: Wiley, 1974.

Ferster, C. B. A functional analysis of depression. *American Psychologist,* 1973, *28,* 857-870.

Ferster, C. B. Behavioral approaches to depression. In R. J. Friedman & M. M. Katz (Eds.), *The psychology of depression: Contemporary theory and research.* New York: Wiley, 1974.

Flowers, J. V. Behavioral analysis of group therapy and a model for behavioral group therapy. In D. Upper & S. M. Ross (Eds.), *Behavioral group therapy, 1979: An annual review.* Champaign, Ill.: Research Press, 1979.

Friedman, A. S. Minimal effects of severe depression on cognitive functioning. *Journal of Abnormal and Social Psychology,* 1964, *69,* 237-243.

Friedman, R. J., & Katz, M. M. (Eds.). *The psychology of depression: Contemporary theory and research.* New York: Wiley, 1974.

Fuchs, C. Z., & Rehm, L. P. A self-control behavior therapy program for depression. *Journal of Consulting and Clinical Psychology,* 1977, *45,* 206-215.

Hamilton, M. A rating scale for depression. *Journal of Neurology, Neurosurgery and Psychiatry,* 1960, *23,* 56-61.

Hammen, C. L., & Krantz, S. Effect of success and failure on depressive cognitions. *Journal of Abnormal Psychology,* 1976, *85,* 577-586.

Hammen, C. L., & Peters, S. D. Differential responses to male and female depressive relations. *Journal of Consulting and Clinical Psychology,* 1977, *45,* 994-1001.

Hammen, C. L., & Peters, S. D. Interpersonal consequences of depression: Reactions to men and women enacting a depressed role. *Journal of Abnormal Psychology,* 1978, *87,* 322-332.

Hathaway, S. R., & McKinley, J. C. *Minnesota Multiphasic Personality Inventory (Revised Manual).* New York: Psychological Corp., 1967.

Hilford, N. G. Self-initiated behavior change by depressed women following verbal behavior therapy. *Behavior Therapy,* 1975, *6,* 703.

Hinchcliffe, M. K., Lancashire, M., & Roberts, F. J. Depression: Defense mechanisms in speech. *British Journal of Psychiatry,* 1971, *118,* 471-472.

Hiroto, D. S. Locus of control and learned helplessness. *Journal of Experimental Psychology,* 1974, *102,* 187-193.

Johnson, W. G. Some applications of Homme's coverant control therapy: Two case reports. *Behavior Therapy,* 1971, *2,* 240-248.

Kanfer, F. H. Self-monitoring: Methodological limitations and clinical applications. *Journal of Consulting and Clinical Psychology,* 1970, *35,* 148-152. (a)

Kanfer, F. H. Self-regulation: Research issues and speculations. In C. Neuringer & J. L. Michael (Eds.), *Behavior modification in clinical psychology.* New York: Appleton-Century-Crofts, 1970. (b)

Kanfer, F. H. The maintenance of behavior by self-generated stimuli and reinforcement. In A. Jacobs & L. B. Sachs (Eds.), *The psychology of private events: Perspectives on covert responses systems.* New York: Academic Press, 1971.

Klein, D. C., & Seligman, M. E. P. Reversal of performance deficits and perceptual deficits in learned helplessness and depression. *Journal of Abnormal Psychology,* 1976, *85,* 11-26.

Lakein, A. *How to get control of your time and your life.* New York: New American Library, 1973.

Lazarus, A. A. Learning theory and the treatment of depression. *Behaviour Research and Therapy,* 1968, *6,* 83-89.

Lazarus, A. A. Multimodal behavioral treatment of depression. *Behavior Therapy,* 1974, *5,* 549-554.

Lewinsohn, P. M. A behavioral approach to depression. In R. J. Friedman & M. M. Katz (Eds.), *The psychology of depression: Contemporary theory and research.* New York: Wiley, 1974. (a)

Lewinsohn, P. M. Clinical and theoretical aspects of depression. In K. S.

Calhoun, H. E. Adams, & I. K. M. Mitchell (Eds.) *Innovative treatment methods in psychopathology.* New York: Wiley, 1974. (b)

Lewinsohn, P. M. The behavioral study and treatment of depression. In M. Hersen, R. M. Eisler, & P. M. Miller (Eds.), *Progress in behavior modification* (Vol. 1). New York: Academic Press, 1975.

Lewinsohn, P. M., & Atwood, G. E. Depression: A clinical-research approach. *Psychotherapy: Theory, Research and Practice,* 1969, *6,* 166-171.

Lewinsohn, P. M., Biglan, A., & Zeiss, A. M. Behavioral treatment of depression. In P. O. Davidson (Ed.), *The behavioral management of anxiety, depression and pain.* New York: Brunner/Mazel, 1976.

Lewinsohn, P. M., & Graf, M. Pleasant activities and depression. *Journal of Consulting and Clinical Psychology,* 1973, *41,* 261-268.

Lewinsohn, P. M., & Libet, J. Pleasant events, activity schedules, and depression. *Journal of Abnormal Psychology,* 1972, *79,* 291-295.

Lewinsohn, P. M., & Shaffer, M. The use of home observations as an integral part of the treatment of depression: Preliminary report and case studies. *Journal of Consulting and Clinical Psychology,* 1971, *37,* 87-94.

Lewinsohn, P. M., & Shaw, D. A. Feedback about interpersonal behavior as an agent of behavior change: A case study in the treatment of depression. *Psychotherapy and Psychosomatics,* 1969, *17,* 82-88.

Lewinsohn, P. M., Weinstein, M. S., & Alper, T. A behaviorally oriented approach to the group treatment of depressed persons: A methodological contribution. *Journal of Clinical Psychology,* 1970, *4,* 525-532.

Lewinsohn, P. M., Weinstein, M. S., & Shaw, D. Depression: A clinical-research approach. In R. D. Rubin & C. M. Franks (Eds.), *Advances in behavior therapy, 1968.* New York: Academic Press, 1969.

Liberman, R. P. A behavioral approach to group dynamics: I. Reinforcement and prompting of cohesiveness in group therapy. *Behavior Therapy,* 1970, *1,* 141-175.

Liberman, R. P., & Raskin, D. E. Depression: A behavioral formulation. *Archives of General Psychiatry,* 1971, *24,* 515-523.

Libet, J. M., & Lewinsohn, P. M. The concept of social skill with special references to the behavior of depressed persons. *Journal of Consulting and Clinical Psychology,* 1973, *40,* 304-312.

Lichtenberg, P. A definition and analysis of depression. *Archives of Neurology and Psychiatry,* 1957, *77,* 519-527.

Lieberman, M. A. Group methods. In F. H. Kanfer & A. P. Goldstein (Eds.), *Helping people change: A textbook of methods.* New York: Pergamon Press, 1975.

Lubin, B. Adjective checklists for the measurement of depression. *Archives of General Psychology,* 1965, *12,* 57-62.

Mahoney, M. J. The self-management of covert behavior: A case study. *Behavior Therapy,* 1971, *2,* 575-578.

Mahoney, M. J., & Thoresen, C. E. *Self-control: Power to the person.* Monterey, Calif.: Brooks-Cole, 1974.

Maier, S. F. Failure to escape traumatic shock: Incompatible skeletal motor responses or learned helplessness? *Learning and Motivation*, 1970, *1*, 157-170.

Meichenbaum, D. H. Cognitive modification of test anxious college students. *Journal of Consulting and Clinical Psychology*, 1972, *39*, 370-380.

Meichenbaum, D. H., & Cameron, R. The clinical potential of modifying what clients say to themselves. *Psychotherapy: Theory, Research and Practice*, 1974, *11*, 103-117.

Melges, F. T., & Bowlby, J. Types of hopelessness in psychopathological process. *Archives of General Psychiatry*, 1969, *20*, 690-699.

Miller, W. R., & Seligman, M. E. P. Depression and the perception of reinforcement. *Journal of Abnormal Psychology*, 1973, *82*, 62-73.

Miller, W. R., & Seligman, M. E. P. Depression and learned helplessness in man. *Journal of Abnormal Psychology*, 1975, *84*, 228-238.

Miller, W. R., & Seligman, M. E. P. Learned helplessness, depression and the perception of reinforcement. *Behaviour Research and Therapy*, 1976, *14*, 7-17.

Miller, W. R., Seligman, M. E. P., & Kurlander, H. M. Learned helplessness, depression and anxiety. *Journal of Nervous and Mental Disease*, 1975, *161*, 347-357.

Overmier, J. B. Interference with avoidance behavior: Failure to avoid traumatic shock. *Journal of Experimental Psychology*, 1968, *78*, 340-343.

Overmier, J. B., & Seligman, M. E. P. Effects of inescapable shock upon subsequent escape and avoidance learning. *Journal of Comparative and Physiological Psychology*, 1967, *63*, 23-33.

Rehm, L. P. Assessment of depression. In M. Hersen & A. S. Bellack (Eds.), *Behavioral assessment: A practical handbook*. Oxford: Pergamon Press, 1976.

Rehm, L. P. A self-control model of depression. *Behavior Therapy*, 1977, *8*, 787-804.

Rehm, L. P., Fuchs, C. Z., Roth, D. M., Kornblith, S. J., & Romano, J. A comparison of self-control and social skill treatments of depression. *Journal of Consulting and Clinical Psychology*, in press.

Rehm, L. P., & Kornblith, S. J. Behavior therapy for depression: A review of recent developments. In M. Hersen, R. M. Eisler, & P. M. Miller (Eds.), *Progress in behavior modification* (Vol. 7). New York: Academic Press, 1979.

Reisinger, J. J. The treatment of "anxiety-depression" via positive reinforcement and response cost. *Journal of Applied Behavior Analysis*, 1972, *5*, 125-130.

Robinson, J. C., & Lewinsohn, P. M. An experimental analysis of a technique based on the Premack Principle for changing the verbal behavior of depressed individuals. *Psychological Reports*, 1973, *32*, 199-210.

Rose, S. D. *Group therapy: A behavioral approach*. Englewood Cliffs, N.J.: Prentice-Hall, 1977.

Rush, A. J., Beck, A. T., Kovacs, M., & Hollon, S. Comparative efficacy of cognitive therapy and pharmacotherapy in the treatment of depressed outpatients. *Cognitive Therapy and Research,* 1977, *1,* 17-38.

Rush, A. J., Khatami, M., & Beck, A. T. Cognitive and behavior therapy in chronic depression. *Behavior Therapy,* 1975, *6,* 398-404.

Seligman, M. E. P. Depression and learned helplessness. In R. J. Friedman & M. M. Katz (Eds.), *The psychology of depression: Contemporary theory and research.* New York: Wiley, 1974.

Seligman, M. E. P. *Helplessness: On depression, development and death.* San Francisco: Freeman, 1975.

Seligman, M. E. P., & Beagley, G. Learned helplessness in the rat. *Journal of Comparative and Physiological Psychology,* 1975, *88,* 534-541.

Seligman, M. E. P., Klein, D. C., & Miller, W. R. Depression. In H. Leitenberg (Ed.), *Handbook of behavior modification and behavior therapy.* Englewood Cliffs, N.J.: Prentice-Hall, 1976.

Seligman, M. E. P., & Maier, S. F. Failure to escape traumatic shock. *Journal of Experimental Psychology,* 1967, *74,* 1-9.

Seligman, M. E. P., Maier, S. F., & Geer, J. The alleviation of learned helplessness in the dog. *Journal of Abnormal and Social Psychology,* 1968, *73,* 256-262.

Seligman, M. E. P., Rosellini, R., & Kozak, M. Learned helplessness in the rat: Reversibility, time course, and immunization. *Journal of Comparative and Physiological Psychology,* 1975, *88,* 542-547.

Seward, J., & Humphrey, G. L. Avoidance learning as a function of pretraining in the cat. *Journal of Comparative and Physiological Psychology,* 1967, *63,* 338-341.

Shaffer, J. B. P., & Galinsky, M. D. *Models of group therapy and sensitivity training.* Englewood Cliffs, N.J.: Prentice-Hall, 1974.

Shaw, B. F. Comparison of cognitive therapy and behavior therapy in the treatment of depression. *Journal of Consulting and Clinical Psychology,* 1977, *45,* 543-551.

Shipley, C. R., & Fazio, A. F. Pilot study of a treatment for psychological depression. *Journal of Abnormal Psychology,* 1973, *82,* 372-376.

Skinner, B. F. *Science and human behavior.* New York: Free Press, 1953.

Taylor, F. G., & Marshall, W. L. Experimental analysis of a cognitive-behavioral therapy for depression. *Cognitive Therapy and Research,* 1977, *1,* 59-72.

Vasta, R. Coverant control of self-evaluations through temporal cueing. *Journal of Behavior Therapy and Experimental Psychiatry,* 1976, *7,* 35-38.

Weissman, M. M., & Paykel, E. S. *The depressed woman: A study in social relationships.* Chicago: University of Chicago Press, 1974.

Wolpe, J., & Lazarus, A. A. *Behavior therapy techniques.* New York: Pergamon Press, 1966.

Yalom, I. D. *Theory and practice of group psychotherapy.* New York: Basic Books, 1970.

Chapter 11

Behavioral Group Therapy with Chronic Day Hospital Patients: Problems, Pitfalls, and Recommendations

Gary A. Williamson
Francisco X. Barrios

Abstract

A behavioral consultation experience in a fledgling day hospital program is described. The consultation demonstrated that highly successful behavioral techniques such as token reinforcement and response cost can fail in a particular setting through design error, staff resistance, patient countercontrol, and/or weak system support. It was learned that successful behavioral consultants must know the consultee's system, their own role in that system, and patient attitudes toward the proposed treatment approach before implementing an intervention program.

As any well-read student of psychology knows, the only applied clinical research projects ever published in journals are the ones which yield positive results. There may be an exception or two here or there, but the general trend for years has been to emphasize experimental results over methodology. Behavior therapists have not been immune to this virus. Witness the table of contents of any volume of *Behaviour Research and Therapy,* the *Journal of Applied Behavior Analysis,* or *Behavior Therapy.* As practitioners, we have become obsessed with positive results, even though the scientist in us knows better (Greenwald, 1975).

For those interested only in positive results, this chapter will be a disappointment. As can be inferred from the title, the project described herein obtained negative results. The purpose of recounting some of the frustrations of behavioral group therapy with chronic patients is to encourage well-meaning but inexperienced behavioral consultants to proceed cautiously in planning and implementing intervention programs. The chapter is organized into five parts. First, there is a brief introduction to the population of chronic psychiatric patients and to the setting in which the project took place. Second, there is a detailed account of the method used to plan and carry out the project itself. Third, the results of the project are presented in graphic form. Fourth, a post mortem interpretation of the results is given. Finally, three caveats for future behavioral consultants are issued.

Project Population and Setting

It is generally agreed that the population of chronic psychiatric patients residing in inpatient settings has decreased considerably in the past 25 years (Paul & Lentz, 1977). This trend usually has been attributed to the introduction of psychoactive drugs, the return of the moral treatment philosophy so prevalent in the last century, the decentralization of state hospitals, and the emergence of the community mental health movement (Nietzel, Winett, MacDonald, & Davidson, 1977).

This trend toward decreased institutionalization, however, has not been an unmixed blessing. For one thing, as Paul (1969) pointed out, long-term patients were not participating in this decrease. Another drawback appears to be the fact that many psychiatric patients were and are being released into the community without adequate preparation for dealing with the environmental stressors that may have helped to hospitalize them in the first place. Thus, the stage appears to be set for a revolving door phenomenon, characterized by high rates of recidivism that have been estimated at anywhere up to 70%.

This rather discouraging state of affairs has persisted, even in the face of massive commitment by government and mental health professions in the 1960s to the idea of community-based mental health treatment (Goldston, Note 1). One possible reason for this apparent failure of community-based treatments to keep psychiatric patients out of the hospital for longer periods of time may be that far from constituting a unitary approach to psychological intervention, the community mental health movement has been plagued by many of the same conflicts which have bedeviled traditional clinical psychology and psychiatry. As Zax and Specter (1974) have pointed out, the community mental health label has been used to cover a wide spectrum of activities that share little in common except the name. At one end of the spectrum are individuals who talk of *community psychiatry* and who view it as a subspecialty of psychiatry requiring medical training for its practice (Bernard, 1964). According to proponents of this viewpoint, the illness model of behavior disorders is quite valid, and community psychiatry is distinguished from traditional psychiatry only in its acknowledgment of social forces as relevant etiologic agents in the development of individual behavior disorders. At the other end of this spectrum are those persons, usually trained as social scientists, who call this field *community psychology* and who directly challenge the basic assumptions of traditional mental health delivery systems. According to these individuals (Denner & Price, 1973; Iscoe, 1971), community psychology should be characterized by: (a) a preference for seeking out persons in need of mental health services, rather than waiting for them to seek out the professional; (b) a strong preference for eschewing intrapsychic processes as the relevant antecedents of behavior disorders in favor of social, interpersonal, and cultural variables; and (c) the abandonment of political and moral neutrality and the adoption of an advocate role in order to help disadvantaged individuals change the social conditions that contribute to their distress, rather than helping them adjust to these situations.

The mental health delivery system in which our project was nested is not exempt from the above-mentioned ideological and political pressures. It is a state-funded/county-administered operation responsible for public sector mental health care in and around Houston, Texas, and its avowed goal is the provision of mental health services exclusively to high risk clients, who are in imminent danger of rehospitalization. Partially as a result of this goal, the predominant mode of treatment in this system is psychopharmacological, while psychologically based interventions have been either deemphasized due to the "unsuitable" nature of the clientele, or have consisted of watered-down versions of insight-orientated psychotherapy.

The project took place in a predominantly Mexican-American, lower socioeconomic class neighborhood near the Houston Ship Channel, and the facility was a converted two-story boarding house which was never completely converted. The mental health unit is only one of more than a dozen service agencies housed under the same roof, including such diverse entities as a medical clinic, an office of the state rehabilitation commission, and a city police substation. However, unlike these latter facilities, the mental health unit is underutilized by the neighborhood population (Acosta, 1977), and so the primary recipients of mental health services are poor blacks and whites.

The mental health unit during 1977-1978 consisted of two parts. One part offered individualized chemotherapy and psychotherapy. The other part was a rudimentary day hospital that emphasized group interaction of clients, along with medication. Most clients of the mental health unit participated in the individualized program, leaving only approximately 20 clients for the day hospital component. Total staff for the unit included one psychiatrist, two psychiatric residents, a master's-level psychologist, one registered nurse, and five aides. The day hospital received the assistance of only 5 of the 10 staff members, and that much only on a part-time basis.

Prior to the present project, the day hospital had been operating for about 6 months. The program incorporated recreational activities, occupational therapy, movies, and group psychotherapy. Three groups of six to eight patients met for group therapy roughly three times per week. One group was composed of chronic patients who had been out of the hospital for some time and who were stabilized on medications. The second group consisted of patients who were not only stabilized but were also attempting to reintegrate themselves into the community via vocational rehabilitation or employment. The third group was made up of chronically psychotic patients who had recently been released from a state psychiatric facility and had not yet settled back into community life.

Given the policies of the mental health system, the inadequate facilities in which to operate, the poor staff-to-patient ratios, and the difficulty of working with chronic patients, a new unit administrator sought consultation on innovative ways of treating his patients.

Establishing the Consultation Program

The unit administrator, a psychiatrist with a faculty appointment at a nearby medical school, informally called on the second author to help improve the quality of care at the fledgling day hospital of his mental health unit. The second author, a new Ph.D. with solid behavioral training, was eager to accept this responsibility but had barely enough time to provide adequate service. The first author, meanwhile, was looking for an externship to complement his VA hospital internship and offered to assist in the consultation. Several discussions later, we began formulating plans for intervention at the mental health unit.

Planning the Intervention

The first step in formulating concrete plans for intervention was to search the literature for demonstrated successful approaches with chronic patients. Partly because of our behavioral bias and partly because of the extant literature, we discovered that the use of operant conditioning principles with this population has become well documented in recent years (Atthowe & Krasner, 1968; Ayllon & Azrin, 1968; Kazdin, 1977).

However, we also found that behavioral changes obtained in the hospital setting often are not transferred to the natural environment upon discharge (Atthowe, 1973). One reason for this phenomenon, we discovered, was the failure of the mental health system to provide continuity of care for the recently discharged chronic psychiatric patient (Nietzel et al., 1977). Our reading suggested that a day hospital could provide high quality continuity of care to patients by means of behavior therapy.

Our decision to recommend the group treatment modality was motivated by two major considerations. First, most psychiatric problems are social-interactional in nature, and group treatment provides an opportunity for observing problem behaviors and practicing more adaptive behaviors in a protected setting (Rose, 1977). Second, group therapy seems to be less expensive than individual treatment in terms of staff time and salaries (Upper & Ross, 1977). In short, the literature suggested that behavioral group therapy might be an efficacious and economical treatment strategy with the target population at the mental health unit.

The second step in planning to intervene at the mental health unit was to provide rudimentary training in behavior modification principles. Having recently stepped out of the classroom ourselves, we assumed that everyone in the mental health field had a burning desire to learn alternative methods of patient care. While this assumption did not prove fatal to the project, in effect it ignited a slow fuse which eventually detonated two months later. As we in time learned, the staff at large had only minimal interest and motivation to study behavior modification. Nevertheless, we distilled the content of Kazdin's (1975) *Behavior Modification in Applied Settings* and taught the staff basic principles in three inservice training sessions. Session 1 involved the presentation of concepts of reinforcement, punishment, schedules, and charting. Session 2 was comprised of applications of operant methods to the problem of anxiety. Session 3 included actual case presentations of patients under the care of the mental health unit. At the conclusion of the third inservice training session the staff was informed by the administration that in coming weeks the consultants would be setting up a treatment program which utilized the principles that had just been taught. At this point we did not realize that the staff had been coerced into acceptance of the project. All we knew was that one staff member was really excited about the possibilities of an alternative treatment approach for her rather regressed patients. The rest of the staff appeared almost indifferent to the intervention.

Designing the Treatment Package

Selection of participants. It was the motivated young therapist around whom we chose to build the intervention. For one thing, she was receptive to our ideas; for another, she was assigned to the most difficult patients at the mental health unit. If we could effect change with this handful of regressed patients, we thought, we would have a firm foundation on which to build larger interventions. On a conscious level, we wanted to keep the study small because we were thinking in terms of a demonstration project. Unconsciously, however, perhaps we were keeping it small because we already sensed resistance. In either event, our next contact was with the enthusiastic therapist.

During this visit it became apparent that the major source of frustration for the therapist was the ingroup behavior of her patients. She felt as though she had no control over them and could not even begin to conduct group psychotherapy. As consultants, we suggested that she invite us to one of her group therapy sessions so that we could observe her difficulties in vivo. What we encountered upon joining the group for one session was, on one hand, an unruly and self-centered covey of psychiatric patients and, on the other hand, a relatively inexperienced and permissive group therapist. The combination of the two was hardly conducive to group effectiveness. Patients walked in and out of the group at will, sat in stuporous silence, started private conversations, belched out loud, or monopolized group time. Meanwhile, the therapist asked the patients what was on their minds today and how that made them feel.

Goals and target behaviors. After empathizing with the therapist in the debriefing period following the group session, we talked with her in terms of goals she had for the sessions. It was clear that she wanted to increase certain desirable behaviors in her patients, while simultaneously decreasing undesirable behaviors. In effect, she wanted to teach her patients how to behave in group therapy sessions and assumed that once patients behaved like group members, therapy would occur. We decided not to challenge this latter assumption and instead focused on helping the therapist to increase desirable and decrease undesirable patient behaviors in group therapy sessions. With our assistance, she was able to identify the following behaviors to be increased by each group member:

1. Labeling personal feelings.
2. Discussing personal-life problems.
3. Soliciting feedback from others.
4. Presenting a clean, neat appearance.
5. Attending group therapy sessions.

The behaviors to be decreased by each member were:

1. Interrupting another speaker in group.
2. Leaving the group meeting early.
3. Changing the subject prematurely.
4. Bumming cigarettes from others.
5. Remaining silent for much of the session.

Treatment techniques. Once the target behaviors were chosen, the next step was to select treatment techniques to modify those behaviors. The techniques of token reinforcement and response cost were chosen because of the widespread success of token economies in recent years (Kazdin, 1977; Stahl & Leitenberg, 1976). Part of the appeal of token systems is that they provide direct feedback to patients. However, such feedback in the context of a group therapy session could prove disruptive. The compromise solution to this dilemma in our case was to provide immediate nonverbal feedback on behavior during the therapy sessions and to delay the rewards until later. Feedback during the sessions was given on a large blackboard on which were listed patient names and the 10 target behaviors. A trained observer checked off a behavior each time it was displayed during therapy and added up the tallies at the end of the session. As a result, patients received immediate individualized feedback via the blackboard, as well as summary feedback via the points earned for behaviors exhibited.

The immediate feedback component of the treatment technique required a person with well-developed observation skills. A quick survey of available on-site staff revealed that the most likely candidate for the job was the day hospital bus driver. The observer was given written operational definitions of the target behaviors and was trained in behavior recording in vivo during a prebaseline period of 2 weeks in group therapy sessions. Once the observer and the group therapist reached a level of 80% agreement on the occurrence of the 10 monitored behaviors, the observer was considered trained. Consensus was determined in a post-session debriefing meeting which was supervised by the first author. Pleased as we were with the reliability, dependability, and enthusiasm of the trained observer, we decided to train a mental health aide as a backup observer to cover for the bus driver in the event of illness or unforeseeable circumstances. We utilized an analogous procedure to train one of the other group therapists as our backup group therapist as well. As the project unfolded, we came to appreciate our own contingency planning.

Patient preparation. At the same time in which the observers were being trained, the patients were also being prepared for the new treatment plan. Patients were requested to complete the Reinforcement Survey Schedule (Cautela & Kastenbaum, 1967) so that we could ascertain what specific objects would be rewarding. Reinforcers such as cigarettes, candy, privileges, and games were identified by the patients in this manner. Then, day hospital staff, along with the consultants, determined the point values necessary to obtain these rewards. Point values were generally ascribed on the basis of the difficulty and/or expense involved in providing the reinforcer. At this time point values were also assigned to the target behaviors on the basis of their relative desirability.

Quite predictably, once the patients completed the Reinforcement Survey Schedule, they asked if there was a change in the wind. Unfortunately, the consultants were not onsite when the question was asked, so day hospital staff, rather than delay an answer, gave a crude explanation of the upcoming intervention. The immediate effect of this explanation was to quiet the patients. However, over the next 2 weeks, they kept asking about what they would lose with the new system and if the change would be a permanent one. The introduction of the observer to the group also occurred during this 2-week period, giving the patients further evidence of the change to come. In retrospect, it seems that the day hospital patients were not adequately prepared for these preintervention changes. A better strategy would have been to give the formal introduction and explanation of the token system prior to the administration of the reinforcer survey. We thus might have reduced much anticipatory anxiety.

Evaluation strategy. The final step in designing the treatment package was to build in an evaluation technique. We elected to employ an ABAB intensive research design in this project, with each participant serving as his own control (Hersen & Barlow, 1976). We chose this design for several reasons: the small number of participants, difficulty in obtaining matched controls, and our personal preference for graphic analysis. Graphic analysis, in particular, was important because we were more interested in clinical significance than in statistical significance, and we thought it was important to forcefully demonstrate the intervention's effectiveness (Kazdin, 1975). The first baseline period was instituted for seven therapy sessions over 3 calendar weeks. (Group sessions met only three times per week because of the limited staff at the day hospital.) During this period the observer sat in the group and charted the 10 target behaviors of each patient on a clipboard. No feedback was given to the patients at this time. Following the baseline period, the point system was introduced for 10 sessions, and patients were allowed to exchange points earned for objects they had rated as rewarding. Points were fined for the display of undesirable behavior in these sessions. During each therapy session in the treatment period, tallies of each patient's behaviors were recorded on a large blackboard as they occurred. After each session a patient's total points were registered in a bank book. Reinforcers were sold at the clinic store each day at the conclusion of the therapy session.

After the 10-session intervention period, a reversal period of 7 sessions was instituted. The observer returned to her chair and clipboard, recorded the target behaviors for each patient, and gave no feedback on performance. On completion of the second baseline period, a second treatment period of 10 sessions was initiated. In this period the token reinforcement and response cost paradigm employed earlier was utilized. The total duration of this experimental phase of the project was determined by the holiday and vacation schedule of the day staff and the consultants.

Troubleshooting the program. Two problems arose in the course of the ABAB design. One problem was anticipated: the absenteeism of participating staff. During the second baseline period the reliable, dependable bus driver became unreliable because of family illness and continued to have periodic absences for the remainder of the project. The group therapist also missed several sessions in the latter half of the experimental phase. Fortunately, trained substitutes were available for both of these personnel, but the validity of their performance was not monitored.

The second problem in the project was not anticipated. As originally conceived, the program called for each patient to earn points for desirable behaviors and to be fined points for undesirable behaviors. However, certain patients displayed so much undesirable behavior that they could never work themselves out of debt. To remedy this situation, we decided that each patient's daily point total could not fall below zero. In this way, each therapy session presented new prospects for earning rewards. Subsequent data analyses were computed on this zero-based system.

For the purpose of formal analysis the experimental phase of the project ended after the tenth session of the second treatment period. Although the subjective impression of the day hospital staff was that the treatment was working, we were eager to see the data for ourselves. The results of the project from a prebaseline practice phase through the second treatment period are presented below.

RESULTS

The behavioral results for the total group of seven patients who completed all experimental phases were analyzed in three separate ways. First, a simple frequency count of the number of on-target and off-target behaviors was computed. Second, in order to take into account the differential weights assigned to various behaviors, we tallied the total number of points earned and fined in the token system. Finally, the individual participants' point totals across the duration of the intervention were computed to obtain an estimate of individual differences. All results are represented in terms of the five phases of the project: Prebaseline (Practice), Baseline I, Training I, Baseline II, and Training II.

Behavioral Frequencies

Figure 11.1 depicts the daily frequency of target behaviors across different phases of the project. As can be seen, there were meaningful gains in the number of on-target behaviors during Training I, while off-target behaviors continued essentially at Prebaseline (Practice) level. In the reversal from Training I to Baseline II, off-target behaviors increased dramatically and then dropped back to Baseline I level. On-

Figure 11.1. Daily Frequency of Target Behaviors

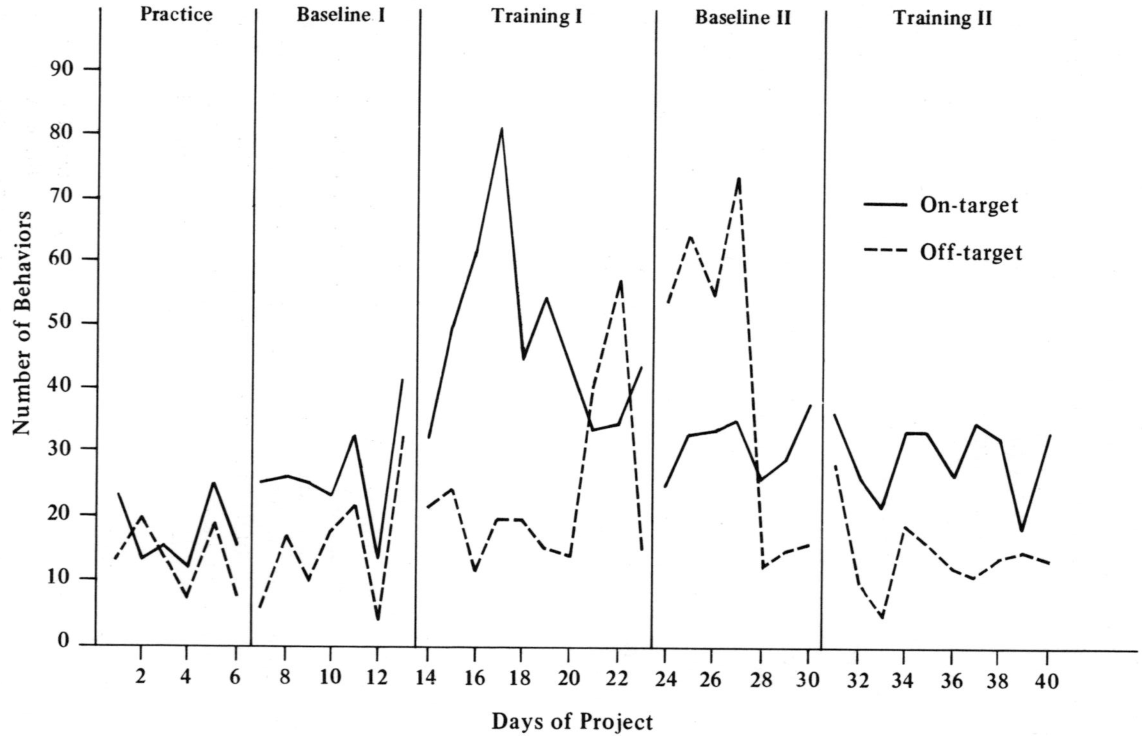

target behaviors, meanwhile, immediately decreased to Baseline I level and remained there. The final treatment period showed little change in either on-target or off-target behaviors from Baseline I. In short, a simple frequency count of target behaviors revealed that the treatment package did not achieve functional control over the target behaviors.

Instead, it appears that the behavioral treatment increased the variance in both on-target and off-target behaviors across the ABAB design. This suggests that the treatment had some influence on the patients' behaviors, if not functional control over them. Not only did the introduction of the token system correlate with an increase in on-target behavior, but also the removal of the tokens correlated with an increase of off-target behavior. Certainly when reinforcement is removed, a decrease in on-target behavior is expected. However, we also observed an increase in off-target behavior, which was entirely unexpected. It is tempting to infer that the patients, in protest of the loss of reinforcement, pulled a job-action during group therapy. However, if this was the case, they carried the protest too far, for when the token system was reestablished, there was essentially no change in the frequency of on-target or off-target behaviors. Whatever the interpretation given, one thing is certain: there was more behavior emitted (both on-target and off-target) during Training I and Baseline II than at any other time in the project.

Points Earned and Fined

The results in terms of points earned and points fined are presented in Figure 11.2. These data, which correct for weighted value of various target behaviors, show a pattern similar to that of the simple frequency data. Points earned increased markedly during Training I, and points fined increased sharply during Baseline II. In addition, points fined during Training II were roughly equivalent to points fined during Baseline I. The only difference between the two forms of the data analysis was that in the weighted analysis, points earned finally stabilized at a level significantly above baseline, while in the simple frequency analysis on-target behaviors stabilized right at baseline level. This would seem to indicate that the group members were beginning to engage more readily in those behaviors which were weighted more heavily by staff at the outset of the program. Unfortunately, the termination of data collection prevents us from knowing whether this trend continued.

Individual Differences

The third form of data analysis for the project involved the inspection of individual differences in performance. These results indicated that individual variables such as age and ethnicity appeared to be relevant

Figure 11.2. Mean Number of Points Earned and Fined in Experimental Phase

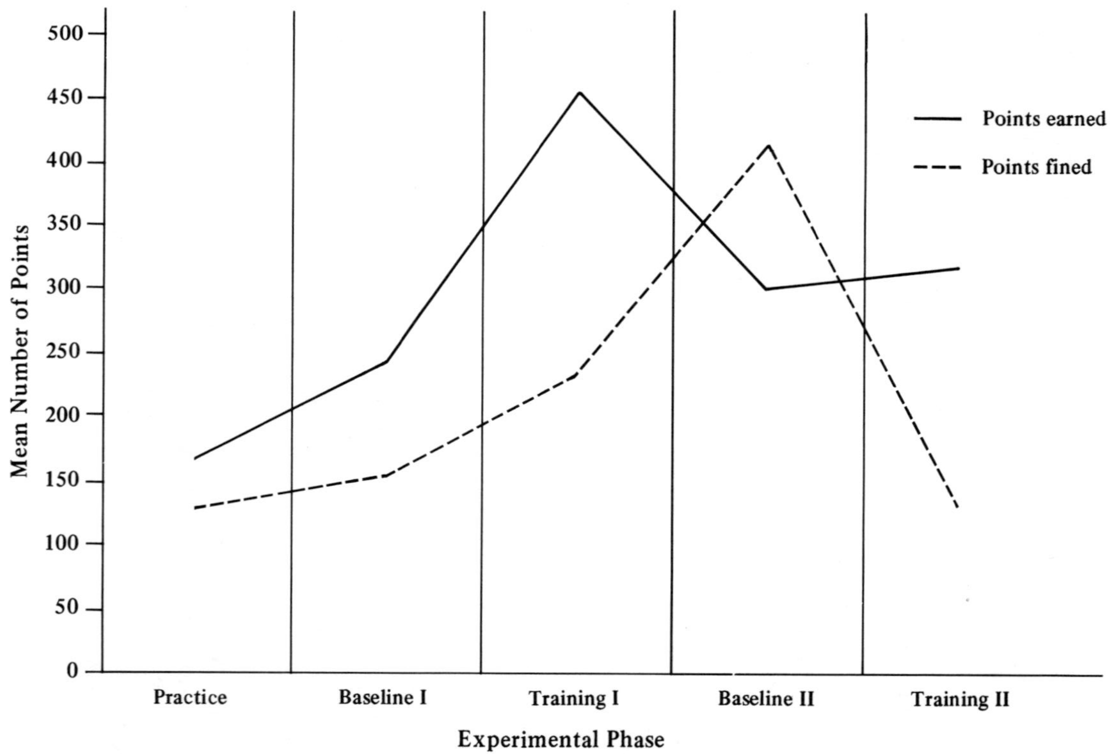

in predicting the degree of responsiveness to this program. For example, one of the subjects, who was the only Mexican-American in an otherwise all-black group, manifested minimal responsiveness throughout the study, perhaps as a result of difficulty in understanding and communicating with the therapist and other group members. Similarly, another member, who was about 10 years older than the rest of the group members, accounted for a disproportionate amount of on-target behavior for the majority of the sessions. Finally, two relatively young participants accounted for a good deal of the disruptive behavior observed in Baseline I. These findings would seem to argue that the criteria of being labeled schizophrenic and recently released from a state hospital are apparently insufficient or irrelevant in attempting to select a group of clients for similar treatment. Thus, it would appear that by accepting the clinic's established guidelines for separating clients into groups, we inadvertently fell victim to the client uniformity myth warned against by Kiesler (1966).

Discussion

Clearly, the subjective impressions of the staff that the treatment was working were not entirely accurate. True, there was a clinically significant increase in on-target behaviors during the first training session and a predictable decrease in these behaviors when rewards were not provided. However, on-target behaviors did not increase as expected in the second training session; therefore, functional control of the target behaviors by the behavioral intervention was not achieved. It was not possible empirically to demonstrate the efficacy of behavioral group therapy in this project. The reasons for failure to achieve experimental control are varied, but they can be clustered into four categories: design error, staff resistance, patient countercontrol, and weak system support. The purpose of discussing these reasons for failure is twofold: first, by doing so we hope to relieve our own frustrations over negative results, and second, we may prevent other investigators from inadvertently falling into the same traps. In this section, then, the four sources of error variance will be discussed and caveats for future investigators will follow.

Design Error

The category of design error includes all mistakes made by the consultants during the project. One of these mistakes may have been the brevity of treatment periods. Group sessions were conducted only three times per week, and each session lasted for less than 1 hour. Moreover, the actual treatment periods themselves ran for just over 3 weeks. Perhaps the relatively short and interrupted treatment periods of the project provided insufficient time for new learning to occur. As consultants, we could have insisted on more frequent therapy sessions and longer treatment periods.

Another error we made was in selling the project to the day hospital staff. In fact, we broke every one of Bellack and Franks' (1975) rules for behavioral consultation. First, we tried to teach theoretical underpinnings instead of on-the-spot application of techniques. Second, we used a didactic approach as opposed to a case conference presentation model of education (with one exception). Third, we neglected to actively model or demonstrate the techniques we taught the staff. And fourth, we challenged the staff's ideology by using behavioral jargon instead of their own language. In short, we read Bellack and Franks (1975) too late.

The other major error made by the consultants was to train participating staff inadequately. At the onset of the project we spoon-fed behavior modification techniques to the staff and did not require them to practice or role play the techniques under supervision. As the project progressed, more than the usual amount of skill-maintenance activities was necessary. It became apparent during the second treatment period that the staff was still in the skill-acquisition phase. Meanwhile, we were vainly trying to maintain skills in staff who had not yet acquired them.

Staff Resistance

Unquestionably, staff resistance was present in various forms during the project: absenteeism, haphazard data collection, tardiness for consultation sessions, complaints about work load, failure to adhere closely to therapy procedures, etc. According to Hersen and Bellack (1978), staff resistance to consultation can result from inadequate staff preparation, conflicting goals of staff and consultants, incomplete training of staff, and/or ineffective staff consultation once a program is underway. Staff resistance in our project could have been generated by any of these four situations. For example, although we were aware from the beginning of the administration's ultimate goal of employing behavioral techniques in the day hospital, the staff was not informed of this decision until the third inservice training session we conducted. As a result, the staff was inadequately prepared for the consultation and may have harbored different goals from those of the consultants and the administration.

In addition, the staff seemed to accept implicitly numerous quasi-psychodynamic concepts about the necessary and sufficient elements of psychotherapy, yet they never verbalized any negative or skeptical attitudes about behavior therapy. Perhaps a covert conflict between staff and consultants over the type of therapy being applied manifested itself in the form of staff resistance. Finally, staff resistance may have resulted from incomplete staff training or from ineffective staff consultation once the program was underway.

Patient Countercontrol

This category of explanation for negative results includes both the unwitting and insidious behaviors of patients that undermine therapy. The patients in this project were not involved in the identification of target behaviors of the therapy group. They had no voice concerning the length of treatment periods or when the reversal from treatment to observation should take place. Thus, they may have felt like the proverbial guinea pigs in an experiment, and they may have displayed anger about this by increasing off-target behaviors and decreasing on-target behaviors in the second half of the program. The results in such a case would be the informal job action mentioned earlier in the chapter.

Three other precipitants of patient resistance may have operated during the token program. First, there may have been interference from psychotic symptomatology during group sessions. Although the observable manifestations of hallucinations, delusions, and thought disorder decreased during therapy sessions, perhaps these symptoms persisted on a covert level. If so, the acquisition and maintenance of desirable target behaviors could have been impaired. Second, medication also could have impaired learning. Throughout the program, dosage levels were titrated on some patients and new medications were prescribed for others. Few patients remained on a constant dose of a specific medication for the full program. Consequently, an unstable chemotherapy regimen may have altered patients' abilities to increase desirable and decrease undesirable target behaviors. Third, differences in patient value systems may have contributed to the program's outcome. As already noted, the patients in the project constituted a culturally heterogeneous group. The values of middle-aged black males are decidedly different from those of young Mexican-American males; the values of lower-class males are considerably different from those of middle-class males; and the values of chronic schizophrenic patients differ from those of day hospital staff members and those of consulting psychologists. It may be that patients somehow resented the imposition of our middle-class target behaviors on their particular behavior repertoires.

Were we to select the target behaviors today, we would utilize the following approach. First, we would identify the predominant ethnic background and age of group members. Second, we would call in a paraprofessional staff member or indigenous community member without emotional problems to advise us on the appropriate behaviors of the predominant ethnic group. The advisor might even be utilized as a cotherapist in the therapy group. Third, during group sessions we would make an attempt to differentiate white middle-class social skills from street behavior. Patients could then be taught to determine when one of these styles is more effective than the other and could use role playing to practice appropriate behaviors in each style.

Weak System Support

The level of explanation for negative results involving system support has to do with the participation in the project of the on site administration. Hersen and Bellack (1978) have observed that consultation failure can usually be traced to poor administrative support and staff resistance. They also point out that consultation typically is initiated without the staff's initial interest, recommendation, or approval. This results in a happy administration which sires an innovative program, and an unhappy staff which undermines it. Clearly, in our case, the administration made a unilateral decision to bring in consultants. This was the last time the administration provided support of this project. We, the consultants, were not permitted to have any authority or clinical responsibility for the project. In fact, once the token program was designed it had to be scrutinized and evaluated by three administrative committees before it could be implemented. The status quo programs of the day hospital were not subjected to such examination.

Another example of weak system support pertains to the noncontingent reinforcement in other facets of the day hospital program. The administration of the clinic decided that behavioral group therapy could be offered only 3 hours per week. As a result, patients obtained behavioral treatment only 9% of the time available each week. In such a system it is unlikely that maladaptive behaviors will be decreased and new/adaptive behaviors will be acquired. Learning may not have occurred in this program because there was too much interference from stimuli in the rest of the day hospital.

Finally, poor administrative support manifested itself in the patient population which was made available for the day hospital program. There were many patients under the care of the mental health unit, but only the most regressed patients participated in the day program. Many of these patients had been prematurely discharged from inpatient status at the state hospital. Interestingly, many of the patients who received individual therapy at the unit would have been good candidates for the day hospital. They could have been models for the more disturbed patients. Unfortunately, the administration did not view such integrated patients as appropriate for a day hospital program, and so whenever one joined the therapy group, he was soon discharged. This was very frustrating for all treatment staff concerned.

Suggestions for Future Consultants

One way in which we can exorcise our frustrations about the results of this project is to provide warning to other well-meaning behavioral consultants. Attention to these warnings is not guaranteed to prevent consultation problems or even to provide positive treatment results. Rather, the advice is offered to maximize the chances of success while preserving the consultants' own mental health during a project.

First and most important, behavioral consultants need to specify the nature of their relationship to the consultees from the beginning. Consultants need to know whether they are wanted, and by whom. They also need to be aware of their authority and responsibility in the consultees' system and must have a clear understanding of the amount of backup and support that the administration will provide. To agree to consult in a system in which these parameters are not explicit is to ask for trouble. Consultants must know their roles in the system from the consultees' point of view.

Second, consultants should not attempt to take on a challenge by trying to implement a behavior therapy project in an essentially hostile system. Contrary to academic fantasies, the mental health network has not "gone behavioral"; clinic staffs are not eagerly waiting for behavioral consultants to stand up and speak "the truth." For many mental health workers behavior therapy poses a two-pronged threat. It is an ideological threat to the more dynamic and client-centered therapy dispensed in many community clinics, and it is a pragmatic threat to many staff members because of extra record keeping and reward giving required. Given a choice between the status quo, which many staff feel is not so bad, and a behavioral program which mandates more effort, most consultees will opt for the status quo. Before implementing a behavioral consultation project, then, it is important to know the system's attitude toward behavior therapy.

Third, consultants need to involve the patients in planning the project, and they should not rely solely on staff opinions about the kind of population with which they will be working. Sitting down eyeball-to-eyeball with the patients and getting to know them is the best way to find out what is really important to them and what their values are. It is important to explore any short-term goals that patients may have and assess the feasibility of these goals. Feeling out patients about different forms of behavioral treatment and about the length of treatment and baseline periods is also essential. The information obtained from these encounters will affect not only the nature of the intervention but also its success. In short, it is best to involve the patients in designing the intervention.

Having read the foregoing material, one might wonder why we, as consultants, were not aware of the potential pitfalls in our project before beginning the intervention. Our answer is twofold. First, although both of us graduated from rigorous APA-approved graduate programs and internships, we had little formal training and limited experience in community consultation. In a sense, we were insufficiently prepared to perform this task and got in over our heads. Second, most of our reading and thinking about behavioral consultation occurred *after* the failure of the health unit intervention. This resulted in good copy for a book chapter but had no impact on the project itself. We owe our current competence in behavioral group therapy to a successless consultation in an unreceptive system. In the process we learned the meaning of the words of Sir William Osler, the great teacher of physicians, who said, "It is always best to do a thing wrong the first time" (Strauss, 1968).

Reference Notes

1. Goldston, S. E. *Concepts of community psychiatry: A framework for training* (U.S. Public Health Service Publication No. 1319). Bethesda, Md.: U.S. Department of Health, Education, and Welfare, 1965.

References

Acosta, F. X. Ethnic variables in psychotherapy: The Mexican American. In J. L. Martinez (Ed.), *Chicano psychology*. New York: Academic Press, 1977.

Atthowe, J. M. Behavior innovation and persistence. *American Psychologist*, 1973, *28*, 34-41.

Atthowe, J. M., & Krasner, L. Preliminary report on the application of contingent reinforcement procedures (token economy) on a "chronic" psychiatric ward. *Journal of Abnormal Psychology*, 1968, *73*, 37-43.

Ayllon, T., & Azrin, N. H. *The token economy: A motivational system for therapy and rehabilitation*. New York: Appleton-Century-Crofts, 1968.

Bellack, A. S., & Franks, C. M. Behavioral consultation in the community mental health center. *Behavior Therapy*, 1975, *6*, 388-391.

Bernard, V. W. Education for community psychiatry in a university medical center. In L. Bellak (Ed.), *Handbook of community psychiatry and community mental health*. New York: Grune & Stratton, 1964.

Cautela, J. R., & Kastenbaum, R. A reinforcement survey schedule for use in therapy, training, and research. *Psychological Reports*, 1967, *20*, 1115-1130.

Denner, B., & Price, R. H. (Eds.). *Community mental health: Social action and reaction*. New York: Holt, Rinehart & Winston, 1973.

Greenwald, A. G. Consequences of prejudice against the null hypothesis. *Psychological Bulletin*, 1975, *82*, 1-20.

Hersen, M., & Barlow, D. H. *Single case experimental designs: Strategies for studying behavior change*. New York: Pergamon Press, 1976.

Hersen, M., & Bellack, A. S. (Eds.). *Behavior therapy in the psychiatric setting*. Baltimore: Williams & Wilkins, 1978.

Iscoe, I. Professional and subprofessional training in community mental health as an aspect of community psychology. In G. Rosenblum (Ed.), *Issues in community psychology and preventive mental health.* New York: Behavioral Publications, 1971.

Kazdin, A. E. *Behavior modification in applied settings.* Homewood, Ill.: Dorsey Press, 1975.

Kazdin, A. E. *The token economy: A review and evaluation.* New York: Plenum Press, 1977.

Kiesler, D. J. Some myths of psychotherapy research and the search for a paradigm. *Psychological Bulletin,* 1966, *65,* 110-136.

Nietzel, M. T., Winett, R. A., MacDonald, M. L., & Davidson, W. S. *Behavioral approaches to community psychology.* New York: Pergamon Press, 1977.

Paul, G. L. The chronic mental patient: Current status–future directions. *Psychological Bulletin,* 1969, *71,* 81-94.

Paul, G. L., & Lentz, R. J. *Psychosocial treatment of chronic mental patients: Milieu vs. social learning programs.* Cambridge: Harvard University Press, 1977.

Rose, S. D. *Group therapy: A behavioral approach.* Englewood Cliffs, N.J.: Prentice-Hall, 1977.

Stahl, J. R., & Leitenberg, H. Behavioral treatment of the chronic hospital patient. In H. Leitenberg (Ed.), *Handbook of behavior modification and behavior therapy.* Englewood Cliffs, N.J.: Prentice-Hall, 1976.

Strauss, M. B. *Familiar medical quotations.* Boston: Little, Brown, 1968.

Upper, D., & Ross, S. M. Behavioral group therapy: I. Emotional, avoidance, and social skills problems of adults. In M. Hersen, R. M. Eisler, & P. M. Miller (Eds.), *Progress in behavior modification* (Vol. 5). New York: Academic Press, 1977.

Zax, M., & Specter, G. A. *An introduction to community psychology.* New York: Wiley, 1974.

Part Four

Process and Outcome Studies

The two chapters included in this final section are significant because there is a great need for research within behavioral group therapy in almost all aspects—techniques, populations, parametric issues, etc. Chapter 12 by John V. Flowers and his co-workers reports three studies that were done to explicate the functions which group cohesion and member flexibility play on within-group and extra-group change. The work reported here is based on several previous studies by these researchers which have validated the procedures employed for process and outcome measures. One finding which is of particular importance to the clinician involved in running groups is the deterioration effect noted for clients who do not disclose a problem *when no one else in the group discloses a similar problem in the group.* One implication of this finding may be the need for obtaining informed consent as to potential risks of group therapy under conditions of nondisclosure.

Chapter 13 by Nancy B. Cohn, Rochelle J. Mann, John V. Flowers, and Curtis D. Booraem reports a study to change the rate of self-reinforcement among group members. Since low rates of self-reinforcement are common among depressed individuals, their findings are relevant to a problem which is frequently encountered by many group leaders.

Chapter 12

Three Studies toward a Fuller Understanding of Behavioral Group Therapy: Cohesion, Client Flexibility, and Outcome Generalization

John V. Flowers
Curtis D. Booraem

Study 1: The Components of Cohesion in Group Therapy

John V. Flowers, Mary J. Rotheram, Barbara Kenney, and Curtis D. Booraem

Study 2: The Effects of Group Cohesion and Client Flexibility on Therapy Outcome

John V. Flowers, Karen A. Hartman, Rochelle J. Mann, Susan Kidder, and Curtis D. Booraem

Study 3: Generalization and Maintenance of Client Outcome in Group Therapy

John V. Flowers, Bruce Tapper, Susan Kidder, Garrie Wein, and Curtis D. Booraem

Abstract

Three successive studies were performed in an attempt to more fully understand behavioral group therapy. In the first study a set of variables that covary and differ significantly between what the authors label as lower and higher cohesive group sessions were identified and operationalized. In the second study, which employed external ratings of clients' abilities to cope with specified problems, it was demonstrated that clients showed greater improvement on problems disclosed in higher cohesion sessions than on problems disclosed in lower cohesion sessions. Through employing one cohesion variable, role flexibility, as a client rather than session variable, it was also demonstrated that clients who varied more in roles adopted during the group experience improved more than clients who varied less. Since the results of this study clearly indicated that generalization of client improvement from one to another life problem area is not automatic, a third study was performed to inspect the variables involved in generalization and maintenance of client improvement. The third study demonstrated that clients improve most on problems that they disclose and discuss in group therapy, improve less on nondisclosed problems that are like a problem disclosed by another group member, and actually show decreased coping ability on nondisclosed problems that are also not like any other problem disclosed during the group experience. This study also demonstrated that client improvement in each category of problem decreased only about 10% after 3 months.

This chapter reports three studies on behavioral group therapy, each of which is an extension of the findings of the previous study. Since the authors and findings are specific to each of the three studies, researchers should indicate the specific study (e.g., Study 3) they are citing and should cite the chapter without such a specific study designation only if referring to this introduction or to the discussion.

The first of the trilogy was designed to identify some of the components of group cohesion (Goldstein, Heller, & Sechrest, 1966; Yalom, 1975). Previous investigations, whether the studies were inferential (Liberman, 1970; Ribner, 1974) or correlational (Certner, 1973; Gruen, 1965; Query, 1964; Yalom & Rand, 1966), have tended to measure group cohesion by a single measure and have tended to treat group cohesion as though it were a single variable. Study 1 was designed in the belief that cohesion is a construct which can be understood best by identifying and operationalizing a network of variables that covary systematically over group sessions. Without such component analysis, the effects of group cohesion would be difficult to understand, and it would be very difficult to train group therapists in the behaviors necessary for systematic manipulation of group cohesion. Thus, Study 1 reported here was performed in preparation for further research on the effects of cohesion and the components of cohesion on client in-group and out-of-group behavior.

The results of Study 1 led to the design of two follow-up studies, one reported here as Study 2 and one reported elsewhere (Flowers, Booraem, & Hartman, in press). In the Flowers et al. (in press) study the authors found that clients improved more on problems which were disclosed in higher cohesion sessions or groups than on problems disclosed in lower cohesion sessions or groups. A potential weakness with that study is that it was conducted with a student population. The study reported here (Study 2) employed a community population with a greater range of age and problems to determine if the finding was also true for a more representative clinical population. Study 2 also extends the research by examining the effects of a single component of cohesion, role flexibility, on client outcome. Since role flexibility is a client variable, as well as a session variable, what was being inspected was whether the clients who contribute most to the group's cohesion are also the clients who improve the most in terms of outcome. The particular component of role flexibility was chosen as the initial component of cohesion to be inspected in terms of client outcome because, of the cohesion variables identified, it seemed to have the best in- to out-of-group generalization possibilities.

While Study 3 can be more adequately introduced after the reader has read the results of Study 2, there are two issues that the reader should be aware of even prior to reading Study 2. While Study 1 operationalizes cohesion so that Study 2 can be performed, and while Study 2 demonstrates that cohesion is important in terms of client outcome, the reader should realize that Study 2 does not demonstrate any maintenance of client improvement over time. Thus, Study 3 examines the maintenance issue not examined in Study 2. Secondly, the reader should be aware that the methodology used to assess client improvement in Study 2 presumes that coping ability is problem-specific and that generalization does not occur. The reader should be aware that if coping generalization does occur, the findings of Study 2 are even stronger, since such generalization weakens the statistical test used to find client improvement. While this is a very safe research design, significant findings on Study 2 would raise the question of under what conditions do clients generalize their improvement to other areas in their life besides the specific area dealt with in the group. Since the findings of Study 2 are significant, Study 3 does examine the issue of generalization of coping skills.

STUDY 1
THE COMPONENTS OF COHESION IN GROUP THERAPY

Method

Subjects. The subjects were 16 students, 9 females and 7 males, with a mean age of 27.3 years, enrolled in a laboratory class on group process. The subjects were randomly assigned to one of two groups (A or B) of eight students each. Each group met separately on different days for 20 one-hour sessions (4 demonstration and 16 training) over a 10-week period.

Therapists. The first four demonstration sessions were conducted by two graduate students (the second and third authors of Study 1) experienced in group therapy techniques who thereafter participated as members in both groups. None of the data reported, however, involved these experienced graduate students. After Session 4, the trainees were assigned as leader pairs and conducted every fourth group session. Thus, during the training phase, each trainee was a leader 4 times and was a nonleading participant 12 times. All sessions were videotaped for future data analysis by the senior author. All data reported are from Training Sessions 5 through 20.

Procedure. To provide content for the quasi-therapeutic experience, each subject completed a problem card (Goodman, 1969) before every session. Each person listed two problems, one of which would be easier and one of which would be harder to disclose in a group situation. In the group each subject was free to choose which problem he wished to disclose, to disclose another problem not yet written on a problem card, or to disclose no problem.

In each session subjects and therapists were given 20 red and 20 blue tokens (2 in. × 2 in. pieces of stiff construction paper) and were instructed to give these tokens along with statements they made to other group members. A blue token was to be given with each statement intended as positive (e.g., a compliment) and a red token with each statement intended as negative (e.g., a criticism). All tokens received by a subject were deposited in an empty container for later count. Each group member had a can for this purpose. Each subject's tokens were numbered so that the following information was available after each group session: the total number of red and blue tokens each subject distributed, to whom the subject gave them, how many of each type the subject received, and from whom the subject had received them. Otherwise the groups were conducted as described elsewhere (Flowers, 1979).

Objective ratings. The objective rating of an individual's role during each session was determined (from the tokens given and received) in the following manner:

1. Subject input role was determined by the percentage of blue tokens received. When a subject's percentage of blue tokens received was one standard deviation above the group mean for that session, that performance was labeled *positive input,* i.e., the subject got proportionately more positive feedback than the average group member that session. When a subject's percentage of blue tokens received was one standard deviation below the group mean for that session, the performance was labeled *negative input,* i.e., the subject got proportionately more negative feedback than the average group member for that session. Input

for subjects within one standard deviation of the group mean was labeled as balanced for that session.

2. Subject output role was determined by the proportion of blue tokens to total tokens *given.* With the same guidelines as above, a subject's output was labeled as positive, negative, or balanced for each session.

Role flexibility. Comparison of a subject's input and output rating from a session yielded three possibilities: (a) that the subject had the same ratings on both input and output (reciprocal), (b) that the subject had totally opposed ratings on input and output (i.e., positive input and negative output or negative input and positive output), or (c) that the subject's input and output were neither reciprocal nor totally opposed (flexible).

Cohesion criterion variable. Employing a criterion suggested by Krumboltz and Potter (1973), we rated all group sessions in terms of other members' attending to the speaker during the session. All group sessions were videotaped, and six trained undergraduate raters recorded the data as follows. An event was defined as any statement lasting 5 seconds or longer. After 5 seconds of speech by any single group member, the videotape was stopped, and each subject was rated as either:

1. Speaking.
2. Attending (eyes or body obviously directed toward the speaker).
3. Not attending (no eye contact or body obviously directed away from the speaker).
4. Unratable (subject having his back to the camera, being excluded from the picture, or showing doubtful attending behavior).

The six raters were divided into two groups of three raters each. Mean reliability (percentage of agreement) was 92.6%, with a range over sessions of 85% to 99%. Interrater reliability between the two groups was checked at Sessions 1, 2, 6, and 10 and averaged 92.4%. This insured that once trained, the two groups did not differ in their category definitions.

Problem disclosure. For the problem disclosure rating another six undergraduates rated the same videotape, scoring every subject's statement longer than two words as either a self-disclosure of a problem or another type of statement. Again, with two groups of three raters each, the interrater reliability (percentage of agreement) was 93.4%, ranging from 85% to 98% over sessions. Checking for consistency in category definitions was done by calculating reliability between groups at Sessions 1, 2, 6, and 10; it was 93.7%.

Satisfaction and trust. The subject, at the end of each session, completed a questionnaire listing those group members she trusted or did not trust during that session. The subject also rated her satisfaction with the group session on a Likert-type, 5-point scale.

Results

Each group had 16 one-hour sessions on videotape for a total of 32 hours of data. Two of these sessions were lost due to technical difficulties; hence, 30 one-hour sessions were rated for attending-to-the-speaker behavior or for frequency of problem disclosures. Employing a split-half procedure on the percentage of members attending-to-the-speaker, we found that subjects attended to the speaker 72.0% in the 15 higher attending sessions and 62.3% in the 15 lower attending sessions (t (28) = 7.76, $p < .001$). The data also indicated that Group A had less overall attending behavior (64.6%) than Group B (69.3%, t (28) = 2.39, $p < .05$). Thus, the criterion variable did discriminate both sessions and groups. This latter finding allowed for a post hoc comparison of Group A (lower attending) with Group B (higher attending) as well as a comparison of lower and higher attending sessions.

Employing attending as the first indicator of cohesion, we found that in the 15 higher cohesion sessions, the subjects disclosed 4.21 problems per session, as compared with 2.21 problems per session in the lower cohesion sessions (t (28) = 3.86, $p < .001$). In the higher cohesion group (B), the subjects disclosed 3.86 problems per session, as compared with 2.46 problems per session in the lower cohesion group (A) (t (28) = 2.88, $p < .01$).

It was hypothesized that there would be a higher percentage of negative feedback in higher cohesion sessions. With the percentage of red tokens used as the measure of negative feedback, a trend was noted in the predicted direction. Subjects were negative on 19.3% of their nonneutral messages in the lower cohesion sessions, as compared to 22.8% of their nonneutral statements in the higher cohesion sessions (t (28) = 1.63, $p < .10$). However, the hypothesis was supported when the groups were compared. Subjects in the lower cohesion group were negative on 17.2% of their nonneutral statements, as compared to a 23.8% rate in the higher cohesion group (t (28) = 2.45, $p < .01$).

It was also hypothesized that while there would be more negative feedback in higher cohesion sessions and groups, there would be a higher frequency of negative roles, i.e., group members who were one standard deviation from the group mean in terms of negative input or output, in lower cohesion sessions. A mean of 1.66 subjects were negative in terms of their output in lower cohesion sessions, as compared to 1.06 subjects per session in the higher cohesion sessions (t (28) = 2.30, $p < .05$). The finding was identical in terms of the number of subjects who received negative input: a mean of 1.66 subjects per session were given negative input in lower cohesion sessions, as compared to 1.06 subjects per session in the higher cohesion sessions (t (28) = 2.32, $p < .05$).

In terms of subject flexibility, the reader should remember that a subject could have identical, totally oppositional, or flexible input versus output roles each session. *Identical* input-output roles meant that the subject was positive both in terms of input and output, balanced in terms of input and output, or negative in terms of input and output. *Oppositional* meant that the subject was positive in terms of input but negative in terms of output, or negative in terms of input and positive in terms of output. *Flexible* was defined as any other combination, such as positive in terms of input and balanced in terms of output, or balanced in terms of input and either positive or negative in terms of output, etc.

It was hypothesized that more subjects would have flexible input versus output roles in higher cohesion sessions and that more subjects would have either identical or oppositional input versus output roles in lower cohesion sessions. Results indicated that in lower cohesion sessions oppositional or reciprocal roles were adopted 110 times, while flexible roles were adopted 37 times. In higher cohesion sessions oppositional or reciprocal roles were adopted 68 times, while flexible roles were adopted 60 times ($\chi^2 = 14.11$, $p < .001$). In the lower cohesion group (A), reciprocal or oppositional roles were adopted 106 times, and flexible roles were adopted 40 times. In the higher cohesion group (B), reciprocal or oppositional roles were adopted 56 times, and flexible roles were adopted 73 times ($\chi^2 = 24.10$, $p < .001$).

The final two hypotheses postulated that there would be greater self-reported satisfaction and a greater number of members rated as trusted in higher cohesion sessions than in lower cohesion sessions. Mean session satisfaction in lower cohesion sessions was 2.18, as compared to 3.96 in higher cohesion sessions (t (28) = 4.23, $p < .001$). The mean number of other group members listed as trusted was 2.44 in lower cohesion sessions, as compared to 4.68 in higher cohesion sessions (t (28) = 3.91, $p < .001$). In the lower cohesion group (A), mean satisfaction was 2.56, and the mean number of members trusted was 2.97, as compared with a mean satisfaction of 3.58 and a mean number of members trusted of 4.18 in the higher cohesion group (B) (t (28) = 2.86, $p < .01$ and t (28) = 3.41, $p < .001$, respectively).

STUDY 2
THE EFFECTS OF GROUP COHESION AND CLIENT FLEXIBILITY ON THERAPY OUTCOME

As stated previously, the results of Study 1 led to the design of the present study in which the major experimental question was "Do clients improve more on issues discussed in cohesive sessions than on issues discussed in less cohesive sessions?" A second experimental question examined was "Do clients who demonstrate more cohesive behavior (defined in the present study as role flexibility) improve more than clients who demonstrate less role flexibility in the group sessions?"

Method

Subjects. The subjects were 16 community members, 7 males and 9 females, with a mean age of 34.6 years, enrolled in a group therapy course offered to the community through a university. All members enrolled in the class with the understanding that participation in group therapy sessions would be a major part of their course work. All subjects were naive to the experimental hypotheses. Subjects were randomly assigned to two groups (A and B), and each group met for 10 one-hour sessions over a 5-week period.

Therapists. The therapists for both groups were two graduate students (the second and third authors of Study 2) experienced in group therapy and supervised by the senior author.

Procedure. The procedure used to conduct the therapy groups was identical to that described in Study 1.

Objective ratings. In addition to the objective ratings of subject input and output described in Study 1, the subjects were also rated on subject activity and subject mode to yield a more thorough assessment of role flexibility.

Subject activity for each session was determined by a count of the total number of tokens a subject gave and received. A subject whose total was at least one standard deviation below the group mean for that session was labeled low in activity, while a subject whose total was at least one standard deviation above the group mean was labeled high for that session. All other subjects were labeled as medium.

Subject mode was determined by the proportion of tokens given to those of both colors both given and received. A subject who gave more than he received (i.e., whose percentage of tokens given to total given and received was at least one standard deviation above the session mean) was labeled as a giver. A subject who received more than she gave by the same calculation was labeled a receiver. All other subjects were labeled as interactors for that session.

Role flexibility. For each session each subject was objectively rated on four variables, and a subject could adopt any of three roles on each variable:

1. Subject activity (high, medium, low).
2. Subject mode (giver, interactor, receiver).
3. Subject input (positive, balanced, negative).
4. Subject output (positive, balanced, negative).

Subjects were rank ordered in terms of who adopted the most to the least number of differing roles in the 10 group sessions.

Eye contact and disclosure. Attending to the speaker and disclosure were rated in vivo by six raters, three for each variable. Interrater relia-

In some cases it is extremely clear how a given component might influence group cohesion and affect client outcome, while in other cases the effects are not so apparent. Increasing attending to the speaker could increase modeling effects and could also increase the amount of attention the speaker perceives when disclosing, thereby reinforcing the disclosing behavior. In terms of affecting the client outside the group, attending to the speaker is often a skill taught by assertion trainers (Booraem & Flowers, 1978) in order to improve social communication.

Increasing disclosure (within limits) gives the group the material with which to work and operationally demonstrates the disclosing member's trust of the group. Outside of group, appropriately increased disclosure can lead to more intimate and presumably more fulfilling relationships (Altman & Taylor, 1973).

It is our contention that understanding the beneficial effects of an increased proportion of negative feedback also requires that one take into account that there is also a decreased frequency of group members with negative input and output roles in a higher cohesion session. In a higher cohesion session negative feedback is spread more democratically among group members. Moreover, negative feedback is employed as help-intended communication, which behaviorally means that negative statements are delivered with accompanying positive feedback for the purpose of teaching discrimination to both the giver and receiver. Thus, negative feedback is increased in a higher cohesion session because it is safer and because members want to tell others what they agree and disagree with, what they think will work, and what will not. There are fewer negative input and output roles because feedback is from all members and is spread to all members. Negative feedback is also used with positive feedback, thus reducing the possibility of any one member getting negative feedback totally out of proportion to the group. An additional way in which the increase of negative feedback may help the client is in terms of desensitization to out-of-group criticism.

The component of increased role flexibility is one of the most intriguing found in Study 1. Group therapists often talk informally about client flexibility and about group therapy's beneficial effect on such flexibility. However, research on this variable is nonexistent. From our perspective, flexibility means the ability to engage in a greater number of appropriate behavior patterns in the same time period. Study 2 demonstrates the relationship of in-group flexibility to client improvement. We are currently postulating that this relationship exists as a function of generalization and that increased in-group flexibility leads to increased flexibility in out-of-group behavior, allowing the client to employ more alternatives with which to cope with life's problems. Obviously, this hypothesis about the link of in-group behavior and client change requires further research to be supported or refuted.

The increase in the percentage of client-client interaction in higher cohesion sessions could help the group process by spreading the punishment and reinforcement power throughout the group. It could affect outcome by providing more sources of assessment, alternative generation, and intervention.

The components of trust and satisfaction are actually self-report measures that indicate cohesion rather than components that can be varied to influence cohesion. It is not surprising that there is more satisfaction and more trust in a higher cohesion session or group than in a lower session or group, given the components identified above.

These are not the only components of a cohesive group session. They comprise an initial list which will help lead to an understanding of how to better conduct effective group therapy, and an understanding of why such therapy is effective in terms of client out-of-group change.

The second point of discussion comes primarily from Studies 2 and 3. Group therapy helps clients improve the quality of their lives and helps them do this to a greater extent in more cohesive sessions. Given the goal of improving out-of-group functioning, group therapy works. The methodology used to assess this in Study 2 also suggests that the generalization of skills the client learns to employ with relation to the disclosed problems is neither automatic nor assured in relation to the rest of the client's life. The results of Study 3 clarify this. Generalization of improvement gained in group therapy does occur; however, it occurs in a very specified manner. Client generalization in terms of coping ability is (a) most probable if the client herself has disclosed and dealt with a similar problem, (b) less probable and/or less strong if the client has not disclosed a similar problem but another group member has so disclosed, and (c) least probable and/or strong if no similar problem has been disclosed. This is encouraging because, while generalization effects are not as strong as direct intervention effects, they do occur and are predictable.

Another encouraging aspect of the results of Study 3 is that while client improvement assessed 3 months following the completion of group was slightly reduced (about 10%) as compared to improvement assessed immediately after the termination of group, the reduction was not significant, and the generalization effects were still present in the delayed assessment. In terms of methodology, we would encourage the delayed assessment employed here as opposed to the traditional follow-up assessment where the effects of asking the same question twice are unknown.

The results of the control group in Study 3 are important methodologically and also lead to some interesting speculation. The control group members, who received no therapy, were rated as slightly improved on their coping ability over all problems that they listed. This could be due to rater bias or to the more optimistic notion that people solve some of their problems without benefit of therapeutic intervention. If the expectation is that people should or do usually cope better with their problems simply as a result of the passage of time, this could also be a regression toward the mean. The mean of problem coping may not be no-change and may actually be solution.

More important methodologically is the fact that there were no differences in ratings for problems that the control subjects thought they would have disclosed in the group and problems that they probably would not have disclosed. If the control subjects correctly identified the problems that they would have disclosed, had there been a therapy group, disclosure is not a selection bias, i.e., clients are not merely disclosing problems in the group that will improve irrespective of group intervention. Given the present findings, this is extremely important. The findings of Studies 2 and 3 strongly suggest that group therapy helps clients resolve their problems if they disclose the problems and deal with them in the group. If the clients are disclosing problems on which they are going to improve anyway, the group is superfluous in terms of improvement. The findings of the control group suggest this is not the case.

It is noteworthy that the control subjects were rated at higher coping levels on one specific set of problems than the experimental subjects. On a small percentage of problems (13% in Study 3) that were not disclosed and were unlike any disclosed in the group, the external raters judged that there had been a deterioration of coping ability over the time the group was conducted; the 3-month lag time assessment demonstrated the same deterioration of coping ability. For the control subjects all problems were both not disclosed and unlike any other disclosed since these subjects were not in a group, yet this alone did not cause the external raters to judge a deterioration in coping ability. It is our present hypothesis that nondisclosed problems, in a group with a relatively normal range of problem disclosures, may consist of uniquely difficult problems that are resistant to change without systematic intervention, and that since neither the problem nor any problem like it is disclosed, no such intervention occurs. A second and equally strong possibility is that the experience of a group in which a nondisclosed problem is unlike any other person's disclosure may discourage the client in terms of shared commonality and self-efficacy and that the group therapy experience may be counterproductive for that client in terms of that problem. For example, Group A had no sexual problem disclosures, yet two members did list sexual problems that they did not disclose. Both members were rated by the external raters as being worse in coping ability after the group experience on those problems than prior to the group. In Group B there was a disclosed sexual problem, and the two nondisclosed sexual problems were rated as improved by the external raters 3 months after the termination of the group.

Behavioral group therapy is a powerful clinical tool that can be taught, conducted systematically, and assessed. The methods employed in the present studies can, for the most part, be employed in clinical group therapy settings rather than only in traditional research settings. There is a crucial need for more practicing behaviorally oriented therapists to run groups, conduct research, and publish their results.

References

Altman, I., & Taylor, D. A. *Social penetration: The development of interpersonal relationships.* New York: Holt, Rinehart & Winston, 1973.

Booraem, C. D., & Flowers, J. V. A procedural model for the training of assertive behavior. In J. M. Whiteley & J. V. Flowers (Eds.), *Approaches to assertion training.* Monterey, Calif.: Brooks/Cole, 1978.

Certner, B. C. Exchange of self-disclosure in same sexed groups of strangers. *Journal of Consulting and Clinical Psychology,* 1973, *40,* 292-297.

Flowers, J. V. Behavioral analysis of group therapy and a model for behavioral group therapy. In D. Upper & S. M. Ross (Eds.), *Behavioral group therapy, 1979: An annual review.* Champaign, Ill.: Research Press, 1979.

Flowers, J. V., Booraem, C. D., Brown, T. E., & Harris, D. E. An investigation of a technique for facilitating patient-to-patient interactions in group therapy. *Journal of Community Psychology,* 1974, *2,* 39-42.

Flowers, J. V., Booraem, C. D., & Hartman, K. A. Client improvement on higher and lower intensity problems as a function of group cohesiveness. *Psychotherapy: Theory, Research and Practice,* in press.

Goldstein, A. P., Heller, K., & Sechrest, L. B. *Psychotherapy and the psychology of behavior change.* New York: Wiley, 1966.

Goodman, G. An experiment with companionship therapy: College students and troubled boy—assumptions, selection and design. In B. G. Guerney (Ed.), *Psychotherapeutic agents: New rules for non-professionals, parents and teachers.* New York: Holt, Rinehart & Winston, 1969.

Gruen, W. A contribution toward understanding of cohesiveness in small groups. *Psychological Reports,* 1965, *17,* 311-322.

Krumboltz, J. D., & Potter, B. Behavioral techniques for developing trust, cohesiveness and goal accomplishment. *Educational Technology,* 1973, *13,* 26-30.

Liberman, R. P. A behavioral approach to group dynamics: I. Reinforcement and prompting of cohesiveness in group therapy. *Behavior Therapy,* 1970, *1,* 141-175.

Query, W. T. Self-disclosure as a variable in group psychotherapy. *International Journal of Group Psychotherapy,* 1964, *14,* 107-115.

Ribner, N. G. Effects of an explicit group contract on self-disclosure and group cohesiveness. *Journal of Counseling Psychology,* 1974, *21,* 116-120.

Yalom, I. D. *The theory and practice of group psychotherapy* (2nd ed.). New York: Basic Books, 1975.

Yalom, I. D., & Rand, K. Compatibility and cohesiveness in therapy groups. *Archives of General Psychiatry,* 1966, *13,* 267-276.

Chapter 13

Methods of Increasing Self-Reinforcement in Group Therapy and the Effects on Therapy Outcome

Nancy B. Cohn
Rochelle J. Mann
John V. Flowers
Curtis D. Booraem

Abstract

An attempt was made to vary self-reinforcement in growth-oriented therapy groups. Tokens were employed to effect self-reinforcement in four groups: a group wherein external tokens prompted self-administered tokens, a group in which external tokens reinforced self-administered tokens, a group wherein tokens monitored only self-reinforcement, and a control group in which no tokens were used. Study results indicated that self-reinforcement was increased in the group in which self-administered tokens were prompted by external tokens. Increased self-reinforcement resulted in a higher frequency of growth goal identification and increased participation and enjoyment of pleasant activities.

The effectiveness of self-reinforcement procedures as a component of self-regulation has been demonstrated in a number of studies (Bandura & Kupfers, 1964; Kanfer & Marston, 1963a, 1963b; Marston & Kanfer, 1963; Marston & Smith, 1968). Specifically, the motivational properties of self-reinforcement (Kanfer & Duerfeldt, 1967) and its effectiveness in maintaining behavior change (Kanfer, Duerfeldt, & LePage, 1969) have clearly been demonstrated. Marston (1968) discussed the relevance of self-reinforcement to psychotherapy, and in a subsequent study Rehm and Marston (1968) demonstrated the effectiveness of self-reinforcement in increasing social instigation and positive self-evaluations within a population of socially anxious college males.

It has been suggested that self-reinforcement deficits and a lack of social involvement are apparent in a depressed client population (Tapper, 1978). It has further been suggested that depressed persons engage in a lower frequency of pleasurable activities than nondepressed persons, that depressed clients perceive the activities they engage in as being less pleasurable than do nondepressed clients, and that depressed clients seem to identify fewer activities as pleasurable than do nondepressed clients (Lewinsohn & Libet, 1972).

In clinical self-reinforcement research (Kanfer, 1975) positive verbal self-evaluations have been employed as a form of self-reinforcement. Booraem (1974) conducted an investigation of the differential effectiveness of external reinforcement versus self-reinforcement in the conditioning of assertive responses. This study employed a technique linking tokens to reinforcing statements and served the purpose of monitoring internal and external evaluations. While demonstrating the power of self-reinforcement in relation to the behavioral change sought, Booraem's study did not assess differential methods of increasing self-reinforcement in therapy sessions.

Flowers (1979) has suggested that analogue procedures are an important intervention employed in the context of group psychotherapy. Specifically, systematically increasing self-reinforcement within the context of group therapy may be a useful intervention in treating depressed clients.

The present study was conducted with nondepressed subjects to determine the differential utility of three self-reinforcement procedures employed within the context of group therapy. The effect of increased self-reinforcement on treatment outcome was assessed. Four groups were conducted, each employing a different training procedure. Training conditions included the following: (a) a group in which self-reinforcement was prompted by external tokens and monitored by self-administered tokens, (b) a group in which self-reinforcement was monitored by self-administered tokens and reinforced by external tokens, (c) a group in which self-reinforcement was monitored by self-administered tokens but was neither prompted nor reinforced by external tokens, and (d) a no-token control group.

It was felt that while the presence of tokens alone in the token-monitor condition created a demand for self-reinforcement, conditions where self-reinforcement was externally validated via prompting or reinforcement by others would demonstrate a higher frequency of self-reinforcement. Therefore, it was hypothesized (Hypothesis 1) that there would be a significant increment in self-reinforcement in both the token-prompt and token-reinforce conditions when compared to the token-monitor condition.

Self-reinforcement procedures were expected to change group members' self-perceptions. As is consistent with research conducted with depressed subjects, a greater frequency and pleasurability of life activities was expected to be demonstrated by subjects who engaged in self-reinforcement for those behaviors and presumably whose self-perceptions were thereby changed. It was hypothesized (Hypothesis 2) that after treatment, subjects in the token-prompt and token-reinforce conditions would report a higher frequency and greater pleasurability of life activities than subjects in either the token-monitor or no-token control groups.

Finally, growth goals can be conceptualized as an abstract subset of behaviors related to pleasant life events. People who experience an increase in pleasant life activities would therby be expected to identify more growth goals than people not participating in a large number of pleasant events. Therefore, self-reinforcement procedures were expected to affect the identification of personal growth goals (Hypothesis 3); i.e., it was hypothesized that subjects in the token-prompt and token-reinforce conditions would identify more personal growth goals during group sessions than subjects in the token-monitor or no-token conditions.

METHOD

Subjects

Subjects were 24 female and 19 male students enrolled in an upper division undergraduate group therapy course. Participation in the study was voluntary. Subjects were randomly assigned to training conditions and were unaware of the experimental hypotheses.

Therapists

The therapists were seven advanced undergraduate and graduate students and the course instructor. The seven students had previously completed a graduate course in group psychotherapy designed to train group leaders. Two therapists were randomly assigned to each different training condition.

Materials

Tokens employed in the training groups were used in the manner developed by Flowers and Booraem (1976). Each group member assigned to a token condition received a set of twenty 3 × 2½-in. rectangles of blue construction paper in a small can. Each group member's token set was identified numerically to facilitate the monitoring of tokens and the differentiation of self-administered and externally administered tokens.

Measures

Dependent measures included the MacPhillamy and Lewinsohn (1974) Pleasant Events Schedule and a Personal Growth Want Sheet, developed for the present study, on which subjects recorded their growth goals. The Pleasant Events Schedule assesses how frequently the subject engages in activities that have been found to be of high frequency among a normative nonclinical sample. The scores could range from 0 to 100 in the version of the scale used in the present study. The scale also assesses how pleasurable the subjects find the activities that they do engage in. This yields information about whether what the subject is doing is actually perceived as pleasurable.

The number of tokens received by each member served as a dependent measure. Self-administered and externally administered tokens were counted for each subject at the end of each session.

Assessment Procedure

All subjects completed the Pleasant Event Schedule prior to the onset of the first training session and once weekly thereafter. Each subject also indicated two growth goals, one short range and one long range, on the Personal Growth Want Sheet prior to each group session (see Figure 13.1). The specified goals could be the same or different each session.

Figure 13.1. Personal Growth Want Sheet

Subject ID Number: ____________________

Date: ____________________

Therapists: ____________________

Instructions: Please fill in two growth goals; space is provided below. One goal should be short range (e.g., achievable over the course of this summer), and one goal should be long range (e.g., achievable over a period of 1 to 5 years).

1. I want (short-range goal): to date five new people by August 31.

2. I want (long-range goal): to design and build a house on the lot I own.

Group Procedure

Training sessions of 60 minutes were conducted twice a week for a period of 5 weeks and a total of 10 sessions. The therapeutic format was identical for all four groups. At the beginning of each group session subjects were informed that the purpose of the group was to discuss and to help them achieve their personal goals. They were also instructed in the specific use of tokens for their respective training condition.

Token-prompt condition. During the first session of the token-prompt group the subjects were instructed that tokens were to accompany all positive statements directed to other group members. All subjects were told to give a blue token to another group member when making a positive statement to that member. Subjects were also instructed that after receiving a positive statement and token from another group member, they could make a positive statement about themselves and give themselves a blue token. Thus, tokens monitored self-reinforcement only following the receipt of external reinforcement. While subjects could choose to positively self-evaluate with both a statement and a token after external reinforcement, they were not required to do so. The cotherapists used tokens in the same manner. Each group member was assigned a number, which was indicated on each of her tokens for the purpose of differentiating self-administered tokens from those administered by others. The group was instructed to challenge any statement that sounded positive that was not accompanied by a token and to question any occasion on which a token was not delivered with an appropriate statement. Tokens received by each subject, whether externally administered or self-administered, were deposited in an empty can for later counting. Tokens not used during the session were collected and kept separate from tokens used for either self-reinforcement or external reinforcement. This provided an accurate count of the token transactions during the group.

Token-reinforce condition. During the first session of the token-reinforce condition group subjects were instructed to accompany all positive self-evaluative statements with a token administered to themselves. Group members were told that they could respond to another member's self-reinforcement with a positive statement, accompanied by a token delivered to the member who had self-reinforced. Positive statements directed to another group member not following that member's self-reinforcement were not accompanied by tokens. Thus, tokens were employed only to reinforce another group member's self-reinforcement. Cotherapists employed the tokens in the same manner as other group members. Instructions on the handling of tokens (i.e., where to place them, what to do with unused ones, etc.) were identical to those given in the first treatment condition.

Token-monitor condition. During the first session of the token-monitor group the subjects were instructed to give themselves a token with any positive self-evaluative statement. Tokens were not given to other members and were not used for any other purpose. Cotherapists also self-administered tokens to monitor self-reinforcement. Instructions for token handling were identical to those in the first two conditions.

No-token control condition. No tokens were used in this treatment group. Data were not collected on either self-reinforcement or external reinforcement.

Data Collection

Following each group training session, all tokens and token cans were collected. In all token conditions the frequency of self-reinforcement was calculated by counting the number of tokens in each subject's can bearing that subject's identifying number. Completed copies of the Pleasant Event Schedule and Personal Growth Want Sheet were collected from subjects in all groups at the end of each group session.

RESULTS

Since the first two sessions of each group were used to train the self-reinforcement procedures employed in the three token groups, only the data from Sessions 3 through 10 were analyzed. Analysis of the self-reinforcement data for the third session indicated no significant differences in the frequencies of subject self-reinforcement in the three groups employing tokens to monitor self-reinforcement. The mean frequency of self-reinforcement for all subjects during Session 3 was 1.52. Data for the final four treatment sessions (7 through 10) of each group indicated that the frequency of self-reinforcement by group members in the token-prompt group (M = 4.00) was significantly higher than by group members in either the token-reinforce group (M = 1.05, t (21) = 8.68, $p < .001$) or in the token monitor group (M = 1.53, t (21) = –3.69, $p < .001$).

There were no between-group differences on the Pleasant Events Schedule prior to the onset of treatment. After treatment termination, the token-prompt subjects participated in significantly more pleasurable activities (M = 65.64) than the subjects in either the token-monitor group (M = 51.11, t (21) = 2.62, $p < .01$) or the subjects in the no-token control group (M = 50.67, t (20) = 2.81, $p < .01$). There was no significant difference between subjects in the token-prompt condition and subjects in the token-reinforce condition. Subjects in the token-prompt group also rated the activities they engaged in as being more pleasurable (frequency x pleasurability, M = 114.82) than subjects in either the token-monitor group ($M = 74.11$, $t\ (21) = 2.50$, $p < .05$) or the no-token control group ($M = 83.67$, $t\ (20) = 2.35$, $p < .05$). There was no significant difference between subjects in the token-prompt group and subjects in the token-reinforce group.

Analysis of the Personal Growth Want Sheet indicated that the subjects in the token-prompt group identified significantly more personal growth goals ($M = 11.5$) than either the subjects in the token-monitor group ($M = 6.1$, $t\ (21) = 4.27$, $p < .001$) or the subjects in the no-token control group ($M = 6.9$, $t\ (20) = 2.79$, $p < .01$). There was no significant difference in goal identification between subjects in the token-prompt group and subjects in the token-reinforce group. The number of growth goals actually achieved during the course of the groups did not significantly differ between groups.

Since the data also seemed to indicate a clear relationship of client and therapist frequencies of self-reinforcement, a post hoc correlation of client and therapist self-reinforcement was conducted. Correlating the mean client and therapist self-reinforcement by session yielded a significant positive correlation ($r = .78$, $p < .001$). A further analysis demonstrated that the therapists in the token-prompt group self-reinforced significantly more ($M = 9.42$) than the therapists in either the token-reinforce group ($M = 2.18$, $t\ (19) = 7.72$, $p < .001$) or the token-monitor group ($M = 4.25$, $t\ (19) = 5.39$, $p < .001$).

DISCUSSION

The group process hypothesis, Hypothesis 1, stated that both the token-prompt and the token-reinforce conditions would increase self-reinforcement when compared to the token-monitor condition. As the data clearly indicated, this hypothesis was only partially confirmed—subject self-reinforcement increased in the token-prompt condition but not in the token-reinforce condition. A post hoc analysis of the data yielded one explanation for the failure of the token-reinforce condition to produce significant increases in self-reinforcement. The session-by-session correlation of the subjects' and the therapists' use of self-reinforcement was highly significant. As results indicated, the therapists in the token-reinforce condition self-reinforced significantly less than the therapists in the token-prompt condition. Thus, the failure of the token-reinforce condition to produce increased self-reinforcement can be explained most parsimoniously by the comparatively low frequency of therapist self-reinforcement, which suggests both a deficit in appropriate therapist modeling and a confounding effect in the experimental design.

The two treatment outcome hypotheses, Hypotheses 2 and 3, were based on the assumption that both the token-prompt and the token-reinforce conditions would demonstrate an increment in self-reinforcement. Hypotheses 2 and 3 were only partially supported—the group that demonstrated an increase in self-reinforcement also demonstrated an increase in the frequency and pleasurability of pleasant events and identified more personal growth goals.

This study suggest that when self-reinforcement of targeted behaviors is increased, subsequent behavioral and cognitive changes occur. While the duration of these particular groups was too brief to determine their impact on the achievement of personal growth goals, goal identification was significantly increased as self-reinforcement of the same was increased. Similarly, significant increases in the participation and enjoyment of pleasant activities were demonstrated as self-reinforcement of those behaviors was increased.

The present study does not indicate that the token-reinforcement procedure per se is ineffective in the production of increased self-reinforcement. The study suggests that increases in observable self-reinforcement by group members can be effected by higher frequencies of observable self-reinforcement by the therapists. Future research investigating the optimum methods of increasing self-reinforcement is indicated. It would also seem productive to further examine the effect of increased client self-reinforcement on therapeutic goals, both in individual and in group psychotherapy.

REFERENCES

Bandura, A., & Kupfers, C. J. The transmission of self-reinforcement through modeling. *Journal of Abnormal and Social Psychology,* 1964, *59,* 87-95.

Booraem, C. D. *The differential effectiveness of external vs. self-reinforcement in the acquisition and performance of assertive responses.* Unpublished doctoral dissertation, University of Southern California, 1974.

Flowers, J. V. Behavioral analysis of group therapy and a model for behavioral group therapy. In D. Upper & S. M. Ross (Eds.), *Behavioral group therapy, 1979: An annual review.* Champaign, Ill.: Research Press, 1979.

Flowers, J. V., & Booraem, C. D. The use of tokens to facilitate outcome and monitor process in group therapy. *International Journal of Group Psychotherapy,* 1976, *26,* 191-201.

Kanfer, F. H. Self-management methods. In F. H. Kanfer & A. P. Goldstein (Eds.), *Helping people change: A textbook of methods.* New York: Pergamon Press, 1975.

Kanfer, F. H., & Duerfeldt, P. H. Motivational properties of self-reinforcement. *Perceptual and Motor Skills,* 1967, *25,* 237-246.

Kanfer, F. H., Duerfeldt, P. H., & LePage, A. L. Stability of patterns of self-reinforcement. *Psychological Reports,* 1969, *24,* 663-670.

Kanfer, F. H., & Marston, A. R. Determinants of self-reinforcement in human learning. *Journal of Experimental Psychology,* 1963, *66,* 245-254. (a)

Kanfer, F. H., & Marston, A. R. Determinants of self-reinforcing responses: An analogue to self-confidence training. *Psychological Reports,* 1963, *13,* 63-70. (b)

Lewinsohn, P. M., & Libet, J. Pleasant events, activity schedules, and depression. *Journal of Abnormal Psychology,* 1972, *79,* 291-295.

MacPhillamy, D. J., & Lewinsohn, P. M. Depression as a function of levels of desire and obtained pleasure. *Journal of Abnormal Psychology,* 1974, *83,* 651-657.

Marston, A. R. Self-reinforcement: The relevance of a concept in analogue research to psychotherapy. *Psychotherapy: Theory, Research and Practice,* 1968, *5,* 29-36.

Marston, A. R., & Kanfer, F. H. Human reinforcement: Experimenter and subject controlled. *Journal of Experimental Psychology,* 1963, *66,* 91-94.

Marston, A. R., & Smith, L. Relationship between reinforcement of another person and self-reinforcement. *Psychological Reports,* 1968, *22,* 83-90.

Rehm, L. P., & Marston, A. R. Reduction of social anxiety through modification of self-reinforcement: An instigation therapy technique. *Journal of Consulting and Clinical Psychology,* 1968, *32,* 565-574.

Tapper, B. J. Assertion training with suicidal and depressed clients. In J. M. Whiteley & J. V. Flowers (Eds.), *Approaches to assertion training.* Monterey, Calif.: Brooks/Cole, 1978.

SUBJECT INDEX

AUTHOR INDEX

SpringerBriefs in Applied Sciences and Technology

Raden Sjarief Widjaja · Hasanudin ·
Yuda Apri Hermawan · Azman Ismail ·
Fatin Nur Zulkipli · Andreas Öchsner
Editors

Marine Technology 3

Springer

Editors
Raden Sjarief Widjaja
Department of Naval Architecture
Institut Teknologi Sepuluh Nopember
Surabaya, Jawa Timur, Indonesia

Yuda Apri Hermawan
Department of Naval Architecture
Institut Teknologi Sepuluh Nopember
Surabaya, Jawa Timur, Indonesia

Fatin Nur Zulkipli
College of Computing, Informatics
and Mathematics
Universiti Teknologi MARA
Machang, Kelantan, Malaysia

Hasanudin
Department of Naval Architecture
Institut Teknologi Sepuluh Nopember
Surabaya, Jawa Timur, Indonesia

Azman Ismail
Maritime Engineering Technology
and Centre for Women Advancement
and Leadership, Malaysian Institute
of Marine Engineering Technology
Universiti Kuala Lumpur
Lumut, Perak, Malaysia

Andreas Öchsner
Faculty of Mechanical Engineering
Esslingen University of Applied Sciences
Esslingen am Neckar, Baden-Württemberg,
Germany

ISSN 2191-530X ISSN 2191-5318 (electronic)
SpringerBriefs in Applied Sciences and Technology
ISBN 978-3-032-05612-2 ISBN 978-3-032-05613-9 (eBook)
https://doi.org/10.1007/978-3-032-05613-9

This Springer imprint is published by the registered company Springer Nature Switzerland AG
The registered company address is: Gewerbestrasse 11, 6330 Cham, Switzerland

Preface

This book explores a variety of research topics in marine technology delving into a wide array of crucial research topics within the maritime industry and its technology, encapsulating the latest research findings presented at the 8th International Conference on Marine Technology (SENTA 2023). It provides the exchange of recent knowledge, experiences, and innovations in the field of marine science and technology encompassing naval architecture and technology; marine system and safety; ocean, coastal, and offshore engineering; shipping, port, and maritime logistics; underwater technology; and advanced technology in maritime industry. This book covers studies ranging from ship design and production technique comprising ship design process, ship hydrodynamics, ship structure and advanced ship material, and ship management and production technology. The book also examines the invention in marine operation, marine machinery and maintenance system, and marine safety including digital technology in marine system and safety. It also addresses the topics on coastal, ocean, and offshore technology starting from coastal/offshore hydrodynamics, ocean energy, mooring line analysis to offshore structure analysis. In addition, shipping, port, and marine logistic researches are also conveyed in this book, especially on shipping operation optimization, port development, green port, and smart shipping development. Moreover, underwater technology and advanced technology in marine industry, i.e., computer vision technology for underwater vehicles, digital technology on ship design and production, and advanced computer technology in port and shipping development, are discovered. Overall, this book offers a cradle for the exchange of ground-breaking ideas, fostering collaboration, and potentially setting the stage for significant developments in the marine technology.

Surabaya, Indonesia — Raden Sjarief Widjaja
Surabaya, Indonesia — Hasanudin
Surabaya, Indonesia — Yuda Apri Hermawan
Lumut, Malaysia — Azman Ismail
Machang, Malaysia — Fatin Nur Zulkipli
Esslingen am Neckar, Germany — Andreas Öchsner

Contents

A Review of Optimizing Ship Productivity During Crude Palm Oil Shipping Using an Articulated Tug and Barge System

Daniel Yust Rizal Maretta and I. Gusti Ngurah Sumanta Buana

1 Introduction

Palm oil, or crude palm oil (CPO), plays a significant role in Indonesia's economy, serving as a primary source of income for job providers and contributing to the country's cash flow [1]. According to the Indonesian Palm Oil Entrepreneurs Association (GAPKI), CPO exports are expected to continue growing at a rate of approximately 4–5% in 2019, despite an anticipated increase in domestic CPO demand due to the B20 program, which blends fuel with 20% biodiesel [1]. However, alongside the increasing demand, several logistical and operational challenges have emerged in optimizing the supply chain and transportation of CPO to ensure its smooth distribution, efficiency, and profitability [2].

In 2021, the most significant production of CPO was estimated to have occurred in the Riau Province, with a production volume of approximately 8.96 million tons, roughly equivalent to around 19.55% of the total production in Indonesia. Meanwhile, the Central Kalimantan province contributed significantly to the production, with approximately 7.28 million tons or about 12.47% of the national total [3]. After the crude palm oil production, it is subsequently transferred to Java for further processing. The transportation phase plays a pivotal role in the supply chain of CPO, encompassing the utilization of several transportation modes, such as trucks and ships. A well-optimized supply chain and transportation system are essential for efficient operations [4].

D. Y. R. Maretta (✉) · I. Gusti Ngurah Sumanta Buana
Faculty of Marine Technology, Institut Teknologi Sepuluh Nopember, Kampus ITS, Surabaya, Jawa Timur, Indonesia
e-mail: daniel.yustrizal@poltekpel-sby.ac.id

I. Gusti Ngurah Sumanta Buana
e-mail: buana@seatrans.its.ac.id

R. S. Widjaja et al. (eds.), *Marine Technology 3*,
SpringerBriefs in Applied Sciences and Technology,
https://doi.org/10.1007/978-3-032-05613-9_1

Currently, CPO shipping in Indonesia is primarily conducted using self-propelled oil barge (SPOB), tug and barge, and tanker ships capable of carrying bulk liquid cargo in a single load [5]. However, there are often challenges during the loading and unloading process, particularly for vessels that lack heating equipment. Heating equipment is essential for re-liquefying CPO during the loading and unloading procedures, especially when the cargo has experienced sedimentation or freezing during the journey [6]. CPO has a crystallization point between 21.12 and 23.55 °C [7]. As a result, frozen CPO cargo requires special handling when unloading at the destination port. Port workers must mix and manually clean solid oil at the barge hold. Thus, the unloading time becomes longer, so the ship's idle time also becomes longer. Within the context of maritime operations, the measure of a ship's productivity is commonly expressed as the proportion of output to input, with a primary objective of optimizing the duration spent at sea while reducing the duration spent at port facilities [8]. The type of sea transportation that the corporation chooses is crucial since it will determine the costs that the company must pay. This research aims to compare the current mode of CPO shipment with the articulated tug and barge to gain the maximum sea time and reduce the cost. The study focuses on the shipment of CPO from the originating port in Central Kalimantan to the destination port in East Java. The research will involve various analyses of operational patterns and related calculations to determine the transportation mode that results in the lowest unit cost.

2 Methodology

2.1 Comparative Method

Within the disciplines of economics and operations, researchers employ the comparative approach as a research technique to examine and contrast various items, occurrences, or examples. This approach entails a methodical examination of two or more entities in order to discern their shared characteristics and disparities and afterward derive conclusions from these comparative analyses [9].

Currently, there are several options used for CPO shipment in Indonesia. The first option is CPO without packaging directly using tanker ships and SPOBs. These modes of transportation were selected by considerations of safety and capacity for accommodating large loads. Additionally, it is common for tankers to be equipped with a heating system. Depending on the available facilities, the cargo can be loaded directly onto a tanker ship or unloaded into storage facilities at the destination port. The loading and unloading equipment used is a pump directly transferred by the pipe structure into the manufacturer.

The other option is using tug and barge. One of the primary reasons is cost efficiency. Tugboats and barges are widely regarded as cost-effective and lucrative assets for businesses due to their relatively low operational expenses. The reason for the lower maintenance costs and less need for workers in the tug and barge mode of transportation is due to its inherent characteristics. According to [10], tugboats and barges have a relatively low required freight rate (RFR), which makes their utilization efficient. Nevertheless, it is imperative to conduct a comprehensive evaluation of the safety aspects associated with the functioning of this particular method of transportation.

Another opinion explained by [5] involves using flexy bags to transport liquids in 20-foot containers. Flexy bags come in various sizes, typically ranging from 10,000 to 24,000 l, depending on the product's density, but can be customized to accommodate specific capacity needs. Using flexy bags, the cargo in the shipping containers can be optimized to carry between 16 and 24 tons, depending on road weight restrictions. Flexy bags are made from a specially formulated polyethylene material approved by the U.S. Food and Drug Administration (FDA) for transporting liquid food products. This solution offers a dynamic, flexible, safe, and cost-effective means of shipping liquids. Additionally, being a single-use solution, the flexy bag eliminates contamination risks and reduces costs associated with returns, cleaning, and demurrage. It is highly durable and reliable in terms of performance, ensuring the safe transit of valuable liquid cargo. As a result, the flexy bag is a favored choice for liquid transportation in today's market [11].

All shipping methods that are explained above will be compared using an articulated tug and barge. The articulated tug barge (ATB) is a distinct maritime vessel characterized by the presence of a strong tugboat that is strategically positioned within a designated notch located at the stern of the barge. This arrangement allows the tug to generate propulsion and control the movement of the barge. The articulated tug and barge features a connecting system that is characterized by an articulated or "hinged" design. This design enables movement along a single axis or plane, specifically in the crucial fore and aft pitch region. The concept being discussed here diverges from that of an integrated tug barge (ITB), wherein the tug and barge are physically linked in a fixed manner, effectively functioning as a single entity.

The ATB principle has advantages in optimizing the tug (pusher) use because the tug (pusher) will continue to operate without stopping while the barge can continue discharging/loading cargo. Shipping data will be compared so that the best formula is found to make decisions regarding future shipments of crude palm oil. Moreover, the implementation of the ATB design facilitates enhanced velocity and improved fuel economy. The ATB has the potential to achieve a notable improvement in speeds, ranging from 35 to 40% when compared to conventional towed units [12]. Additionally, the ATB presents the advantage of improved fuel efficiency, resulting in enhanced savings in fuel consumption. The reason for this phenomenon lies in the ATB design, which enables the tug to capitalize on the displacement of the barge, hence minimizing the energy expenditure necessary for vessel propulsion. The utilization of the ATB in the transportation industry renders it a viable choice from an environmental standpoint.

3 Results and Discussion

3.1 *Port Capacity*

In East Java (generally CPO advanced processed), PT Pelindo III (Persero) Tanjung Perak Branch has constructed a liquid bulk dock at the North Nilam terminal to enhance liquid bulk shipping facilities at Tanjung Perak port. Currently, the port's installed capacity is 2,116,800 tons, while the realized liquid bulk flows in 2012 reached 4,949,968 tons, with a consistent 5% annual growth. The liquid bulk dock features a platform capable of handling tankers up to 40,000 DWT, measuring $27 \times 12.1\text{m}^2$, constructed with reinforced concrete, and equipped with fenders, bollards, and a pool depth of 10.0 m LWS. Additionally, it includes four reinforced concrete-breaking dolphins and two mooring buoys. Despite limited facilities, this new dock at the Nilam terminal has proven to be highly productive for loading and unloading domestic containers, outperforming other terminals in Tanjung Perak, according to the Deputy General Manager PT. Port III Tanjung Perak.

In the Gresik area, several ports have facilities and buildings specialized to handle palm oil commodities. But they built port facilities (jetty) for the berths of ships jutting out to sea to get a deep draft, so that generally oil unloading would be done to tank trucks, and the jetty could load more than five medium-sized ships simultaneously.

3.2 *Maximum Payload Capacity*

The construction of new vessels resulted in different payload capacities. The SPOB has a capacity of 5976 tons, the SPCB has a capacity of 3965 tons or equivalent to 194 TEU (twenty-foot equivalent unit), and the container ship has a capacity of 4957 tons or equivalent to 243 TEU. Conversely, the chartered vessels have slightly different payload capacities. The SPOB has a capacity of 3758 tons, the SPCB has a capacity of 3600 tons or equivalent to 176 TEU, and the container ship has a capacity of 6791 tons or equivalent to 332 TEU. The payload weight of the ATB reaches 4300 tons or equivalent to 330 TEU.

3.3 *Time Calculation*

Based on the given data concerning the distances between locations and the speed of each ship, we can compute both the total sailing time (referred to as sea time) and the duration when the ships are not operational (known as port time). Sea time is derived by summing the time taken to sail to the destination port and return to the originating port.

3.3.1 Sea Time

In the research, [5] mentioned that the load of palm oil from central Borneo to East Java is as follows. Sea time for SPOB amounts to 81.53 h. Meanwhile, for SPCB, it comes to 85.82 h; in the container ship case, it stands at 64.37 h.

In contrast, articulated tug and barges experience longer sea times as they do not wait for unloading periods. Furthermore, ATBs possess several noteworthy regulatory advantages. These include enhanced velocity, diminished fuel usage, the capability to navigate in more severe sea conditions, constant accessibility to the barge, and the elimination of susceptible tow wire connections. Another advantage is the reduction in accelerations experienced by the tug, leading to enhanced to create maximum speed [13].

3.3.2 Port Time

In the transportation operation of CPO using a tanker and general tug and barge, unloading often takes considerable time due to the manual cargo melting process. Consequently, the ships have to remain at the unloading port for an extended period, commonly called idle time at the port. Based on Table 1, the duration required for tank cleaning is seen to be relatively lengthy. During the tank cleaning operation, port workers will perform the task, while the tanker or tugboat will remain idle until the unloading process has been finished. The idle time experienced by these ships might be seen as unproductive.

The estimated unloading and loading times for liquid bulk cargo using port pumps are approximately 150 tons per h. As for container loading and unloading, we assume a rate of about 35 B/C/H (boxes/containers/hour). When utilizing flexy bags, loading is estimated to take around 1 h per bag, and unloading is anticipated to require about 0.5 h per bag or 30 min per bag. The calculation of port time includes the duration for approach time (AT), waiting time (WT), and idle time (IT). We assume that AT, WT, and IT each take 5 h at the origin port, and at the destination port, they are presumed to last 7 h [5].

With the introduction of the ATB system, the barge can be left at the unloading port while the tugboat can promptly return to the loading port, often accompanied by another barge that has completed its loading or unloading process. This allows the tugboat to work continuously without interruption, thus maximizing its sea time, while minimizing the time spent at the port (port time). Therefore, the definition of a productive ship is one that can operate as much as possible with a short time spent at the port.

Table 1 Time sheet of unloading CPO

Vessel	Start discharging	Start cleaning freezing cargo	Finish discharging	Vessel idle time
MT. Celline 2	07/12/2022 (16.40)	08/12/2022 (23.25)	09/12/2022 (20.00)	20 h 35 m
BG. BM VII	07/12/2022 (16.30)	08/12/2022 (20.25)	10/12/2022 (05.45)	1 day 9 h 20 m
BG. Duta bahari II	11/12/2022 (16.20)	12/12/2022 (19.20)	13/12/2022 (11.25)	15 h 5 m
BG. Hamco 03	13/12/2022 (12.05)	14/12/2022 (23.15)	17/12/2021 (09.10)	2 days 9 h 55 m
MT. IBM 1	05/12/2022 (06.25)	07/12/2022 (14.35)	12/12/2022 (08.00)	4 days 17 h 25 m
MT. Koan	11/12/2022 (18.05)	12/12/2022 (22.10)	16/12/2022 (12.55)	3 days 14 h 45 m
BG. Kapuas 11	03/12/2022 (09.22)	05/12/2022 (14.30)	08/12/2022 (16.45)	3 days 2 h 15 m
BG. Luis 8	03/12/2022 (18.00)	05/12/2022 (06.55)	06/12/2022 (14.15)	1 day 6 h 20 m
BG Nusa lease	05/12/2022 (06.05)	06/12/2022 (22.00)	09/12/2022 (03.40)	2 days 5 h 40 m
BG. Samudera palma	08/12/2022 (21.50)	12/12/2022 (10.02)	16/12/2022 (04.55)	3 days 18 h 53 m

Source PT X, Gresik, East Java

3.4 Cost Calculation

Within the maritime sector, the determination of whether to lease or construct a new vessel is a multifaceted undertaking, subject to an array of influential variables. The components encompassed in this analysis comprise the initial capital outlay, ongoing maintenance expenditures, operational costs, voyage-related outlays, and loading and unloading fees.

When evaluating the costs of new-build and rental ships, several cost factors need to be considered, such as the initial investment, maintenance, operational costs, voyage expenses, and loading/unloading expenses. The rental costs for SPOB amount to Rp. 138,147,929,749, for SPCB it is Rp. 406,265,314,362, and for container ships, it is Rp. 536,373,579,011. On the other hand, the new-build costs for SPOB are Rp. 144,281,361,352, for SPCB it is Rp. 423,452,558,942, and for container ships, it is Rp. 534,747,215,850 [5].

However, when compared to the new-build costs of an ATB, which has a total length of 81.94 m and uses a coupler connection system, the estimated total cost for building the integrated tug barge is $3,445,889 or approximately Rp. 42,959,893,905 [14]. Based on this price comparison, it can be concluded that constructing a new ATB is more cost-effective than the other options.

4 Conclusion

In the context of the CPO transportation sector, it is common for ships to experience significant waiting periods as a result of the manual liquefaction process required for the CPO cargo, specifically crude palm oil, which tends to settle. Consequently, the vessels have a prolonged stay at the unloading port, commonly called ship idle time. The issue at hand is effectively resolved using the ATB system, also known as the articulate tug and barge system. This system allows the tugboat to depart from the port without the need to wait for the completion of cargo loading and unloading processes. This strategy optimizes operational efficiency by enabling ships to maximize their sea time duration while minimizing their port stay. Moreover, the construction cost of ATB is comparatively more cost-effective when compared to the construction of other vessel types such as SPOB, SPCB, and containers. The previously mentioned data offer significant insights into the cost-effectiveness of several transport alternatives within the CPO business.

References

1. Diniaty, D., Hanum, F., Hamdy, M.I.: Analisis pengendalian mutu (quality control) CPO (crude palm oil) pada PT. XYZ. J. Tek. Ind. **5**, 92–99 (2019)
2. Suwarno, W.: Kebijakan Sawit Uni Eropa dan Tantangan bagi Diplomasi Ekonomi In-donesia. J. Hub. Int. **8** (2019). https://doi.org/10.18196/hi.81150
3. Badan Pusat Statistik: Statistik Kelapa Sawit Indonesia 2021, Jakarta (2022)
4. Indraswanti, B.I.E., Gunawan, R.: Sunoto: Peran Komoditas Sawit Pada Perekonomian Bengkulu. Converg.: J. Econ. Dev. **5**, 45–61 (2023)
5. Lailiah, I.: Analisis Pola Operasi Dan Perancangan Angkutan Crude Palm Oil (CPO): Studi Kasus Kalimantan Tengah—Jawa Timur (2018)
6. Wulandari, A.I., Arifuddin, A.M.N., Nudi, A.M.A.N.: Analisis Umur Kelelahan Struktur Ruang Muat Kapal Spob Akibat Beban Sloshing. J. Teknol. Terpadu. **10** (2022)
7. Wulandari, N., Muchtadi, T.R., Budijanto, S.: Sugiyono: Sifat Fisik Minyak Sawit Kasar Dan Korelasinya Dengan Atribut Mutu [Physical properties of crude palm oil and their correlations to the quality attributes]. J. Teknol. Ind. Pangan. **XXII** (2011)
8. Lathrop, D.S.: Ship productivity as a measure of maritime performance. Marit. Policy Manag. **6**, 183–186 (1979)
9. Bolbakov, R.G., Sinitsyn, A.V., Tsvetkov, V.Y.: Methods of comparative analysis. J. Phys. Conf. Ser. **1679** (2020). https://doi.org/10.1088/1742-6596/1679/5/052047
10. Audiba Karchely, V., Fadillah, A., Manullang, S.: Analisis Angkutan Crude Palm Oil Multi Trip Dengan Kapal Tongkang. J. Penelit. Transp. Laut. **22**, 79–86 (2021). https://doi.org/10.25104/transla.v22i2.1615
11. Gunawan, I., Vanany, I., Widodo, E., Mulyana, I.J.: Improving traceability system in Indonesian coconut oil company. In: IEEE International Conference on Industrial Engineering and Engineering Management, pp. 51–55. IEEE Computer Society (2019)
12. Intercon: Tug-Barge Couplers—Advantages. https://intercon.com
13. Brown, R. A Study of Energy Efficiency in Articulating Tug and Barge Design (2022)
14. Hidayat, W.: Studi Desain Integrated Tug-Barge (ITB) Pengangkut CPO (crude palm oil) Yang Sesuai Untuk Perairan Sungai Siak, Riau (2015)

Cost and Effects of Vehicle Ship Lashing on Ferry Vessels

Wayan Baniyasa and Setyo Nugroho

1 Introduction

The movement of products and people between different locations is made easier thanks to maritime transportation, which also plays a significant role in connecting islands. The ferry is one sort of ship that is frequently utilized in maritime transportation. Ferries are a highly effective and safe method of moving people, cars, and freight between islands [1].

Aspects of safety and security must be taken into consideration when running a ferry, particularly when loading and securing cargo. Safe and effective lashing can reduce the possibility of cargo damage and keep the ship stable while at sea. However, insufficient or ineffective lashing might have negative effects.

In this article, the author will talk about the costs associated with lashing on ferries as well as the potential outcomes. The author will examine the many aspects of lashing cost, such as the type of cargo, the manner of lashing, and the relevant safety laws and standards. The author will also talk about how insufficient lashing affects the financial health of shipping firms as well as the operational safety of ferries.

The purpose of this study is to present a thorough understanding of the significance of appropriate lashing in preserving ferry vessel safety and operational effectiveness. The goal of this article is to help the marine industry and shipping businesses by analyzing the factors that affect lashing costs and the effects of insufficient lashing.

One cannot overstate how crucial correct lashing is for ferries. During maritime voyage, improper or insufficient lashing can cause harm to the cargo. Poorly secured

W. Baniyasa · S. Nugroho (✉)
Institut Teknologi Sepuluh Nopember, Surabaya, Indonesia
e-mail: snugroho@seatrans.its.ac.id

W. Baniyasa
e-mail: wbaniyasa@gmail.com

R. S. Widjaja et al. (eds.), *Marine Technology 3*,
SpringerBriefs in Applied Sciences and Technology,
https://doi.org/10.1007/978-3-032-05613-9_2

cargo can shift or fall during the voyage, costing the shipping firm money and possibly putting the ship's crew and passengers in danger. Additionally, the price of lashing is a crucial factor for shipping businesses. The company's earnings and ability to compete in the market may be impacted by high lashing expenses. To minimize company spending and preserve operational effectiveness, it is crucial to assess the variables that determine lashing costs, including the type of cargo, the complexity of the lashing, and any applicable restrictions.

Ferry lashing also heavily depends on the applicable safety norms and laws. International organizations like the United Nations Conference on Trade and Development (UNCTAD) and the International Maritime Organization (IMO) have created safety standards and laws governing the stevedoring of cargo aboard ships [2].

With the high delivery of goods or the movement of people from one island to another, it is necessary to improve safety, both for cargo and passengers. One of the problems that often occurs is the overturning of trucks or cargo on the main deck of the ship due to large waves. This will affect the stability of the ship which can later endanger the ship and cause it to sink [3].

To prevent this, the port authority requires lashing for ships and cargo. However, in its implementation in the field there are several obstacles, namely:

1. The vehicle has been lashed but not done properly.
2. The ship does not lash because it takes a long time and crew energy.
3. Lashing takes up the transportation deck which will reduce the ship's capacity.
4. Short voyage distance makes the operator feel that the ship will be safe on the voyage [4, 5].

The author will cite a variety of pertinent and trustworthy sources in this study, including related research as well as rules and regulations from international organizations. These sources will give one a strong knowledge foundation to comprehend the significance of good lashing on ferries. So, let us start looking into this subject and learning about the pertinent components of the cost of lashing a ship as well as the potential impacts.

2 Methodology

This study utilizes a quantitative methodology to analyze the impact of implementing a dedicated lashing crew team for securing vehicles on Ro-Ro ferries. The methodology involves the following steps:

1. Data Collection: Operational data was collected from a major Ro-Ro ferry operator in Indonesia on the following parameters over a 1 month period:
    - Number of 2-wheel and 4-wheel vehicles transported per trip.
    - Average lashing time per vehicle.
    - Current lashing practices and challenges.

2. Lashing Crew Team Concept: A lashing crew team concept was developed consisting of:
   - Five crew members.
   - Salary and overhead costs estimated.
   - Lashing time per vehicle benchmarked from industry data.
3. Cost-Benefit Analysis: A cost-benefit analysis model was built to compare the current lashing practices to the proposed lashing crew team concept in terms of:
   - Lashing crew costs.
   - Potential time savings.
   - Impact on ticket pricing.
4. Operational Impact Analysis: The operational impacts were analyzed in terms of:
   - Loading and unloading time changes.
   - Process efficiency gains.
   - Safety and stability improvements.
5. Implementation Plan: An implementation plan was outlined for rolling out the lashing crew team concept including:
   - Crew hiring and training program.
   - Integration with current port and vessel operations.
   - Change management and communications strategy.

This quantitative approach draws on real-world data, financial modeling, operational analysis, and implementation planning to provide a comprehensive methodology for evaluating the lashing crew team concept. The results will quantify the costs, benefits, and operational impacts to help determine the feasibility and effectiveness of this approach for improving lashing practices on Ro-Ro ferries [6].

3 Results and Discussion

Based on the problems that occur in the field. The author has an idea to hold lashing installation services for ship cargo. This has been applied to the type of Ro-Ro Pure car carrier ship in South Korea. Where when the car is on the ship, the lashing team will automatically lashing the car using the lashing available on the ship.

This alternative will later be applied to all passenger or Ro-Ro ships that carry cargo, especially cars or trucks. That way there will be no more reasons for ferry/crossing service providers not to lash vehicles for various reasons. The results of this research are expected to help choose which lashing option is the most economical and does not interfere with the technical operation of the ship.

In this scenario, author will use simulated calculations using a case scenario in Ujung-Kamal ferry crossing port [7, 8].

3.1 Planning and Calculation

There is a great opportunity in doing the lashing business, considering that the mode of passenger ships that operate daily also transports both two wheels and four wheels according to the following data written in Fig. 1.

From the data, we can see the massive profit potential for lashing business. This is important to determine since without profit potential, there will be no viable cost calculation for lashing plannings. From this data, we can use this to make a calculation to determine how much ticket price change for this lashing service.

From the data in Fig. 2, with a capital of Rp 9,765,067 to procure ship lashing, the following profit can be obtained:

Number of vehicles lashing per trip:

Two wheels: 109

Four wheels: 9

JENIS ANGKUTAN / BULAN	TRIP			RODA - 2			RODA - 4			BARANG		
	UJUNG	KAMAL	JUMLAH	UJUNG	KAMAL	JUMLAH	UJUNG	KAMAL	JUMLAH	UJUNG	KAMAL	JUMLAH
JANUARI	831	834	1665	43687	43601	87288	3860	4577	8437	1944	777	2721
FEBRUARI	707	703	1410	45009	44944	89953	3265	3634	6899	1944	475	2419
MARET	695	691	1386	47892	47781	95673	3547	3893	7440	2049	535	2584
APRIL	609	601	1210	46504	45024	91528	3528	3487	7015	1931	852	2783
MEI	889	889	1778	40842	42659	83501	3604	3702	7306	2090	1083	3173
JUNI	834	833	1667	43952	45247	89199	3855	4532	8387	1527	554	2081
JULI	768	766	1534	46834	47085	93919	3736	4226	7962	1664	96044	97708
AGUSTUS	724	725	1449	46166	44470	90636	3462	3555	7017	1971	935	2906
SEPTEMBER	814	804	1618	47993	47559	95552	3514	3761	7275	2306	1127	3433
OKTOBER	771	768	1539	49008	47203	96211	3533	3585	7118	2283	873	3156
NOVEMBER	670	669	1339	44409	43406	87815	2217	2383	4600	1344	433	1777
DESEMBER	677	679	1356	40601	41898	82499	2552	2661	5213	1413	309	1722

Fig. 1 Kamal-Ujung port production data

Rata - rata jumlah kendaraan dalam satu tahun	742	739	1481	45383	45207	90590	3347	3584	6930
Jumlah kendaraan terbanyak dalam satubulan	889	889	1778	49008	47781	96789	3860	4577	8437
Jumlah kendaraan maksimal dalam satu pelayaran				55	54	109	4	5	9
Kebutuhan lashing tiap kapal				61	59	120	5	6	10
Harga lashing per unit (@ Rp. 75.000)	Rp 75,000			Rp 4,547,987	Rp 4,434,120	Rp 8,982,106	Rp 358,211	Rp 424,750	Rp 782,961
Total harga seluruh lashing	Rp 9.765.067								
Gaji Pekerja Lashing (20 orang)	Rp 4,600,000						Rp 92,000,000 per bulan		
							Rp1.104.000.000	per tahun	

Fig. 2 Business calculations

RODA - 2			RODA - 4			pendapatan lashing	
UJUNG	KAMAL	JUMLAH	UJUNG	KAMAL	JUMLAH	roda 2	roda 4
43687	43601	87288	3860	4577	8437	Rp 174.576.000	Rp 42.185.000
45009	44944	89953	3265	3634	6899	Rp 179.906.000	Rp 34.495.000
47892	47781	95673	3547	3893	7440	Rp 191.346.000	Rp 37.200.000
46504	45024	91528	3528	3487	7015	Rp 183.056.000	Rp 35.075.000
40842	42659	83501	3604	3702	7306	Rp 167.002.000	Rp 36.530.000
43952	45247	89199	3855	4532	8387	Rp 178.398.000	Rp 41.935.000
46834	47085	93919	3736	4226	7962	Rp 187.838.000	Rp 39.810.000
46166	44470	90636	3462	3555	7017	Rp 181.272.000	Rp 35.085.000
47993	47559	95552	3514	3761	7275	Rp 191.104.000	Rp 36.375.000
49008	47203	96211	3533	3585	7118	Rp 192.422.000	Rp 35.590.000
44409	43406	87815	2217	2383	4600	Rp 175.630.000	Rp 23.000.000
40601	41898	82499	2552	2661	5213	Rp 164.998.000	Rp 26.065.000
					total per jenis kendaraan	Rp 2.167.548.000	Rp 423.345.000
biaya lashing					total	Rp 2.590.893.000	
roda 2	Rp 2.000				nilai keuntungan	Rp 1.486.893.000	
roda 4	Rp 5.000						

Fig. 3 Business calculations (2)

Cost for lashing:

Two wheels: IDR 2000

Four wheels: IDR 5000

From the result, we can see that this business calculation is feasible with only a little addition on the ticket price that is not going to give significant financial burden to ferry-RoRo ship users (Fig. 3).

3.2 Effects of Enforced Lashing on Current Situation

Proper and safe lashing is an important aspect of ferry operations to maintain cargo safety and vessel stability during the voyage. However, in practice, problems often arise where the crew is reluctant or unwilling to involve themselves in the lashing process. One common reason is the perception that lashing is time-consuming. In addition, there are also other reasons that may be an inhibiting factor in lashings on ferries. This paper will discuss these and other reasons that may contribute to the crew's inconvenience or reluctance in lashing [9, 10].

(a) Time Required: One of the main reasons for crew reluctance in lashing is the perception that the process is time-consuming. Ferries often have tight departure schedules, and crews can feel burdened by the additional time required for proper lashing. They may think that the time spent on lashing may interfere with departure schedules and reduce operational efficiency.

(b) Resource Limitations: Some ferries may face human resource limitations, especially in terms of the number of crew available. In this situation, the crew may feel burdened with the additional task of lashing which can require extra effort and time. These human resource limitations can be an obstacle in properly carrying out the lashings.

(c) Lack of Awareness and Training: Lack of awareness about the importance of proper lashing and lack of crew training in proper lashing procedures can also lead to reluctance in carrying out this task. Poorly trained crews may not feel confident in performing lashings and prefer to avoid the responsibility.
(d) Level of Difficulty and Complexity: Some types of cargo on a ferry, such as heavy vehicles or dangerous cargo, can require more complicated lashings and require more careful calculations. This level of difficulty and complexity can make the crew feel anxious or uncomfortable in carrying out the lashing, especially if they do not have sufficient knowledge or experience in handling certain types of cargo.
(e) Lack of Attention to Safety: At times, crew members may overlook the importance of proper lashing because they are not fully aware of the risks and consequences that can result from poor lashing. Lack of awareness of the negative impacts that can occur on cargo, vessel stability, and crew and passenger safety can be the reason for insufficient attention to proper lashing [11, 12].

In the face of these issues, it is important for the ferry management to take the following steps:

- Provide adequate training to the crew on the importance of proper lashing, proper lashing procedures, and handling of special cargo.
- Pay attention to the availability of sufficient human resources to engage the crew in lashing without disrupting the departure schedule.
- Increase crew awareness of the consequences of poor lashings through safety education and campaigns.
- Provide adequate equipment and facilities to facilitate efficient and safe lashing.
- Enforce strict policies on proper lashing and make it an integral part of the ship's safety culture.
- By addressing these issues, it is expected that the crew will be more engaged and responsible in proper lashing. It is important to remember that proper lashing is a long-term investment in the safety of the vessel, the cargo, and the safety of everyone involved in the ferry journey [13, 14].

4 Conclusion

Based on the results of the discussion, there should be no problem with lashing because, from the calculation above, we can see that it is a feasible business model. Problems with lashing on ferries include crew reluctance due to perceived time consumption, limited resources, lack of awareness and training, level of difficulty, and lack of attention to safety. Efforts are needed to provide training, raise awareness, and overcome these barriers to ensure proper and safe lashing.

Acknowledgement The authors gratefully acknowledge to the Transportation Policy Agency and Ujung-Kamal Ferry RoRo port for providing data and information to the author. Thanks to Institut Teknologi Sepuluh Nopember (ITS) for the guidance from the lecturers in compiling this article.

References

1. Brambilla, M., Martino, A.: Research for TRAN committee-the EU maritime transport system: focus on ferries. European Union (2016)
2. IMO/ILO/UNECE Code of Practice for Packing of Cargo Transport Units (CTU Code) | UNECE. https://unece.org/transport/intermodal-transport/imoilounece-code-practice-packing-cargo-transport-units-ctu-code
3. CSS-Code: Code of Safe Practise for Cargo Stowage and Securing—Res. A.714(17)—Netherlands Regulatory Framework (NeRF)—Maritime. https://puc.overheid.nl/nsi/doc/PUC_2092_14/3/
4. Embankment, A.: Revised Guidelines for the Preparation of the Cargo Securing Manual (2020)
5. Det Norske Veritas: Cargo Securing Manual for M/S Test Vessel (2001)
6. Suyono, J.: Analisa Biaya Lashing dan Pemilihan Metode Pelashingan pada Kapal Roro di Indonesia. J. Ilm. Tek. Ind. **17**(1), 49–56 (2018)
7. Pramitha, D., Pramono, R.: Determination of optimal lashing method for Ro-Ro vessels: case study of Indonesian domestic routes. J. Mar. Eng. Technol. **19**(2), 109–117 (2020)
8. Hafiz, M.A., Kurniawan, F.: Analysis of cargo lashing on roll-on/roll-off ship with a case study of ferry ship in Indonesia. IOP Conf. Ser.: Mater. Sci. Eng. **1188**(1), 012052 (2021)
9. Code of Safe Practice for Cargo Stowage and Securing (CSS Code). https://www.imo.org/en/OurWork/Safety/Pages/CSS-Code.aspx
10. Transport of Dangerous Goods. United Nations, New York and Geneva (2019)
11. International Association of Ports and Harbors (IAPH): Guidelines for Safe Access to Ports and Ships. IAPH, Tokyo (2016)
12. Lee, J.S., Kim, H.J., Lee, S.J.: Analysis of lashing forces for vehicles on Ro-Ro vessels during ship sailing. Ocean Eng. **139**, 59–68 (2017)
13. Maritime and Coastguard Agency (MCA): Code of Safe Working Practices for Merchant Seafarers. MCA, Southampton (2018)
14. International Union of Marine Insurance (IUMI): Cargo Securing Manual: Guidelines for Industry Best Practice. IUMI, Zurich (2021)

Analysis of Competition Between Land Transportation and Sea Transportation in the North Kolaka Regency

Bayu Saputra Syah, I. Gusti Ngurah Sumanta Buana, and Kusuma Dhipanusa

1 Introduction

Indonesia is the biggest archipelago country in the world with the possession of 13,466 islands. This country is surrounded by the ocean, whose size reached 5.8 million km^2 consisting of 3.1 million km^2 of territorial waters and 2.7 million km^2 of the Economic Exclusive Zone [1]. One of the strategic regions in Indonesia is the North Kolaka Regency, Southeast Sulawesi Province. Besides having 3391 km^2 of land and small islands, the North Kolaka Regency also has extensive sea that stretches among the Gulf of Bone, covering more than 12,376 km^2.

Based on its geographic location, the North Kolaka Regency has a prospect for strategic development. This is because the region is located at the Trans Sulawesi road as a regional economic line. Not only that, the North Kolaka Regency also has a direct boundary with South Sulawesi which is regarded as the most advanced and developed region in the Indonesian East region. The availability of transportation then becomes the main requirement in supporting regional development, especially for regions with great potential resources [2]. Transportation also can reduce time and make the costs more affordable [3].

The development of transportation becomes more important to be implemented in the North Kolaka Regency. This is influenced by the increased number of its inhabitants. In 2025, the predicted number is 168,802 people. It will increase significantly in 2030 because the number is predicted to reach 185,550 people [4]. The increased

B. S. Syah (✉)
Department of Marine Transportation Engineering, Sepuluh Nopember Institute of Technology, Surabaya, Indonesia
e-mail: bsaputrasyah@gmail.com

I. G. N. Sumanta Buana · K. Dhipanusa
Rotterdam Mainport Institute, Rotterdam University of Applied Science, Rotterdam, The Netherlands
e-mail: buana@seatrans.its.ac.id

R. S. Widjaja et al. (eds.), *Marine Technology 3*,
SpringerBriefs in Applied Sciences and Technology,
https://doi.org/10.1007/978-3-032-05613-9_3

number of inhabitants will influence the increased demand for transportation. Thus, this study aims to analyze the transportation condition in the North Kolaka Regency by focusing on land and sea transportation. Therefore, this study will fill the gap of knowledge in previous research. In the previous study, there is no explanation or analysis regarding the comparison of these two transportation modes in the region. This study then provides a new insight by comparing land transportation and sea transportation in the North Kolaka Regency.

2 Methodology

This study used a qualitative research method that consists of searching, collecting, and analyzing non-numerical data. By applying the qualitative method, a more comprehensive understanding of certain phenomena can be obtained [5]. In this context, we used land transportation and sea transportation in the North Kolaka Regency as our unit of analysis. The authors then used a literature review design by reading, reviewing, and examining the published materials concerning the transportation condition in the North Kolaka Regency [6]. The authors then conducted several steps in formulating this study, namely (1) collecting various data and information from different sources; (2) reviewing and analyzing the content by reading them carefully; (3) developing and writing the findings into the study; and (4) developing conclusions and some recommendations both for policymakers and future research.

3 Results and Discussion

3.1 Condition of Land Transportation

Southeast Sulawesi Province is supported by a land transportation network consisting of several kinds of roads, such as 1293 km of national road, 1487 of provincial road, and 3785 of regency or city road [3]. Road transportation facilities are then including private transportation and public transportation. For public transport, it consists of intercity bus transportation inter-province (AKAP), intercity bus transportation in province (AKDP), transportation services for the isolated region (*perintis* transportation), and tourism transportation.

There are some reports regarding the condition of land transportation in Southeast Sulawesi, including the North Kolaka Regency, namely (1) public transportation to the tourism location is very limited; (2) AKAP is only available at night; (3) AKDP is quite well organized; (4) public transportation (angkot) is available but only developed in first-tier cities; (5) rural public transportation (angkudes) is available but not well developed and not recorded properly; (6) some travel transportations still have black plates; (7) costs of car rental transportation is around IDR 400,000–500,000 per day; and (8) taxis are only available in first-tier city [3].

However, the government of the North Kolaka Regency tried to improve the dissemination of land transportation by creating a program policy. In the context of operationalization and maintenance of land transportation mode, the government has appointed local government owned enterprises (BUMD), village government, or local organizations as responsible parties [7]. The government also tried to put more operational locations of land transportation in the productional areas which have road access and other supported transportation facilities [8]. In order to optimize the land transportation performance, the government then added some new routes to increase land transportation services so that society can feel satisfied and comfortable [9]. In 2022, the Department of Transportation in Southeast Sulawesi also implemented the rate adjustment for public transportation in the province, especially for economic class, in several regions. The rate has increased by 12–13% from the previous rate [10].

3.2 Condition of Sea Transportation

Besides land transportation, the society of the North Kolaka Regency also usually uses sea transportation, especially when they want to go into Siwa Regency. This is because the North Kolaka Regency and Siwa Regency are separated by the sea, so that society needs to use crossing transportation to achieve their destination. In order to facilitate the crossing, the government has provided the port of Tobaku which is supplemented by three units of ferry motorboats, namely KMP Merak, KMP New Rose, and KMP New Camelia.

Each day, there are three departure and arrival trips at port of Tobaku. Each ferry has different capacity, but on average, these ferries can carry 194 adult passengers, 15 children, 22 class II vehicles, 30 class IV vehicles, 6 class VI vehicles, and 1 class VII vehicle within one day. This information can be seen in more detail in Fig. 1. The high number of passengers or vehicles using sea transportation is because of the effectiveness provided by the ferry motorboats to achieve the destination. The location also has a direct border with South Sulawesi which is known as the most developed area in the region. Not only that, the port of Tobaku is also located in the middle of the North Kolaka Regency and not far away—only 5 km away—from the regency capital, Lasusua.

Ministry of Transportation then will make the port of Tobaku as the best example for crossing port in the Southeast Sulawesi Province in 2024. Not only the Port of Tobaku, but the North Kolaka Regency also has the port of Watunohu as a collector port. There are several commodities that can be found at the port, such as coconut, coffee bean, pepper, nutmeg, chocolate, and sago. The number of passengers is also projected to increase each year [2].

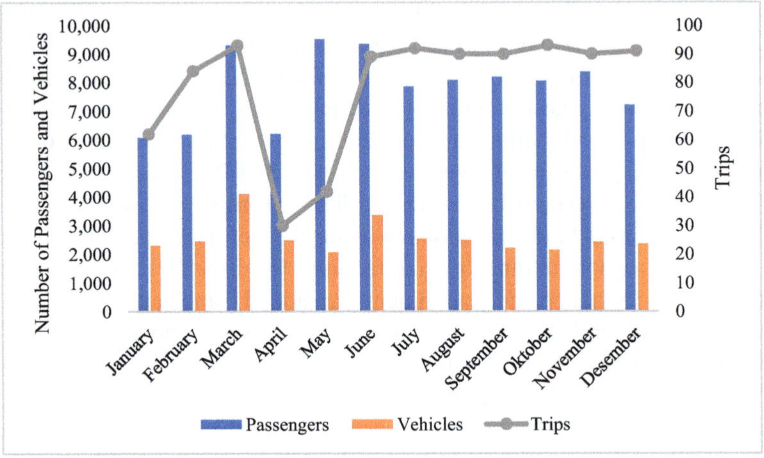

Fig. 1 Monthly production 2021. *Source* Land Transportation Management Center Region XVIII Southeast Sulawesi [11]

3.3 Competition Between Land Transportation and Sea Transportation: Who Will Be Winning?

We need to remember that Sulawesi Island is the development center in East Indonesia. The Sulawesi economic corridor has been created to make Sulawesi a market gate from East Asia, Australia, and America (see Fig. 2). However, this market gate cannot exist if no supports are coming from sea transportation. The use of ferries or ships then became more crucial to connect Sulawesi, particularly the North Kolaka Regency, with other regions and countries. The ship offers a huge advantage because of its flexibility to load various types of passengers, such as people, vehicles, and commodities [12]. This explains why the number of passengers in sea transportation is higher than in land transportation. In 2021, there were more than 90,000 passengers using sea transportation in port of Tobaku. Meanwhile, the number of passengers who used land transportation from the North Kolaka Regency to Siwa is less than 70,000 people [3].

Sea transportation has proven its capability to provide more benefits than land transportation. Different from land transportation which only provides transportation that is connecting between cities within Sulawesi, sea transportation has an important role as a national connectivity supporter to connect Sulawesi with other main islands, such as Kalimantan, Sumatra, Java, and Papua. Not only that, but sea transportation also supports local or intraregional connectivity by connecting islands within Sulawesi and becoming an alternative route of land transportation [13]. As the most strategic port in the North Kolaka Regency, port of Tobaku has become the best choice for society. This port is located not far away from the capital center; thus, it is easier to access because of its strategic location. Port of Tobaku then plays an essential role in supporting economic growth in local society.

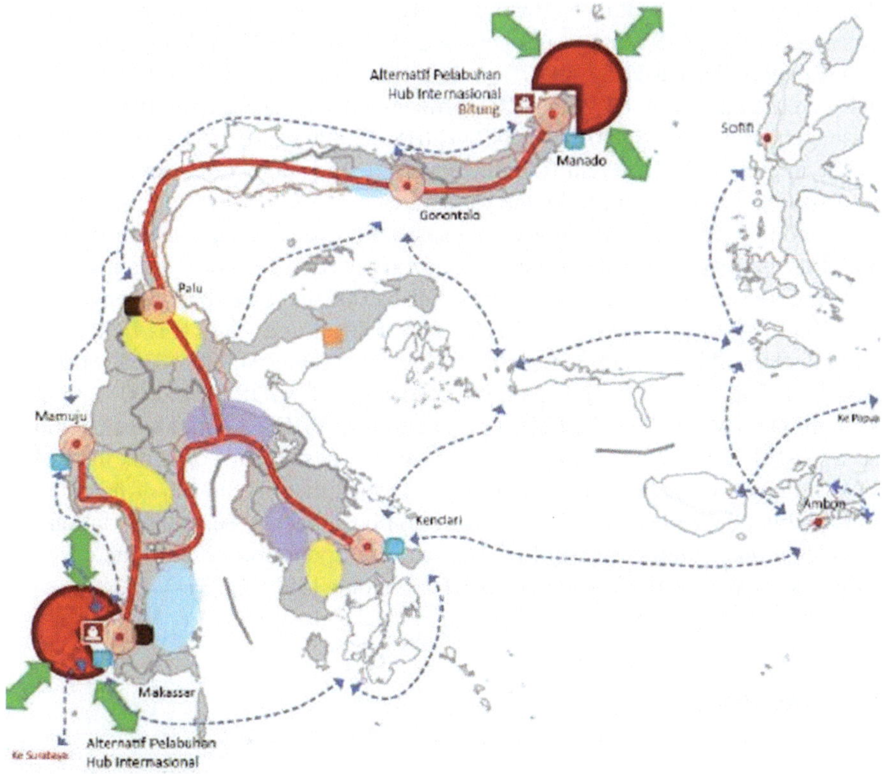

Fig. 2 Sulawesi economic corridor. *Source* Misliah et al. [12]

Furthermore, there is also the port of Watunohu which plays an important role in the North Kolaka Regency. Several strengths possessed by port of Watunohu that cannot be found in the land transportation are (1) it plays as a primary connection for coastline area with Luwu Regency in South Sulawesi province; (2) it becomes a port for product accumulation to increase export value between islands; (3) it has a function to be an intermediary area to support the fluency of distribution of people, commodities, and services to the other regions that will encourage the development of economy, social, politics, and cultures; (4) the volume of loading and unloading activities are increased every year both for national purpose and export; and (5) port of Watunohu becomes the regional crossing destination from South Sulawesi province, namely port of Siwa, port of Palopo, and port of Malili [2].

People usually choose sea transportation over land transportation because of its effectiveness. For instance, if the port of Siwa becomes their destination, people will use the ferry or ship because the distance for the North Kolaka–Siwa route is 32 miles at sea and it will take three hours to cross. The route for sea transportation is simpler and closer than land transportation because if we used land transportation, it would take more time and become less efficient. This is explaining why the number of passengers using the Port of Tobaku is massive. For departure, there were 71,119 passengers in 2017, 95,087 passengers in 2018, and 97,875 in 2019. The number of passengers decreased by 49,070 in 2020 because of the COVID-19 pandemic and increased again by 81,941 in 2021. Meanwhile, for the arrival, the number of passengers was 42,317 in 2017, 40,349 in 2018, 42,890 in 2019, 51,270 in 2020, and 83,438 in 2021 [12]. Different from the departure, the number of passengers during the arrival was not influenced by the pandemic.

Due to less time consumed by sea transportation, the costs are also more affordable. For the Siwa–Lasusua crossing line in the port of Tobaku with a distance of 52 miles, an adult passenger only needs IDR 51,000 and an infant passenger needs IDR 5000. For the vehicles, class I is IDR 58,000 per unit, class II is IDR 103,000 per unit, class III is IDR 209,000 per unit, class IV is IDR 794,000–831,000 per unit, class V is IDR 1,532,000–1,418,000 per unit, class VI is IDR 1,810,000–1,854,000 per unit, class VII is IDR 2,776,000 per unit, class VIII is 3,056,000 per unit, and class IX is IDR 3,751,000 per unit [12]. Moreover, in sea transportation, the carbon footprint has been reduced in several ways. One of them is the use of newer solutions in ship power plants which can reduce combustion [14]. Sea transportation also provides more economic feasibility than land transportation because the costs are lower and more affordable with the society's income [15] (Fig. 3).

4 Conclusion

The North Kolaka Regency has become a strategic region in Southeast Sulawesi province because it is located at the Trans Sulawesi road and has a direct boundary with South Sulawesi as the most developed region in the Indonesian East region. In order to access South Sulawesi from the North Kolaka Regency or vice versa, there are two transportation modes that can be used, namely land transportation through Trans Sulawesi road and sea transportation through port of Tobaku. However, sea transportation is better than land transportation in the North Kolaka Regency because the use of ferries or ships becomes more crucial to connect the North Kolaka Regency with other regions that are located across the sea.

This study then provides two recommendations for the government of the North Kolaka Regency to improve the effectiveness of its sea transportation. First, the ports' facilities need to be improved by changing the broken tools with the new tools and adding new facilities to complement the ports' function. Second, if the number of passengers is continuously increased, the government needs to increase the number of ferries or ships. Thus, the increased number of passengers will not become a problem anymore and the needs of these passengers can be fulfilled. Meanwhile, for future research, this study can become an additional reference to give new insight regarding the effectiveness of sea transportation over land transportation in the North Kolaka region.

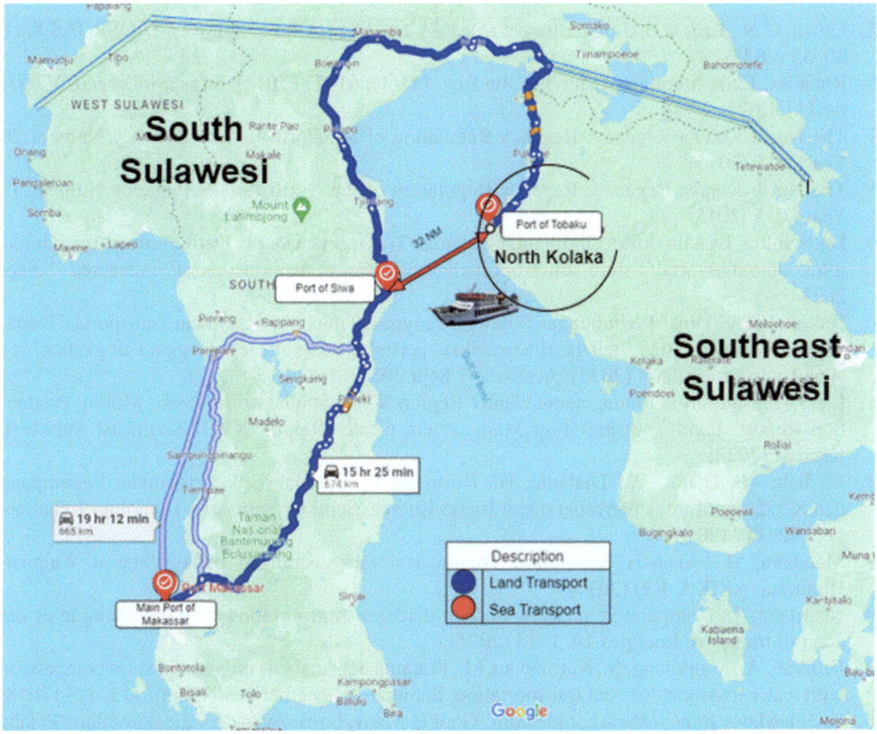

Fig. 3 Tobaku–Siwa crossing route. *Source* Google Map [16]

Acknowledgement The authors gratefully acknowledge to the Directorate of Sea Transportation and Department of Transportation in the North Kolaka Regency and information to the author. Thanks to Institut Teknologi Sepuluh Nopember (ITS) for the guidance from the lecturers in compiling this article.

References

1. Coordinating Ministry of Maritime and Investment Affairs: Rencana Strategis Deputi Bidang Koordinasi Infrastruktur dan Transportasi 2020–2024, pp. 1–43 (2020)
2. Mahmud, M.: Analisis Kelayakan dan Strategi Pengembangan Pelabuhan Watunohu sebagai Pelabuhan Pengumpul di Regency Kolaka Utara. J. Unitek **15**(2), 229–237 (2022)
3. Kresnanto, N.C., Risdiyanto, R., Marlina E.: Studi Analisa Peran dan Potensi Sektor Transportasi Darat dalam Mendukung Pengembangan Pariwisata di Wilayah Pulau Sulawesi. Technical Report, Ministry of Transport the Republic of Indonesia (2015). https://doi.org/10.13140/RG.2.2.34954.72642
4. Japari, D.M.F., Jinca, M.Y., Azmy, M.F.: Selection of airport location for the north Kolaka regency based analytical hierarchy process (AHP). J. Teknol. Dan Keselam. Transp. **3**(1), 1–8 (2020)

5. Ugwu, C.N., Eze, V.H.U.: Qualitative research. Int. Digit. Organ. Sci. Res. IDOSR JAS **8**(1), 20–35 (2023)
6. Rhoades, E.: A literature reviews. Volta Rev. **111**(1), 61–71 (2011). https://doi.org/10.17955/tvr.111.1.677
7. The North Kolaka Regency: Regency Regulation of the North Kolaka Regency Number 20 Year 2015 (2015)
8. The North Kolaka Regency: Regency Regulation of the North Kolaka Regency Number 19 Year 2015 (2015)
9. DetikSultra: Dishub Sultra Optimalkan Layanan Transportasi Darat. DetikSultra. https://detiksultra.com/advertorial/dishub-sultra-optimalkan-layanan-transportasi-darat/. Accessed 7 Sept 2023
10. Hakim, I.S.A.: Dinas Perhubungan Sulawesi Tenggara Tingkatkan Layanan Transportasi Darat. Telisik Indonesia. https://telisik.id/news/dinas-perhubungan-sulawesi-tenggara-tingkatkan-layanan-transportasi-darat (2023). Accessed 7 Sept 2023
11. Land Transportation Management Center Region XVIII Southeast Sulawesi: Montly Production Report. Land Transportation Management Center Region XVIII Southeast Sulawesi, Kendari (2022)
12. Misliah, M., Djafar, W., Djalante, H., Rosmani, R.: Karakteristik Pergerakan Penumpang dan Kendaraan Lintas Penyeberangan Bajoe-Kolaka. Seminar Sains dan Teknologi Kelautan, pp. 112–119 (2018)
13. Mandaku, H.: Analisis Kebutuhan Transportasi Penyeberangan pada Lintasan Waipirit-Hunimua. ARIKA **4**(2) (2010)
14. Neumann, T.: Comparative analysis of long-distance transportation with the example of sea and rail transport. Energies **14**, 1–13 (2021)
15. Fadillah, A., Manullang, S., Kurniawan, M., Pratama, P.: Analysis of land transport comparison with water transport for coal transportation. Seminar Master 2018 PPNS, pp. 115–121 (2018)
16. GoogleMaps: https://www.google.com. Google Maps, https://www.google.com/maps/Pelabuhantobaku-pelabuhansiwa/ (2023). Accessed 7 Sept 2023

Experimental Study of Forecasting Methods for the Passenger Terminal Development Plan of the Nabire Port

Lely Mareta Lestari, Mauritz Halomoan Manontang Sibarani, and Yunika Fitrianti

1 Introduction

Ports are places for intra-and inter-mode transportation, especially the interface for the smooth movement between sea and land transport. Therefore, it is essential to make an appropriate port development plan that meets port technical requirements, environmental sustainability, social spatial planning, shipping safety, economics, and finance [1, 2]. Indonesia, as an archipelagic country, has many potential regions dispersed throughout its hinterland connected by seaports. Thus, the maritime transportation system serves a crucial purpose [3]. Maritime transportation, particularly in archipelagic countries, significantly impacts people's lives through cargo, passengers, and livestock.

Nabire Regency is a town in the Indonesian province of Central Papua, at the western end of New Guinea. Nabire Regency is the Capital of the Central Papua Province after its proliferation in 2022 [4]. Nabire Regency has several ports for commodity inlet-outlet. The main port in the regency is Nabire Port, located in Samabusa Village, Teluk Kimi District, Nabire Regency, Central Papua Province. The Nabire Port is one of the pioneer ship hub ports with a significant number of passengers, and it has more than 100,000 passengers annually [5]. There are ten different passenger ships that call at Nabire Port each year, for a total of 230 calls, with most of them are liner passenger ships having a fixed schedule [5]. However,

L. M. Lestari (✉)
Faculty of Marine Technology, Institut Teknologi Sepuluh Nopember, Surabaya, Jawa Timur, Indonesia
e-mail: lelymareta@gmail.com

M. H. M. Sibarani
Merchant Marine Higher Education Marunda, Jakarta, Indonesia

Y. Fitrianti
Ministry of Transportation of Indonesia, Jakarta, Indonesia

R. S. Widjaja et al. (eds.), *Marine Technology 3*,
SpringerBriefs in Applied Sciences and Technology,
https://doi.org/10.1007/978-3-032-05613-9_4

passenger terminal facilities such as waiting room areas, service areas, and public areas in Nabire Port are only designed to accommodate less than 100 passengers at once. The passenger terminal's current area is 450 m^2, resulting in the inability to accommodate over 100 passengers simultaneously [5]. It is also reported that some passengers were waiting outside the passenger terminal, especially in the container yard, due to insufficient passenger terminals. The condition does not comply with safety procedures as stated in the Ministerial Decree and the ratification regulation. Moreover, the port authority has an obligation to provide and implement safety procedures for passengers [6, 7].

The study uses the Minitab application to conduct a statistical forecasting method to provide an initial perspective on the development plan of the Nabire Port passenger terminal. As well as practical tools for streamlining processes, Minitab offers a vast set of statistics for investigating data and graphs [8]. Passenger data will be forecast for the next five years using data from 2010 to 2022. Forecasting data will be analyzed according to government regulations on passenger terminals to determine the size and development plan of the Nabire Port passenger terminal.

2 Methodology

The research will conduct some forecasting methods to determine the best model to suit for the passenger terminal development plan. The forecasting methods used in the study are as follows:

1. Trend Analysis: The trend analysis is used to fit a general trend model to time series data without seasonality and provide forecasts [8]. With $N = 13$ and fitted trend equation $y_t = 227859 \times (0.9404^t)$;
2. Moving Average: Moving averages are employed to smooth data without trend and offer short-term forecasts [8]. With $N = 13$ and moving average length = 1;
3. Single Exponential Smoothing: Single exponential smoothing can smooth data by calculating exponentially weighted averages and providing short-term forecasts [8]. With $N = 13$ and smoothing constant $\alpha = 0.349149$;
4. Double Exponential Smoothing: Double exponential smoothing calculates dynamic estimates of two components, level and trend. This method can be used as a general smoothing method and to provide short-term data [8]. With $N = 13$ and smoothing constants α (level) = 0.1; γ (trend) = 0.1;
5. Winter's Method: Winter's method calculates dynamic estimates for three components: level, trend, and seasonal. The method can be used for data smoothing and providing medium-range forecasts [8]. With $N = 13$ and smoothing constants α (level) = 0.9; γ (trend) = 0.9; δ (seasonal) = 0.9.

Passenger data from embarking and disembarking at Nabire Port (2010–2022) as stated in Table 1. The data is utilized to determine the optimal forecasting method, focusing on passenger data, which impacts space requirements for facilities

Table 1 Nabire Port passenger data (2010–2022)

No.	Year	Arrival (person)	Departure (person)	Total (person)
1.	2010	78.099	77.127	155.226
2.	2011	82.410	73.000	155.410
3.	2012	91.035	75.425	166.460
4.	2013	89.504	76.476	165.980
5.	2014	167.062	178.628	345.690
6.	2015	91.729	103.499	195.228
7.	2016	86.197	91.285	177.482
8.	2017	85.018	75.515	160.390
9.	2018	85.018	74.515	159.533
10.	2019	77.675	95.440	173.115
11.	2020	29.058	19.085	48.143
12.	2021	29.391	40.014	69.405
13.	2022	67.757	73.431	141.188

Source Port Management Unit Class II of Nabire

like terminals and public spaces. Peak passenger numbers will determine capacity parameters.

Table 1 shows there is a slight trend data with the highest number of passengers in 2014 and the lowest counted in 2020 during the travel restrictions during the COVID-19 pandemic. Post-pandemic, passenger traffic data showed promising growth; it returned to previous trends within two years, making it unsuitable for seasonal data. Furthermore, the forecasting data of passengers are analyzed by the Guidelines for Planning Land Facilities at Collector Ports, Regional Ports, and Local Feeder Ports within the Directorate General of Sea Transportation [9].

The research study focused on policy analysis at Nabire Port Passenger Terminal in Central Papua Province, using secondary survey data without user surveys.

3 Results and Discussion

Based on passenger data and ship calls, as stated in Tables 1 and 2, it can be assumed that each ship carries an average of 707 passengers. The pier's length is insufficient for multiple passenger ships to berth simultaneously, causing it to be assumed that they cannot berth simultaneously. Therefore, the spatial requirement calculation variable will use the 707 passengers' assumption as the number of passengers served on each ship. Based on the Decree of the Director General of Sea Transportation regarding Guidelines for Planning Land Facilities at Collector Ports, Regional Ports, and Local Feeder Ports, the minimum spatial requirements and capacity at the Nabire Port passenger terminal were calculated in total as 1.340 m^2 for 300 passengers (Table 3).

The objective of the statistical experiment of the forecasting method is to obtain a method with the smallest accuracy value. As listed in Table 4 of the accuracy value summary of the passengers forecasting analysis method at the Nabire Port, the highest value is moving average method with MAPE = 42, MAD = 50,765, MSD = 3,974,725,213. On the other hand, Winter's method has the smallest value of MAPE = 30, MAD = 45,194, MSD = 2,784,639,766. As a result, this method can be used to forecast the number of passengers at Nabire Port over the next five years.

Table 5 shows passenger traffic forecasting data for the next four years, which has been presented in bold numerical values. The forecasting results indicate that the average number of passengers is below the spatial calculation variable, resulting in a variable of 707 passengers. The comparison of the passenger terminal area and existing capacity toward spatial and capacity requirements based on land facility standards is shown in Table 6.

Table 2 Nabire Port passenger ship calls (2010–2022)

No.	Year	Perintis ships					Pelni ships				Total
		R-72	R-74	R-76	R-78	R-79	Tidar	Labobar	Dobonsolo	Ciremai	
1.	2010	26	26	26	26	22	26	26	26	26	230
2.	2011	26	26	26	26	22	26	26	26	26	230
3.	2012	26	26	26	26	22	26	26	26	26	230
4.	2013	26	26	26	26	22	26	26	26	26	230
5.	2014	26	26	26	26	22	26	26	26	26	230
6.	2015	26	26	26	26	22	26	26	26	26	230
7.	2016	26	26	26	26	22	26	26	26	26	230
8.	2017	26	26	26	26	22	26	26	26	26	230
9.	2018	26	26	26	26	22	26	26	26	26	230
10.	2019	26	26	26	26	22	26	26	26	26	230
11.	2020	26	26	26	26	22	26	26	26	26	230
12.	2021	26	26	26	26	22	26	26	26	26	230
13.	2022	26	26	26	26	22	26	26	26	26	230

Source Port Management Unit Class II of Nabire

Table 3 Passenger's terminal area estimation based on guidelines

No.	Facility	Capacity	Area estimation (m^2)
1.	Departure lobby	300 passengers	150.48
2.	Ticket counter	30 passengers	21.60
3.	Information center	15 passengers	10.8
4.	Departure hall	360 passengers	326.25
5.	Check-in counter	300 passengers	99
6.	Waiting room	150 passengers	90
7.	Arrival lobby	300 passengers	150.48
8.	Toilet departure area	300 passengers	36
9.	Toilet difabel departure area		6
10.	Toilet arrival area	300 passengers	36
11.	Toilet difabel arrival area		6
12.	Janitor		1
13.	Mushola departure area	60 passengers	48
14.	Mushola arrival area	60 passengers	48
15.	Staff room	4 passengers	8.8
16.	Port health room		12
17.	Smoking room		6
18.	Nursery room		6
19.	Shipping service room		12
20.	Port quarantine room		6
21.	Commercial area		32.63
22.	Circulation area		226.96
	Total		**1.340**

Source Guidelines for Planning Land Facilities at Collector Ports, Regional Ports, and Local Feeder Ports within the Directorate General of Sea Transportation, Ministry of Transportation of Re-public Indonesia

Table 4 Accuracy value summary of the passenger forecasting analysis method at Nabire Port

Forecasting method	MAPE	MAD	MSD
Trend analysis	35	48,518	3,974,725,213
Moving average	*42*	*50,765*	*6,422,666,581*
Single exponential smoothing	41	44,846	4,593,721,424
Double exponential smoothing	36	**41,385**	4,079,797,235
Winter's method	**30**	45,194	**2,784,639,766**

Table 5 Nabire Port passenger forecasting data

No.	Year	Passengers
1.	2010	155.226
2.	2011	155.410
3.	2012	166.460
4.	2013	165.980
5.	2014	345.690
6.	2015	195.228
7.	2016	177.482
8.	2017	160.390
9.	2018	159.533
10.	2019	173.115
11.	2020	48.143
12.	2021	69.405
13.	2022	141.188
14.	***2023***	***116.680***
15.	***2024***	***206.414***
16.	***2025***	***118.870***
17.	***2026***	***135.185***

The bold numerical values is the data of passenger traffic forecasting for the forthcoming four-year period

Table 6 Data comparison between existing facility and portland facility standard

Variable	Existing	Standard
Passenger terminal area	450 m^2	1.340 m^2
Passenger terminal capacity	707 person	300 person

4 Conclusion

To conclude, the Nabire passenger terminal needs urgent development to meet passenger needs, especially sea transportation, due to an escalating passenger forecast in five years. The current terminal building is insufficient to accommodate approximately 707 passengers.

According to portland facility standards, the passenger terminal building should have a minimum of 1.340 m^2 for a 300-passenger capacity. In other words, to accommodate approximately 707 passengers each call, it is required to have the passenger terminal area twice the size of the standard area. However, this research can be expanded by conducting a primary survey of users (passengers) to obtain more comprehensive research results that can be used to justify further technical studies. The research suggests using statistical forecasting methods to predict future passenger data and plan accordingly. These findings could be crucial for policy-makers and port authorities in planning for the future development of the Nabire Port.

Furthermore, the research results can be utilized as a foundation for updating the Directorate General of Sea Transportation's Guidelines for Planning Land Facilities for future passenger facilities.

Acknowledgement The author would like to express sincere gratitude to all those who have contributed to completing this research paper, Pak Mauritz for the constructive input, and Kak Yunika for a lot of research ideas. This work would not have been feasible without their counsel and encouragement. To all professors, lecturers, and staff of the Department of Marine Transportation Engineering, ITS for great chances and resources.

References

1. Government of Indonesia: Undang-Undang Nomor 17 Tahun 2008 tentang Pelayaran, Ministry of Transportation, Indonesia (2008)
2. Government of Indonesia: Peraturan Pemerintah Nomor 61 Tahun 2009 tentang Kepelabuhanan, Ministry of Transportation, Indonesia (2009)
3. Jinca, M.Y., Humang, W.P.: Perencanaan dan pengembangan transportasi wilayah kepulauan. Nas Media Pustaka, Yogyakarta (2023)
4. Government of Indonesia: Undang-Undang Nomor 15 Tahun 2022 tentang Pembentukan Provinsi Papua Tengah, Ministry of State Secretariat, Indonesia (2022)
5. Government of Indonesia: Keputusan Menteri Perhubungan Nomor KM 103 Tahun 2019 tentang Rencana Induk Pelabuhan Nabire, Ministry of Transportation, Indonesia (2019)
6. Government of Indonesia: Peraturan Menteri Perhubungan Nomor PM 37 Tahun 2015 tentang Standar Pelayanan Penumpang Angkutan Laut, Ministry of Transportation, Indonesia (2015)
7. Government of Indonesia: Peraturan Menteri Perhuungan Nomor PM 119 Tahun 2015 tentang Standar Pelayanan Penumpang Angkutan Laut, Ministry of Transportation, Indonesia (2015)
8. Minitab: Statistical Modeling. https://support.minitab.com/en-us/minitab/20/help-and-how-to/statistical-modeling/time-series/how-to/time-series-plot/ (2023). Accessed 17 Sept 2023
9. Government of Indonesia: Keputusan Direktur Jenderal Perhubungan Laut Nomor PP.001/3/17/DJPL-2017 tentang Pedoman Perencanaan Fasilitas Darat pada Pelabuhan Pengumpul, Pelabuhan Pengumpan Regional dan Pelabuhan Pengumpan Lokal di Lingkungan Direktorat Jenderal Perhubungan Laut, Directorate General of Sea Transportation Ministry of Transportation, Indonesia (2017)

Unveiling New Dimensions: A Scoping Review and Bibliographic Analysis in Port Resilience with a Focus on Oil and Gas Marine Terminals

Putu Indra Mahatrisna, Raja Oloan Saut Gurning, and Anak Agung Bagus Dinariyana

1 Introduction

The maritime oil and gas industry faces numerous risks, such as geopolitical uncertainties, weather-related events, accidents, and technological failures. Earlier research has shown that shipping networks are vulnerable and susceptible to intentional attacks [1]. Threats to terminal operations have increased in recent years. In the past, natural disasters like typhoons, earthquakes, and tsunamis or human causes like ship collisions and strikes were significant threats and relatively common. However, cyber-terrorism and infectious diseases such as Covid-19 are relatively infrequent and rising as new threats [2, 3].

The concept of resilience was first introduced in ecology by Holling [4]. In marine transportation, the resilience concept pertains to the ability of various components within the maritime system, including ports, shipping lines, vessels, and supply chains, to effectively foresee, plan for, respond to, and recover from various disruptions [5]. This study aims to offer a thorough overview of current knowledge and ascertain frameworks, as well as potential avenues for future research on port

P. I. Mahatrisna (✉) · A. A. B. Dinariyana
School of Interdisciplinary Management and Technology, Institut Teknoogi Sepuluh Nopember, Surabaya, Indonesia
e-mail: pimahatrisna@skkmigas.go.id

A. A. B. Dinariyana
e-mail: kojex@its.ac.id

R. O. S. Gurning
Department of Marine Engineering, Faculty of Ocean Technology, Institut Teknologi Sepuluh Nopember, Surabaya, Indonesia
e-mail: sautg@its.ac.id

R. S. Widjaja et al. (eds.), *Marine Technology 3*,
SpringerBriefs in Applied Sciences and Technology,
https://doi.org/10.1007/978-3-032-05613-9_5

resilience within the oil and gas sector. This will be achieved through the utilization of a combined approach involving a scoping review (ScR) and bibliographic analysis (BA), commonly referred to as "ScoRBA" introduced by Wijaya et al. [6].

2 Methodology

The bibliometric literature review method is valuable for efficiently and effectively assessing the current state of research on a specific topic. It enables the identification of trends and provides insights into potential avenues for future investigation. However, it is important to acknowledge that this method also possesses certain limitations, such as the potential for insufficient and in-depth analysis outcomes. To address existing gaps in knowledge, it is necessary to employ a combination of bibliometric methods and scoping reviews, as pioneered by Arksey et al. [7] and Bradburry-Jones et al. [8] to explore the available literature on a given topic systematically. This study employs three applications to achieve optimal outcomes, such as VOSViewer, OpenRefine, and Mendeley.

2.1 Search Strategy

The analysis of the bibliography begins by formulating an initial use case for the ScoRBA, which encompasses various pieces of information such as: *Research Topic*: Enhancing Port Resilience in Oil and Gas Marine Terminals through a Risk Assessment Approach; *Dataset*: Scopus; *Keyword*: "Port Resilience"; *Time Span*: 2013–2022; *Document Type*: Journal Article; *Language*: English. Approximately 341 relevant papers were obtained and utilized as initial data through a keyword search conducted in the Scopus database. Moreover, the subsequent step involves conducting keyword analyses to identify trending and highly referenced keywords within the scholarly community in recent times.

2.2 Keywords Analysis

With the assistance of the VOSViewer application and the previously obtained Scopus dataset, the author conducted a keyword suitability analysis to determine the most relevant keywords to the current topic and trend. Initially, as shown in Fig. 1 the network visualization image reveals interconnected research areas. Three clustering areas are related to the port resilience keyword based on the search results: the red group relates to urban and community resilience, the blue group relates to natural disasters, and the green group relates to disruption management and sustainability. In this analysis of network visualization, the author chose to focus on the green

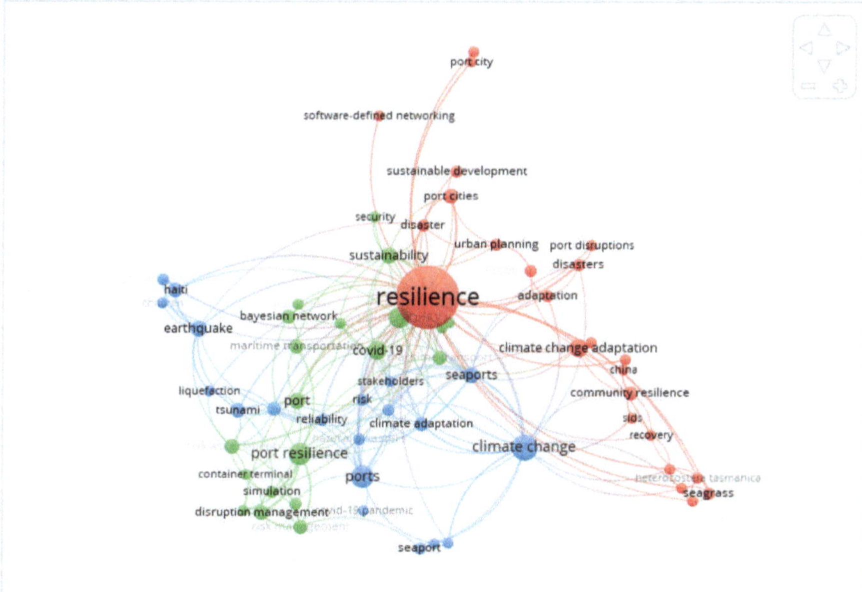

Fig. 1 Keyword network visualization by VOSViewer

research group, which is associated with disruption management and sustainability, because this keyword is closely associated with the operational risk at the oil marine terminal that is the subject of this study. Moreover, in the same application, we use the overlay visualization feature to obtain keyword information that has been cited most frequently by researchers recently, as depicted in Fig. 2. The red dots on the image represent the following keywords: *Covid-19, Bayesian Network, Reliability, and Climate Adaptation.* In this particular case, the author opted to investigate the Bayesian network and reliability keywords for further examination, as they were found to be most relevant to the previously chosen topic.

In addition, after conducting a keyword analysis, relevant papers were then extracted using the *OpenRefine* application and conducted a manual review based on the PAGER framework that consist of relevance patterns, advances, gaps, evidence for practice, as well as the future research recommendation [8], in order to obtain all the literatures that, according to the author, are the most relevant to the topic discussed.

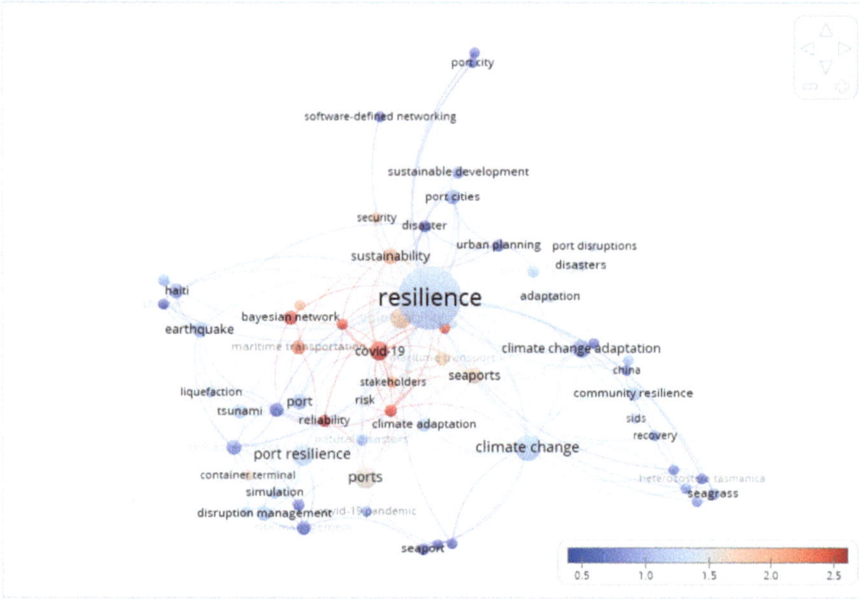

Fig. 2 Keyword overlay visualization by VOSViewer

3 Results and Discussion

3.1 Literature Review

Based on analysis of relevant papers, it has identified articles with the highest number of citations (> 50 citations) that have been published in many journals, including Computers and Industrial Engineering (199 citations), Reliability Engineering and System Safety (> 100 citations), and the International Journal of Sustainable Transportation (> 50 citations). Furthermore, it also identified several authors who have been highly active in contributing to writing high-quality articles that other researchers have widely cited. These authors include Loh H. S. et al., who published five articles (2014–2017); Hossain N. U. I. et al., who published two articles (2019–2020); and Shaw D. R. et al., who published two articles (2017–2019).

From the annual distribution analysis results, the number of publications released throughout the ten years found that the average number of publications was two per year, peaked during 2020–2021 as many as nine publications and then declined in 2022. According to our perspective, this occurrence may be attributed to the significant impact on port operations during Covid-19, which raised many stakeholders' awareness about the need for port resilience for business sustainability.

3.2 Port Resilience Strategy

According to the literature review, resilience in critical infrastructure, particularly port and maritime systems, is a common subject. The researchers aim to develop and refine models, frameworks, and strategies to enhance port resilience that can be categorized into three main areas: (i) the management of risks, (ii) the development of frameworks to enhance resilience, and (iii) the quantitative assessment of resilience [2].

A comprehensive port resilience strategy for the oil and gas sector requires a multidimensional approach, frequently using multiple methodologies. Quantitative methods like mathematical modeling, simulation, and multicriteria decision making (MCDM) can develop and optimize the resilience framework, starting with the conceptual framework and integrating data-driven insights from case studies and empirical research. Bayesian analysis may handle risk analysis uncertainties, and complex network analysis and game theory can reveal system-wide interactions and strategic dynamics. Since they complement each other and provide a holistic understanding of port resilience in the oil and gas sector, the appropriate approach or methods to construct such a framework will depend on the study objectives and port operational complexity.

3.2.1 Vulnerability Assessment

The analysis of port resilience within an integrated maritime network has been the subject of a recent study conducted by Catalayud et al. [9] and Liu et al. [1]. These studies emphasize the importance of adopting a systematic network perspective when assessing the vulnerability of ports, highlighting the need for a maritime network viewpoint in vulnerability assessment. This section outlines the methodologies employed in establishing the proposed vulnerability assessment framework. Once a port is selected for examination within a shipping network, a hierarchical framework for vulnerability assessment is developed. This structure identifies the distinct vulnerability measures and their hierarchical arrangement that define the shipping network. Jiang et al. [10] demonstrate the feasibility of generating comprehensive vulnerability estimates by integrating fuzzy theory and the ER approach, which involves amalgamating several vulnerability assessments.

3.2.2 Design of Resilience Capacities Network

The resilience triangle concept which includes absorptive, adaptive, and restorative elements was introduced by Vugrin et al. [11] and expanded by Francis and Bekera [12] inspired the development of the resilience capacities framework. Subsequently, the framework was refined by incorporating the nine characteristics of resilience proposed by Kim et al. [2], where measurements of absorptive capacity include robustness, redundancy, and visibility among the nine factors. The assessment of adaptive capabilities consists of the dimensions of adaptability, collaboration, agility, and information sharing, whereas the evaluation of restorative abilities focuses on response and recovery metrics. Robustness and redundancy are absorptive capacity-quantifying metrics.

3.2.3 Resilience Analysis

Numerous researchers have employed various qualitative and quantitative approaches and methodologies to assess the level of resilience exhibited by a marine terminal, drawing upon a diverse body of earlier studies. The Delphi technique [13], dynamic Bayesian network [13], decision support system (DSS) with digital twinning [14], and Monte Carlo simulation [15] represent quantitative methodologies within the field. In addition to the methodologies above, scholars have employed many qualitative methods, including the resilience assessment grid (RAG) developed by Djunaidi et al. [16] and the functional resonance analysis method (FRAM) proposed by Qiao et al. [17].

3.2.4 Resilience Index Cost Ratio Analysis

In this evaluation, the total cost is determined by the duration of recovery for each port and its corresponding daily expenses. Its dimensions and economic condition primarily assess the time of recovery for a port following an interruption. According to Jiang et al. [10], ports of larger size and lower levels of economic development often require a longer recovery duration. This is attributed to the increased resource requirements for repairing a larger port after its destruction.

To maximize the benefits of a shipping network throughout the recovery process, it is imperative to consider the resilience and cost of disaster response. The resilience-cost ratio indicator for shipping network performance can be determined by utilizing the resilience index model and the shipping network recovery cost model.

3.2.5 Recovery Strategies

Implementing qualitative strategies to enhance port resilience involves prioritizing recovery efforts based on the strategic significance of various port network elements. The random, degree centrality, closeness centrality, and betweenness centrality-based recovery strategies provide distinctive means of achieving this objective [18]. The random approach ensures that all port areas receive attention, but dealing with critical infrastructure may require greater efficiency. Degree, closeness, and betweenness centrality strategies concentrate recovery efforts on the most interconnected, central, and intermediate elements, ensuring rapid restoration of vital structures.

To effectively enhance port resilience, it is recommended that these strategies be combined, adapted to the specific port's characteristics, and regularly updated as the network evolves. Collaboration with stakeholders and relevant agencies is necessary for a coordinated and thorough response to disruptions, ensuring the port's continued functionality and service delivery.

4 Conclusion

A combined scoping review-bibliographic analysis ("ScoRBA") methodology is used to discover strategies to improve marine terminal disruption resilience. The research examines 341 Scopus journal articles with port resilience keywords from 2013 to July 2022. A keyword appropriateness analysis using the VOSViewer let the author find the most relevant terms for the topic and trend. In the network visualization image with three clustering areas—red, blue, and green—the author focused on the green cluster, which is associated with disruption management and sustainability and operational risk at the oil marine terminal under study.

According to the overlay visualization function, the red dots on the image represent Covid-19, Bayesian network, reliability, and climate adaptation. The author selected to examine the Bayesian network and reliability keywords because they were most relevant to the issue. This study grouped port resilience research into three primary areas: (i) risk management, (ii) framework development to enhance resilience, and (iii) quantitative assessment of resilience [2]. Based on the literature review, we can classify resilience framework construction at a port or marine terminal into vulnerability assessment, resilience capacity network design, resilience analysis, resilience index cost ratio analysis, and recovery strategies.

References

1. Liu, H., Tian, Z., Huang, A., Yang, Z.: Analysis of vulnerabilities in maritime supply chains. Reliab. Eng. Syst. Saf. **169**, 475–484 (2018). https://doi.org/10.1016/j.ress.2017.09.018
2. Kim, S., Choi, S., Kim, C.: The framework for measuring port resilience in korean port case. Sustainability (Switzerland) **13** (2021). https://doi.org/10.3390/su132111883
3. Mou, N., Liao, M., Zhang, H., Peng, P., Liu, X.: Evaluation on location advantages of the ports along the maritime silk road. J. Geo-Inf. Sci. **20**, 613–622 (2018). https://doi.org/10.12082/dqxxkx.2018.170514
4. Holling, C.S.: Resilience and stability of ecological systems. Ann. Rev. Ecol. Syst. **4**, 1–23 (1973). https://doi.org/10.12987/9780300188479-023
5. Notteboom, T., Pallis, T., Rodrigue, J.P.: Disruptions and Resilience in Global Container Shipping and Ports: The COVID-19 Pandemic Versus the 2008–2009 Financial Crisis. Palgrave Macmillan, UK (2021)
6. Wijaya, A., Setiawan, N.A., Ahmad, A.H., Zakaria, R., Othman, Z.: Electroencephalography and mild cognitive impairment research: a scoping review and bibliometric analysis (ScoRBA). AIMS Neurosci. **10**, 154–171 (2023). https://doi.org/10.3934/neuroscience.2023012

7. Arksey, H., O'Malley, L.: Scoping studies: towards a methodological framework. Int. J. Soc. Res. Methodol.: Theor. Pract. **8**, 19–32 (2005). https://doi.org/10.1080/1364557032000119616
8. Bradbury-Jones, C., Aveyard, H., Herber, O.R., Isham, L., Taylor, J., O'Malley, L.: Scoping reviews: the PAGER framework for improving the quality of reporting. Int. J. Soc. Res. Methodol. **25**, 457–470 (2022). https://doi.org/10.1080/13645579.2021.1899596
9. Calatayud, A., Mangan, J., Palacin, R.: Vulnerability of international freight flows to shipping network disruptions: a multiplex network perspective. Transp. Res. Part E: Logistics Transp. Rev. **108**, 195–208 (2017). https://doi.org/10.1016/j.tre.2017.10.015
10. Jiang, M., Lu, J., Qu, Z., Yang, Z.: Port vulnerability assessment from a supply chain perspective. Ocean Coast. Manag. **213**, 105851 (2021). https://doi.org/10.1016/j.ocecoaman.2021.105851
11. Vugrin, E.D., Warren, D.E., Ehlen, M.A., Camphouse, R.C.: A framework for assessing the resilience of infrastructure and economic systems. In: Sustainable and Resilient Critical Infrastructure Systems: Simulation, Modeling, and Intelligent Engineering, pp. 77–116 (2010)
12. Francis, R., Bekera, B.: A metric and frameworks for resilience analysis of engineered and infrastructure systems. Reliab. Eng. Syst. Saf. **121**, 90–103 (2014). https://doi.org/10.1016/j.ress.2013.07.004
13. Hossain, N.U.I., Nur, F., Hosseini, S., Jaradat, R., Marufuzzaman, M., Puryear, S.M.: A Bayesian network based approach for modeling and assessing resilience: a case study of a full service deep water port. Reliab. Eng. Syst. Saf. **189**, 378–396 (2019). https://doi.org/10.1016/j.ress.2019.04.037
14. Zhou, C., Xu, J., Miller-Hooks, E., Zhou, W., Chen, C.H., Lee, L.H., Chew, E.P., Li, H.: Analytics with digital-twinning: a decision support system for maintaining a resilient port. Decis. Support. Syst. **143**, 113496 (2021). https://doi.org/10.1016/j.dss.2021.113496
15. Dhanak, M., Parr, S., Kaisar, E.I., Goulianou, P., Russell, H., Kristiansson, F.: Resilience assessment tool for port planning. Environ. Plann. B: Urban Analytics City Sci. **48**, 1126–1143 (2021). https://doi.org/10.1177/2399808321997824
16. Djunaidi, Z., Tantia, A.A., Wirawan, M.: Analysis of the safety resilience implementation in the maritime industry at public and private companies (a case study in Indonesia). Safety **7** (2021). https://doi.org/10.3390/safety7030056
17. Qiao, W., Ma, X., Liu, Y., Deng, W.: Resilience evaluation of maritime liquid cargo emergency response by integrating FRAM and a BN: a case study of a propylene leakage emergency scenario. Ocean Eng. **247**, 110584 (2022). https://doi.org/10.1016/j.oceaneng.2022.110584
18. Wan, C., Zhao, Y., Zhang, D., Yip, T.L.: Identifying important ports in maritime container shipping networks along the maritime silk road. Ocean and Coast. Manag. **211**, 105738 (2021). https://doi.org/10.1016/j.ocecoaman.2021.105738

Strategic Partnership and Its Effect on Port Performance: Case Study of the New Priok Container Terminal 1 in Kalibaru Terminal

Arya Adhyguna and I. Gusti Ngurah Sumanta Buana

1 Introduction

PT Pelabuhan Indonesia (Persero) or Pelindo is the only state-owned enterprise ("SOE") port operator in Indonesia. As a step in achieving Pelindo's vision as the leader of an integrated and world-class maritime ecosystem, port development is one of the strategies for achieving it. One of Pelindo's biggest achievements in port development is the inauguration of the operation of Kalibaru terminal/New Priok container terminal 1 ("NPCT 1").

Tanjung Priok was originally designed to handle 5 million TEUs of container traffic per year. In 2011, container traffic in this port reached 5.8 TEUs, increased to 6.2 million TEUs in 2012–2013, which means the utilization reached 124% of port design capacity. This indicates that it is necessary to immediately add facilities to accommodate the growth of container flow in the coming year and to reduce logistics costs per TEU so that adequate facilities must be prepared to serve direct call ships with large sizes.

The development of the Kalibaru container terminal aims to accommodate the increase of international cargo at Tanjung Priok port to improve port's performance and achieve optimal capacity of the Tanjung Priok port. Tanjung Priok port handles more than half of the total traffic volume in Indonesia and ranks 25th in the world

A. Adhyguna (✉) · I. Gusti Ngurah Sumanta Buana
Department of Marine Transportation Engineering, Institut Teknologi Sepuluh Nopember, Surabaya, Indonesia
e-mail: arya.adhyguna@gmail.com

I. Gusti Ngurah Sumanta Buana
e-mail: buana@seatrans.its.ac.id

A. Adhyguna
Rotterdam Mainport Institute, Rotterdam University of Applied Science, Rotterdam, The Netherlands

R. S. Widjaja et al. (eds.), *Marine Technology 3*,
SpringerBriefs in Applied Sciences and Technology,
https://doi.org/10.1007/978-3-032-05613-9_6

in 2021 in terms of throughput [1]. The development of the Kalibaru container terminal will put Tanjung Priok's port facilities on par with other world-class ports and provide new network connectivity opportunities to shipping lines. In operation, NPCT 1 is owned by four shareholders, namely Pelindo; Mitsui & Co., Ltd.; PSA International Pte Ltd.; and Nippon Yusen Kabushiki Kaisha (NYK Line). The infrastructure development of NPCT 1 is carried out by Pelindo and the operational facilities (superstructure) and operations are carried out under a strategic partner scheme.

With Pelindo's capabilities at that time, Pelindo decided to look for potential partners to operate together the Kalibaru terminal. Based on the Indonesian regulation concerning Guidelines for State-Owned Enterprises Cooperation [2], it is stated that state-owned enterprises cooperation is carried out with the principles of transparency, independence, accountability, responsibility, benefits, and fairness, and following the provisions of laws and regulations. To fulfill the principles of cooperation, in general, this paper presents a recommendation process for partner selection of cooperation in port development which is carried out in stages starting from the planning process by developing strategies, the process of selecting strategic partners with assessment and negotiations, to the implementation of cooperation.

2 Methodology and Discussion

The successful strategic partnerships should involve high levels of trust, commitment, coordination, and interdependency [3]. Building a strategic partnership requires organizational compatibility and alignment of top management with the strategic partner company or organization, which leads to the need for a comprehensive and accountable selection process [4]. Strategic partnerships relate to long-term relationships that focus on strategic goals to provide added value to customers and benefits to the partnering parties. In this case, strategic partnerships are defined as relationships formed between two or more independent entities to achieve specific goals and benefits, as well as provide a competitive advantage and improve the financial performance of the cooperating parties. A typical terminal awarding procedure consists of three phases [5] (Fig. 1).

As a state-owned company, Pelindo must pay attention to aspects of good governance following existing regulations in conducting strategic cooperation. The tailored strategic partner selection process for NPCT 1 carried out by Pelindo consists of the planning, selection, and post-selection phases as shown in Fig. 2.

Fig. 1 Kalibaru terminal development

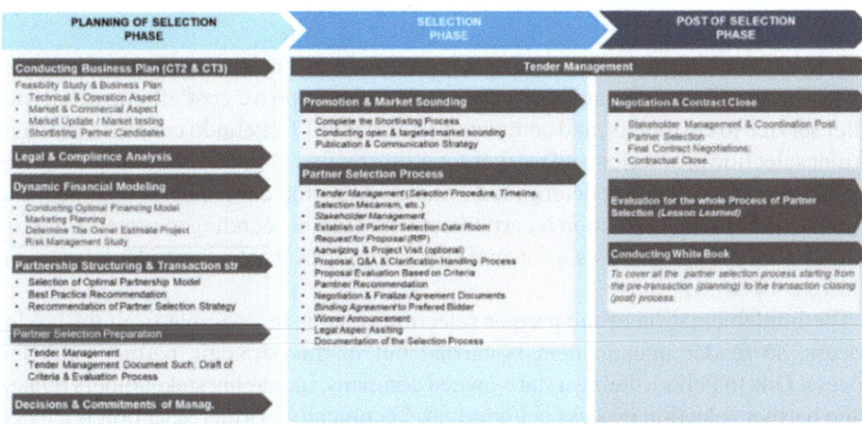

Fig. 2 Strategic partner selection process *Source* Modified by author

2.1 Planning Phase

Strategic planning will focus on adjusting to the internal and external conditions of the company by formulating company strategies and implementing strategies consistently. Strategic planning will build strategic cooperation by maintaining long-term relationships in the hope that company targets can be achieved [6]. In this phase, a business plan review of NPCT 1 was conducted, which contained a business feasibility study, technical and operational analysis, market and commercial analysis, market renewal, and mapping of potential partners (shortlisting). In addition to business studies, it is also necessary to study the legal and compliance aspects because Pelindo is a state-owned company, and the cooperation involves international parties.

Furthermore, the required capital structure, marketing plan, and cost calculation were developed to compile a financial model of the required investment. To mitigate risks that may arise during planning and operations, a risk management study was also conducted at the planning stage of partner selection.

The selection of strategic partners cannot be separated from the tender process, so the planning phase is important in preparing for tender management starting from compiling the required documents and preparing the proposed cooperation agreement documents. In the planning phase, top management decision-making and commitment in terms of partner selection strategies must be obtained before continuing the next process. In the process of selecting partners, Pelindo also coordinates with the Financial and Development Supervisory Agency (BPKP) and the Deputy Attorney for Civil and State Administration (Jamdatun) in order to obtain technical reviews and legal opinions.

2.2 Partner Selection Implementation Process

A company's ability to make the right partner selection aims to implement integration between the company and partners that can improve cost efficiency, giving better service to customers, and company performance [7]. Pelindo conducts strategic partner selection promotion and market sounding by first shortlisting strategic partner candidates close to the predetermined criteria. After the candidate is obtained, the publication of partner selection is carried out with market sounding to promote investment opportunity projects as a form of exploring market interest in the investment to have cooperated.

The implementation of the partner selection stage is inseparable from the tender process, so tender management is carried out in this strategic partner selection process. Due to Pelindo being a state-owned company, managing stakeholders related to the partner selection process is important. Technically, partner selection is carried out from the announcement of the request for proposal (RFP) to prospective partners who have expressed interest, management of proposals received, to the proposal evaluation process to obtain a partner proposal that matches the criteria Pelindo needs in developing NPCT 1. After obtaining recommendations for strategic partner winners, a negotiation evaluation process is carried out by considering criteria on financial, technical and operational, commercial, and other criteria that are considered essential.

In the strategic partner selection process, the evaluation is carried out by scoring each criteria, where the criteria with the largest score becomes the most important thing to consider in the evaluation and assessment process. With the evaluation and negotiation of the assessment of these criteria, a recommendation for the winning candidate is obtained. Prior to the announcement of the winner, a binding agreement is made with the selected strategic partner. If there is no selected partner during the partner selection process, the selection process can be repeated with the prior tender management method.

2.3 Post Partner Selection Phase

After the strategic partner is selected, Pelindo coordinates to fulfill the requirements of the winner and finalizes contract negotiations with the winner of the selected partner. In this contract negotiation, the cooperation scheme in the financial aspects of asset utilization, terminal operations, and concession agreement fees are discussion points that need to be considered between the parties to the cooperation. An evaluation of the entire process of implementing partner selection was performed as a kind of knowledge management to gain lessons learned. The government representative or port authority grants a concession or a franchise to a Pelindo as an operator terminal to finance and build or modernize a specific port facility. Pelindo together with a selected partner to operate the facilities and to obtain revenue from port operations for a designated time period. Pelindo takes all commercial risks during the concession. At the end of the concession period, the government retakes ownership of the improved assets. Arrangements between the government and the operator are set out in a concession contract that may or may not include regulatory provisions.

The final stage was the preparation of a white paper to document from the transaction planning phase to the transaction closing phase. The white paper is prepared by containing the chronology of the company's cooperation or other corporate actions and becomes a crucial need considering the possibility of an audit from the Supreme Audit Agency (BPK) or external auditors. The white paper is prepared in accordance with facts that are carried out systematically, correctly, and based on actual documents that exist or are used as the basis for the implementation of the company's corporate actions. After the partner selection, an important thing that needs to be done is the monitoring and evaluation process of the cooperation with strategic partners to be able to identify its effect on operational performance.

3 Strategic Cooperation and Operational Performance

The hypothesis that strategic cooperation affects operational performance is proven [3]. Strategic cooperation carried out by the company can improve the company's operational performance. Research shows that the collaboration developed by the company with its strategic partners will increase the role of partners in the company's business processes. Strategic cooperation will show innovation in company activities to create efficiency and effectiveness in reducing company operational costs and improving operational performance. The presence of strategic partners in the development and operation of NPCT 1 has a good impact on Pelindo. This is evident from the operation of NPCT 1 in 2016, and the throughput realization from 2017 to 2022 as shown in Fig. 3. The decline in 2020 was due to the COVID-19 pandemic which also affected the flow of goods worldwide. Based on this data, it can be concluded that strategic cooperation between Pelindo and international companies can bring a wider market for NPCT 1.

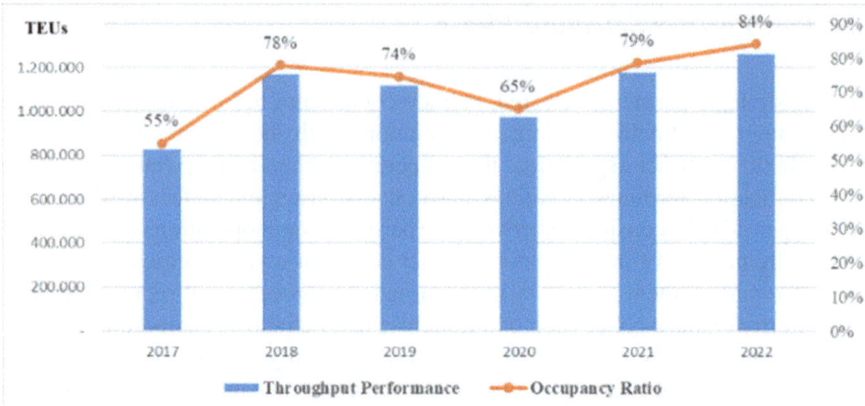

Fig. 3 Operational performance of NPCT1 in 2017–2022. *Source* Financial Report of PT Pelabuhan Indonesia

The cooperation structure carried out by Pelindo with its strategic partners in the development and operation of NPCT 1 and has a good impact on port operational performance. Pelindo's strategic cooperation in NPCT 1 by forming a joint venture with four international companies with an ownership structure, with Pelindo's ownership of 51%. With this international cooperation, the performance of export and import activities at Tanjung Priok port in 2021 grew positively. The total international throughput (exports and imports) in the first quarter of 2021 experienced a growth of 9%, and in January 2022 international throughput growth of 17% compared to the same period in the previous year. In December 2022, the Port of Tanjung Priok got the top 20 the best port performance was published by the United Nations Conference on Trade and Development [8].

With strategic cooperation with global companies, it is possible to open direct shipping (direct call) from Tanjung Priok port. This direct call will make the travel time and cost of shipping goods from Jakarta to the destination country more efficient because there is no need to transit in Singapore. In 2018, the CMA CGM Tage made a direct call from Tanjung Priok port to Los Angeles, United States. The 10,000 TEUs capacity ship offloaded 1905 TEUs and loaded 2385 TEUs. The direct call saved costs of around 20% of US$300 per 20 feet container with time efficiency also reduced from 31 days due to transit to 23 days. The opening of the market at Tanjung Priok port and especially at NPCT 1 has had a positive impact on Pelindo's operations. This is in line with Tarigan and Siagian's research which states that strategic cooperation can have a positive effect on operational performance [3].

In the implementation of NPCT 1 operations, the cooperation scheme used is build, operate, and transfer (BOT) between the NPCT 1 joint venture company and Pelindo. In this cooperation scheme after the cooperation period is completed, all assets will be returned to Pelindo. In the BOT cooperation, Pelindo gets revenue in the form of upfront fees for asset utilization as well as infrastructure and land leasing fees from NPCT 1. In addition, Pelindo also receives dividends from its subsidiary PT

Table 1 Concession fee—contribution

Period	Gross revenue (Rp Million)	*Penerimaan Negara Bukan Pajak* (*PNBP*) non-dividend—concession fee—contribution (Rp Million)
2016 (from Aug 2016)	95,190.35	475.95
2017	1,381,234.74	6905.17
2018	1,947,232.30	9736.16
2019	1,721,708.89	8608.17
2020	1,478,827.32	7394.14
2021	1,755,815.07	8779.07
2022	2,033,548.35	10,167.75

IPC terminal Peti Kemas which is a representative of Pelindo Group in the formation of the NPCT 1 joint venture. During the operation of NPCT 1 since August 2016, NPCT 1 also contributed non-tax state revenue (PNBP) to the state in the form of a concession agreement fee of 0.5% of NPCT 1's gross revenue, with a total each year as in Table 1. Until 2022, NPCT 1 has contributed to PNBP amounting to Rp 52 billion with the highest contribution in 2022 amounting to Rp 10,167 billion. This shows that the operation of NPCT 1 can contribute additional revenue to Pelindo and the state. In reaching agreements related to upfront fees and infrastructure and land leases, Pelindo goes through a negotiation process with partners to obtain costs that can benefit the partnering parties. The negotiating process occurs during partner selection and after the partner has been selected, as specified in the cooperation agreement.

4 Conclusion

The Kalibaru terminal is part of the support for the government's program to increase economic productivity and competitiveness in the international market. The commercial operation of the Kalibaru terminal is supported by the operational readiness of the container terminal in terms of infrastructure, superstructure, and information systems at the berth, yard, and gate. The operational readiness is supported by the cooperation between Pelindo and its strategic partners in the operation of Kalibaru terminal, especially NPCT 1. With the achievement of throughput that has been obtained from the operation of NPCT 1 which almost fulfills its capacity, this cannot be separated from the cooperation carried out by Pelindo with international strategic partners that can bring new market shares for NPCT 1. However, the new of market also creates competition between terminals in Tanjung Priok to NPCT 1. The emergence of this competition can have an impact on Pelindo, which must continue to improve operational performance to continue to get the existing market that already exists and also expand the market even wider.

To encourage direct calls of large-capacity vessels at Indonesian ports, an increase in the volume of export cargo also needs to be considered. Large-capacity ships will make direct calls if the volume of cargo at the port is high, thus meeting the minimum capacity of the ship. From an economic perspective, large-capacity vessels also have a high value because the handling costs will be lower with more cargo that can be transported.

Continuous improvement related to service quality and standardization is also an important benchmark for the success of NPCT 1 to gain a wider market in the midst of open competition between terminals because each shipping company (shipping line) is given the freedom to choose a terminal that becomes a place to unload and load their containers. With so many terminals, it makes it easier for service users to choose a more open terminal destination; this can be an opportunity for each terminal to compete fairly by improving its operational performance and services. Although the strategic partner selection process has included commercial aspects so that partners can bring the volume of ship traffic and ship networks that can increase export and import flows by expanding the market reach of NPCT 1, improvements in terms of the quality of facilities and services must still be developed. On the other hand, this open competition can increase the terminal's competitiveness to become a world-class terminal and increase the level of customer confidence to use port services.

References

1. Bansard International: Global Top 30 Container Ports in 2022, Bansard International. Accessed 16 Jan 2024 [Online]. Available: https://www.bansard.com/en/news/global-top-30-container-ports-2022 (n.d.)
2. Minister of Indonesia State Owned Enterprises: Guidelines for State-Owned Enterprise Cooperation, Indonesia (2021)
3. Tarigan, Z.J.H., Siagian, H.: The effects of strategic planning, purchasing strategy and strategic partner-ship on operational performance. Uncertain Supply Chain Manag. (2021)
4. Yuen, K.F., Thai, V.V.: The influence of supply chain integration on operational performance: a comparison between product and service supply chains. Int. J. Logist. Manag. **28**(2), 444–463 (2017)
5. Notteboom, T., Pallis, A., Rodrigue, J.-P.: Port Economics, Management and Policy (2022)
6. Amrollahi, A., Rowlands, B.: OSPM: a design methodology for open strategic planning. Paper presented at the 22nd Americas Conference on Information Systems (AMCIS 2016), San Diego, California, Aug 2016
7. Piprani, A.Z., Mohezar, S., Jaafar, N.I.: Supply chain integration and supply chain performance: the mediating role of supply chain resilience. Int. J. Supply Chain Manag. **9**(3), 58–73 (2020)
8. UNCTAD: Indonesia Enters the Top 20 Countries with the Best Port Performance. Directorate General of Sea Transportation. Accessed: 05 Oct 2023 [Online]. Available: https://hubla.dephub.go.id/btkp/page/news/read/13051/indonesia-masuk-20-besar-negara-dengan-performa-pelabuhan-terbaik

Experimental Study of Forecasting Methods for the Passenger Terminal Development Plan of the Sri Bintan Pura Port

Ahmad Taufik

1 Introduction

Based on PP No. 61 of 2009 concerning ports, what is meant by "sea port" is a port that can be used to serve sea transportation activities and/or crossing transportation located at sea or on the river [1]. A port is a place consisting of land and/or waters with certain limits as a place of government activities and business activities used as a place to lean, get on and off passengers and/or load and unload goods, in the form of terminals and ship berths equipped with shipping safety and security facilities and port support activities as well as a place for intra- and intermodal transportation movements [2].

Sri Bintan Pura port is located on the West coast of Bintan Island with coordinates 0° 55′ 55″ N and 104° 26′ 06″ E. Administratively, the Sri Bintan Pura port is included in the Tanjungpinang Kota village area, Tanjungpinang Kota district, Tanjungpinang city—Riau Islands Province [3]. Port activities that are generally served by the port of Sri Bintan Pura are boarding and dropping passengers both from domestic and from abroad. Passengers who come to Tanjungpinang have a high number every year. In 2017, passengers at Sri Bintan Pura port reached 465,881 foreign passengers and 1,234,390 domestic passengers, then in 2018 passengers at Tanjungpinang port (Sri Bintan Pura) reached 405,002 foreign passengers and 1,499,763 domestic passengers [4].

As a form of commitment to provide the best service to passenger terminal service users, PT Pelabuhan Indonesia (Persero) Regional I has carried out the development and arrangement of airport-standard passenger terminals at the Sri Bintan Pura passenger terminal in 2018. Furthermore, to align the arrangement of the Sri Bintan Pura passenger terminal with the RTRW of Tanjungpinang city and the RTRW of

A. Taufik (✉)
Institut Teknologi Sepuluh Nopember, Surabaya, Jawa Timur, Indonesia
e-mail: ahmadtaufikppu2017@gmail.com

R. S. Widjaja et al. (eds.), *Marine Technology 3*,
SpringerBriefs in Applied Sciences and Technology,
https://doi.org/10.1007/978-3-032-05613-9_7

Riau Province which plans the construction of the Gurindam ring road, a revision of the Tanjungpinang port master plan is needed [5].

In May 2018, a discussion meeting was held between PT Pelabuhan Indonesia (Persero) Regional I, KSOP Class II Tanjungpinang port, ministry of transportation, in order to accommodate the proposed development of Sri Bintan Pura port that had been carried out by PT Pelindo I. It was agreed to refine the Tanjungpinang port master plan document which had been prepared in 2014 and stipulated in accordance with KM 62 of 2019 concerning the Tanjungpinang port master plan [6].

Hinterland Tanjungpinang Port Batu Anam terminal and Sri Bintan Pura terminal consist of several regencies/cities around the Riau Islands Province. This port is a gateway both for people who will carry out activities in Tanjungpinang city and for tourists (domestic and foreign). Based on RIPPARDA data, Tanjungpinang city foreign tourist visits are still dominated by Singapore with 56,133 people and Malaysia with 11,261 people [7].

The road network to Tanjungpinang port (Sri Bintan Pura) has been connected to the primary arterial road network (central government authority), environmental roads, and secondary collector roads in addition to connectivities to activity centers and other transportation nodes such as the Fisabilillah airport, which has been connected to good road network conditions.

The condition of the pond waters of Sri Bintan Pura port has a depth of −2 to −3 meters LWS. This is not an obstacle to ship movement or mooring because the Sri Bintan Pura port only serves passenger ships with an average weight of 100 GT with a draft of 1.5 m; besides that the topographic conditions at Sri Bintan Pura port are in accordance with the measurement results, including sloping conditions, which are on a slope of 0% to 100 m toward the southeast of Tanjungpinang city. Further to tidal conditions, the nature of the tides is mixed, leaning to double daily. The average water ride at full tide is 124 cm, and at dead tide 56 cm. The low tide (Zo) is 130 cm below DT. The current measurement results show that the maximum current speed at the time of measurement is 0.30 m/s. While the minimum current is 0.00 m/s. Based on visual observations of water conditions at the time of measurement, it appears that the sea surface is calm. When compared with tidal data, it can be seen that the current at the survey site is very small. The tides that occur during measurements have a maximum tidal ride (at spring) of about 207 cm. With conditions like this, the tidal currents generated on average will be small. While at the time of neap tide riding the tide only reached 1.63 cm.

Sri Bintan Pura port has six piers. In addition to the pier at Sri Bintan Pura port also has two domestic and two international pontoon piers, the depth of the port pool on the pier face is at −3 to −4 mLWS while for domestic and international pontoons the port pool is at a depth of −2 to −3 mLWS.

The specifications of the ships docked at Sri Bintan Pura port are passenger ships with a capacity of 400 people; some ships docked at the port include MV Oceana Baruna Ferry with Callsign number YB3702 and MSSI number 525020809 and Nusantara Belt ship48. The specifications of ships docked at Tanjungpinang port Batu Anam terminal are dominated by cargo ships. Several ships docked at the port include KM Sejahtera 20 and KM Intan Samudera 6.

2 Methodology

2.1 *Analysis Methods*

The method of analysis and projection is used to analyze and project the growth in the number of loading and unloading goods and the number of passengers getting on and off. Analysis and projections of the growth in the number of loading and unloading of goods, the number of passenger fluctuations, and the number of ship visits are based on data on the development of the number of loading and unloading of goods and the number of passenger fluctuations that occurred at the port during the period 2005–2018 end. The input data used is taken from the recapitulation data of port operational activities. The data obtained for 12 years is considered sufficient to be used as a basis for consideration of projections of loading and unloading goods and passengers [8].

In forecasting goods and passengers, there are several variables that are interrelated with one another. To see the relationship between variables that affect the growth forecast of the number of loading and unloading goods and the number of passengers, a correlation test analysis was done using statistical test models made in the Microsoft Excel software [9].

To see the forecast growth in the number of loading and unloading goods and the number of passenger fluctuations, forecasting is carried out using the linear regression and the polynomial lung regression with the following conditions [10]:

The method flow is as follows:

(a) Data Collection

Gather historical data on the number of loading and unloading goods and the number of passengers at Sri Bintan Pura port over the relevant time period. Ensure that the data is complete and accurate.

(b) Data Preprocessing

- Perform data preprocessing, including cleaning the data by handling missing or invalid values.
- Split the data into two separate groups: one for the number of loading and unloading goods and another for the number of passengers.

(c) Model Selection

- Choose an appropriate regression model for each data group, then choose between linear regression and polynomial lung regression based on the data characteristics and model assumptions.
- Linear regression is suitable for simple linear relationships between time and the number of loading and unloading goods and passengers.
- Polynomial lung regression is appropriate if the relationship is more complex and cannot be well explained by a linear model.

(d) Model Training

For each data group, train the regression model using historical data. The model will learn the relationship between time (independent variable) and the number of loading and unloading goods and passengers (dependent variable).

(e) Model Evaluation

Evaluate the performance of your model using metrics such as mean absolute error (MAE), mean squared error (MSE), or root mean squared error (RMSE) to assess how well the model fits the training data.

(f) Forecasting and Prediction

Once the regression model is trained and considered adequate, use the model to make forecasts for the future growth in the number of loading and unloading goods and passenger fluctuations based on the given time frame.

(g) Validation and Monitoring

Continuously monitor and validate the model as new time-based data becomes available. If the model no longer fits the actual data, consider updating or improving the model.

(h) Interpretation of Results

Interpret the forecasting results to make decisions regarding terminal development and the addition of mooring points based on the identified trends.

2.2 Data Sources

Sri Bintan Pura port operational data is obtained from PT Pelabuhan Indonesia I (Persero) regional branch I, where port activities are intended for domestic and foreign passengers, existing data from 2005 to 2018 tends to fluctuate from year to year. Domestic passengers still dominate the port through Sri Bintan Pura port where in 2018 it reached 1,499,763 people while foreign passengers reached 405,002 people.

3 Results and Discussion

3.1 Passenger Traffic Analysis and Projections

After determining three analysis methods, the next step is to determine three scenarios, namely pessimistic, moderate, and optimistic scenarios. The scenario is taken based on the growth rate of the three projection methods in the second step [11], namely

1. Linear regression has a growth rate of 0.10% and is used as a pessimistic scenario.
2. Multiple linear regression has a growth rate of 2.12% and is used as a moderate scenario.

3. The arithmetic method has a growth rate of 4.00% and is used as an optimistic scenario.

From the three scenarios that have been analyzed, the selection of projection methods to be used is carried out with the following considerations and assumptions:

1. Pessimistic method: PT Pelabuhan Indonesia 1 (Persero) does not carry out any policy related to the development of Sri Bintan Pura port (do nothing);
2. Moderate method: PT Pelabuhan Indonesia 1 (Persero) makes policies related to the development of Sri Bintan Pura port by considering the trend of traffic up and down passengers time series 15 years back (do something);
3. Optimistic method: PT Pelabuhan Indonesia 1 (Persero) benefits from external policies both on a national, provincial, and municipal scale that are strategic in nature such as the determination of national strategic areas, development of industrial estate centers, and other policies that can directly affect the traffic up and down passengers significantly at Sri Bintan Pura Port;
4. In addition to the above considerations, the R^2 value is also seen as a statistical benchmark for determining selected scenarios. The value of R^2 which is getting closer to 1, it can be said that the data is valid and can be accounted for.

From the four considerations above, the following conclusions are obtained to choose a projection scenario as follows:

Highlighting the dynamic nature of passenger traffic in the Sri Bintan Pura port area by analyzing historical passenger growth data, which has provided a solid foundation for understanding the trends and patterns that influence demand for terminal facilities. This insight is invaluable to terminal operators, as it allows them to make informed choices about terminal infrastructure expansion and upgrades. In addition, the study confirms the need to adopt adaptable and flexible development plans that can respond to changing passenger needs over time.

The recommendation to add mooring points based on passenger growth data emphasizes the importance of aligning infrastructure investments with actual demand. Docking points are a critical element of port facilities, and their availability directly affects port efficiency and capacity. Using data-driven forecasting methods not only justifies the need for additional berth points but also provides a rationale for their strategic placement. This approach not only ensures efficient port operations but also contributes to sustainable development by avoiding overinvestment or inefficient use of resources.

This experimental study demonstrates the power of data-driven forecasting in shaping the passenger terminal development plan of Seri Bintan Pura port. The findings emphasize the need for flexibility and adaptability in infrastructure planning and highlight the importance of aligning investments with actual demand to optimize port efficiency and sustainability. This research can be a valuable reference for terminal operators, port authorities, and policymakers as they face the challenges of managing and developing transportation infrastructure in a dynamic and changing environment.

3.2 Analysis and Projections of Goods Traffic

After determining three analysis methods, the next step is to determine three scenarios, namely pessimistic, moderate, and optimistic scenarios. The scenario is taken based on the growth rate of the three projection methods, namely

1. The polynomial lung method has a growth rate of 0.51% and is used as a pessimistic scenario.
2. The multiple interest method has a growth rate of 2.85% and is used as a moderate scenario.
3. The arithmetic method has a growth rate of 4.00% and is used as an optimistic scenario.

From the three scenarios that have been analyzed, the selection of projection methods to be used is carried out with the following considerations and assumptions:

1. Pessimistic method: PT Pelabuhan Indonesia I (Persero) does not carry out any policy related to the development of Tanjungpinang Port Batu Anam terminal (*do nothing*).
2. Moderate method: PT Pelabuhan Indonesia I (Persero) makes policies related to the development of Tanjungpinang port Batu Anam terminal by considering the trend of loading and unloading goods time series 15 years back (*do something*).
3. Optimistic method: PT Pelabuhan Indonesia I (Persero) benefits from external policies on a national, provincial, and municipal scale such as the establishment of national strategic areas, development of industrial estate centers, and other policies that can directly affect the traffic of loading and unloading goods significantly at Tanjungpinang Port Batu Anam terminal.
4. In addition to the above considerations, the R^2 value is also seen as a statistical benchmark for determining selected scenarios. The value of R^2 which is getting closer to 1 can be said that the data is valid and can be accounted for.

From the four considerations above, the following conclusions are obtained to choose a projection scenario as follows:

It can be concluded that the selected scenario is a moderate scenario; this is because the development trend is in accordance with the time series data of the past 15 years, where the highest loading and unloading data occurred in 2012, which was 379,000 tons with an average of 262,858 tons per year, while when viewed based on projection data using the moderate method in 2039 (long-term development) loading and unloading of goods amounted to 510,101 tons with an average of 392,396 tons per year.

3.3 Analysis of the Origin of Passenger Destinations

The origin of passenger destinations through Tanjungpinang port (Sri Bintan Pura) consists of several areas in Riau Islands Province as well as passengers coming from abroad.

4 Conclusion

From the analysis carried out, the Sri Bintan Pura port facility needs plan is as follows:

4.1 Dock Length Requirement Plan

Analysis of the need for additional moorings at Sri Bintan Pura port can be seen by considering several variables including the number of passengers in one year, the number of ship visits, ship service time, and the value of BOR plans at Sri Bintan Pura port in existing conditions, and the dock can still serve ships optimally.

Up to the long term, both domestic docks and international docks do not require addition; this is because the number of existing moorings has been able to serve the flow of passengers up and down to the long term and has been disciplined in ship service time which only reaches 30 min per ship.

4.2 Terminal Area Development Needs Plan

The increase in the number of passengers at each stage of development requires an increase in terminal area to remain optimal, while the calculation of terminal area is adjusted to the needs of facilities at domestic and international terminals according to the guidelines for land facility planning at collecting ports (PP), regional feeders (PR), and local feeders (PL) of the ministry of transportation.

Apart from that, one can also make new innovations, such as digitizing the security system by adding CCTV facilities or the like, and if possible, adding free Wi-Fi facilities or a place to play games on board in the passenger terminal.

Acknowledgement The following support is acknowledged: Department of Marine Transportation Engineering, Institut Teknologi Sepuluh Nopember, Surabaya, Directorate General of Higher Education, Research, and Technology, Ministry of Education, Culture, Research, and Technology of the Republic of Indonesia.

References

1. President of The Republic of Indonesia: Peraturan Pemerintah Republik Indonesia Nomor 61 Tahun 2009 tentang Kepelabuhanan. Database Peraturan Perundang-undangan. https://peraturan.go.id/id/perpres-no-61-tahun-2009 (2009). Accessed 8 Sept 2023
2. Wikipedia: Undang-Undang Republik Indonesia Nomor 17 Tahun 2008. https://id.wikisource.org/wiki/Undang-Undang_Republik_Indonesia_Nomor_17_Tahun_2008 (2008). Accessed 08 Sept 2023
3. Mustafa, R., Akhirman, A., Nurhasanah, N.: Pengaruh persepsi manfaat, persepsi kemudahan, persepsi kepercayaan, daya tarik promosi dan pengetahuan produk terhadap minat menggunakan e-money card di Pelabuhan Sri Bintan Pura PR. Pelindo 1 cabang Tanjungpinang. Undergraduate Thesis, Universitas Maritim Raja Ali Haji (2019). http://repositori.umrash.ac.id/id/eprint/246
4. Engky, S.: Analisis Faktor-Faktor yang Mempengaruhi Waktu Tunggu Kapal di Pelabuhan Sri Bintan Pura Tanjung Pinang. Undergraduate Thesis, Universitas Maritim AMNI Semarang (2021). http://repository.unimar-amni.ac.id/3667/
5. Langley, C.J., Novack, R.A., Gibson, B., Coyle, J.J.: Supply Chain Management: A Logistics Perspective. Cengage Learning, Australia (2020)
6. Novita, N., Martaleli, B., Alena, U.: Perbandingan tingkat akurasi metode fuzzy time series model markov chain dan metode fuzzy time series jasim untuk memprediksi jumlah penumpang kapal (Studi kasus: Pelabuhan Sri Bintan Pura dalam Negeri). Undergraduate Thesis, Universitas Maritim Raja Ali Haji (2019). http://repositori.umrah.ac.id/1251/
7. Martel, N., Seneviratne, P.N.: Analysis of factors influencing quality of service in passenger terminal buildings. Transp. Res. Rec. **1273** (1990). https://trid.trb.org/view/348928
8. Setti, J.R., Hutchinson, B.G.: Passenger-terminal simulation model. J. Transp. Eng. **120**(4), 517–535 (1994). https://doi.org/10.1061/(ASCE)0733-947X(1994)120:4(517)
9. Uyanık, G.K., Güler, N.A.: Study on multiple linear regression analysis. Procedia-Soc. Behav. Sci. **106**, 234–240 (2013). https://doi.org/10.1016/j.sbspro.2013.12.027
10. Nimon, K.F., Oswald, F.L.: Understanding the results of multiple linear regression: beyond standardized regression coefficients. Organ. Res. Methods **16**(4), 650–674 (2013). https://doi.org/10.1177/1094428113493929
11. Grégoire, G.: Multiple linear regression. Eur. Astron. Soc. Publ. Ser. **66**, 45–72 (2014). https://doi.org/10.1051/eas/1466005

Comparison of Ship-Owned Service Quality Between PT. PELNI and PT. DLU by Using the Fuzzy C-Means Method

Nur Ilman Habil, Rendi Ferdillah, and Tri Achmadi

1 Introduction

Due to increased activity and rising public demand for shipping, competition is inevitable. The Directorate General of Sea Transportation of Indonesia predicts 17.82 million domestic ship passengers in 2022, up 18.85% from 2021s 15 million [1]. After aircraft and railroads, Indonesians utilize ships most. Increased passenger numbers aboard these ships have forced corporations to compete to offer community services and infrastructure and ensure passenger, freight, and crew safety and comfort. PT. PELNI (Indonesian National Shipping) is a state-owned shipping firm founded on April 28, 1952, by Decree of the Minister of Transportation No. M2/1/2. It provides trustworthy and professional ship transportation services. PT. PELNI runs passenger, fast, and freight ships on domestic routes and stops at over 94 ports throughout the archipelago [2]. The work culture of PT. PELNI supports its service aims. PELNI, such as public accessibility to services, increasing marine logistics enterprises, and ensuring firm value growth via creativity, innovation, digitization, and human resource development, to achieve sustainable growth [3]. Additionally, PT. PELNI contributes to environmental development and community services [3]. Besides PT. PELNI, PT. Dharma Lautan Utama (PT. DLU) provides shipping services to promote economic progress and human welfare [4]. PT. DLU continues to improve its fleet and human resources to provide the best service to the community in terms of timely loading capacity and safety and security services in accordance with International Maritime Organization (IMO) rules ratified by the government of Indonesia.

N. I. Habil (✉) · R. Ferdillah · T. Achmadi
Department of Sea Transportation Engineering, Institut Teknologi Sepuluh Nopember, Kampus ITS Sukolilo, Surabaya, Jawa Timur, Indonesia
e-mail: nurilmanhabil@gmail.com

T. Achmadi
e-mail: triachmadi@seatrans.its.ac.id

R. S. Widjaja et al. (eds.), *Marine Technology 3*,
SpringerBriefs in Applied Sciences and Technology,
https://doi.org/10.1007/978-3-032-05613-9_8

PT. DLU provides far above economy class services (beyond economy class) using economy fares. The fleet is maintained by the PT Adiluhung Saranasegara Indonesia shipyard since 2007, and human resources are improved by providing systematic, continuous, and sustainable training [4]. Other than that, PT DLU prioritizes ship facilities and infrastructure and customer service to satisfy the community. DLU improves ship facilities such as ship condition, ship security, and ship stewardess service to meet ferry transportation users' service and safety criteria to deliver this finest service [5]. PT. PELNI and PT. DLU's constantly updated and upgraded facilities and services make many customers choose these two shipping service firms. In 2022, particularly until August, PT. PELNI carried 2.76 million passengers, up 191% from 2021 [2]. This research divides PT. PELNI and PT. DLU's service into two categories, satisfactory and unsatisfactory, based on crew service, customer service, security, cleanliness, length of trip, and playground availability to keep kids entertained. Fuzzy C-means classifies the service quality. Fuzzy C-means clustering groups data points by membership into clusters. This approach groups more data with greater accuracy than standard clustering algorithms and can group outliers and overlapping instances [6].

2 Methodology

2.1 *Fuzzy C-Means*

When clustering yields groupings with high similarity within the same group but low similarity across clusters, it is good [7]. Each data point is clustered into a cluster with a particular degree of membership using fuzzy C-means clustering [8]. The fuzzy C-means clustering (FCM) element of the hard K-means approach employs a fuzzy clustering model to group all data into classes or clusters with degrees of membership between 0 and 1 [9]. The FCM analysis method has numerous steps [10]:

1. Number of clusters to be produced ($1 < c < N$), weighing exponent ($m > 1$), initial goal function, maximum iteration = 0 and initial iteration ($t = 1$).
2. Data generation with random numbers μ_{ik}, where $i = 1, 2, \ldots, c$ and $1 \leq k \leq N$ as an initial matrix element.
3. Formula for calculating kth cluster center

$$v = \frac{\sum_{k=1}^{N} u_{ik}^{m} x_k}{\sum_{k=1}^{N} u_{ik}^{m}}. \tag{1}$$

4. Member distance to cluster center with

$$D_{ik}^2 = \|x_k - v_i\|^2 = (x_k - v_i)^T A(x_k - v_i). \tag{2}$$

5. tth iteration objective function calculation

$$J_{FCM}(X, U, V) = \sum_{i=1}^{c}\sum_{k=1}^{N}(u_{ik})^m D_{ik}^2(x_k - v_i). \tag{3}$$

6. Calculates new membership function value $\mathbf{U}_{t+1}$

$$u_{ik} = \left[\sum_{j=1}^{c}\left(\frac{D(x_k, v_i)}{D(x_k, v_j)}\right)^{\frac{2}{m-1}}\right]^{-1}. \tag{4}$$

7. Compare U matrix membership values. If $\|\mathbf{U}_{t+1} - \mathbf{U}_t\| < \varepsilon$ or $t >$ maximum iteration, it is considered convergent. But, if $\|\mathbf{U}_{t+1} - \mathbf{U}_t\| \geq \varepsilon$ back to step 3.

2.2 Clustering Assumption Testing

To categorize using the selected clustering technique, traditional clustering assumptions must be satisfied. The clustering assumptions are:

1. Multivariate Normal Assumptions

   A hypothesis-driven Q-Q plot between square distances and chi-square distribution quartile values tests multivariate normality,

   H_0: data normally distributed multivariate.
   H_1: data not normally distributed multivariate.

   A straight Q-Q plot with a p-value > 0.05 fails to reject H_0, supporting multivariate normal assumptions [11].
2. Data Sufficient Assumptions

   The Kaiser-Meyer-Olkin (KMO) tcchniquc analyzes if all study data is sufficient to support the hypothesis [12].

   H_0: the amount of data is sufficient to be analyzed.
   H_1: the amount of data is not enough to be analyzed.
3. Multicollinearity Assumptions

   The multicollinearity test detects significant independent variable correlation [13]. A variance inflation factor (VIF) above 10 indicates multicollinearity [13].

2.3 Validity Index

The validity index helps to choose the best number of clusters among the numerous that were created.

1. Silhouette coefficient (SC)

   Silhouette shows the quality and strength of clustered objects [7]. The number of clusters with the highest silhouette value is the number of clusters selected.
2. Xie Beni Index

   The ratio of the total variation within a cluster can be measured using the Xie Beni index [14]. The best number of clusters can be determined by the smallest Xie Beni value among all the number of clusters formed [15].

2.4 Data Source

This research employed original data from a survey of 55 PT. PELNI and PT. DLU travelers and shippers. The study factors include cleanliness (X_1), security (X_2), comfort (X_3), rapid response (X_4), and service quality (X_5). The service quality variable (X_5) has two categories: 1 for poor service and 2 for excellent service. This study employed enough samples to reflect the research population and meet the minimal sample size for data analysis, 30. This research included 55 samples from PT. PELNI and PT. DLU surveys, not shipping business surveys.

3 Results and Discussion

3.1 Data Characteristics

This research presents descriptive statistics to summarize data properties. Table 1 lists the numerical variable data features of PT. PELNI and PT. DLU in this investigation. According to Table 1, PT. PELNI has a lower average of each variable than PT. DLU, which has a more heterogeneous variance. Figure 1 shows categorical data on service quality from PT. PELNI and PT. DLU. Figure 1 shows that PT. PELNI outperforms PT. DLU in service quality.

3.2 Clustering Assumption Testing

The following assumptions must be satisfied before clustering:

1. Multivariate Normal Assumptions

   Hypothesis:
   H_0: Data normally distributed multivariate.
   H_1: Data not normally distributed multivariate.

Table 1 Descriptive statistics

	Variable	Mean	Variance	Min	Max
PT. PELNI	X_1	75.270	179.130	50.000	99.000
	X_2	69.980	132.830	50.000	88.000
	X_3	72.070	148.620	51.000	90.000
	X_4	68.490	161.290	50.000	90.000
PT. DLU	X_1	71.58	197.10	50.00	98.00
	X_2	74.20	183.79	50.00	92.00
	X_3	75.15	200.20	51.00	96.00
	X_4	70.16	150.55	50.00	90.00

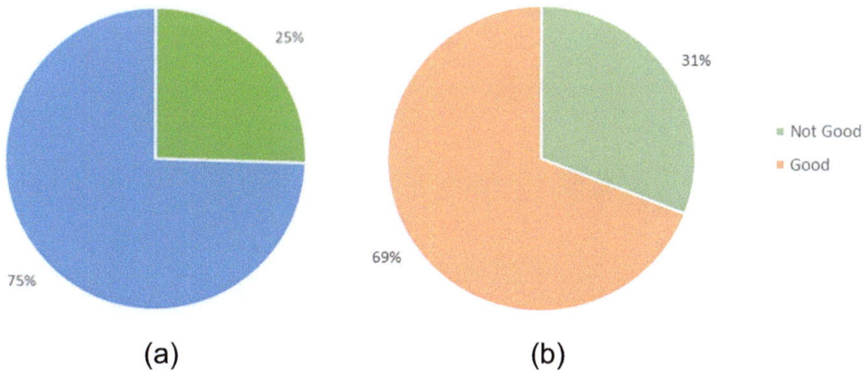

Fig. 1 Characteristics of service quality **a** PT. PELNI and **b** PT. DLU

The QQ-plot for each maritime transportation service firm variable is shown in Fig. 2. The QQ-plot between the Mahalanobis distance and the cluster distances from PT. PELNI and PT. DLU confirms the multivariate normal assumptions. To support QQ-plot findings, PT. PELNI has a correlation coefficient of 0.957 (p-value < 0.001), whereas PT. DLU has 0.967 (p-value < 0.001). The data shows a strong correlation coefficient (p-value < 0.05), rejecting H_0 and meeting the multivariate normal distribution assumption.

2. Data Adequacy Assumptions

   Table 2 shows the KMO MSA test results for each shipping service firm to assess data adequacy. According to Table 2, PT. PELNI's KMO MSA value was 0.514 and PT. DLU's was 0.523. If the KMO value was > 0.5, factor analysis was suitable or the data was adequate to analyze.

3. Multicollinearity Assumptions

   Table 3 shows each variable's VIF value to identify multicollinearity. According to Table 3, the VIF values for each variable and firm were less than 10; hence they were not related. This investigation satisfied multicollinearity.

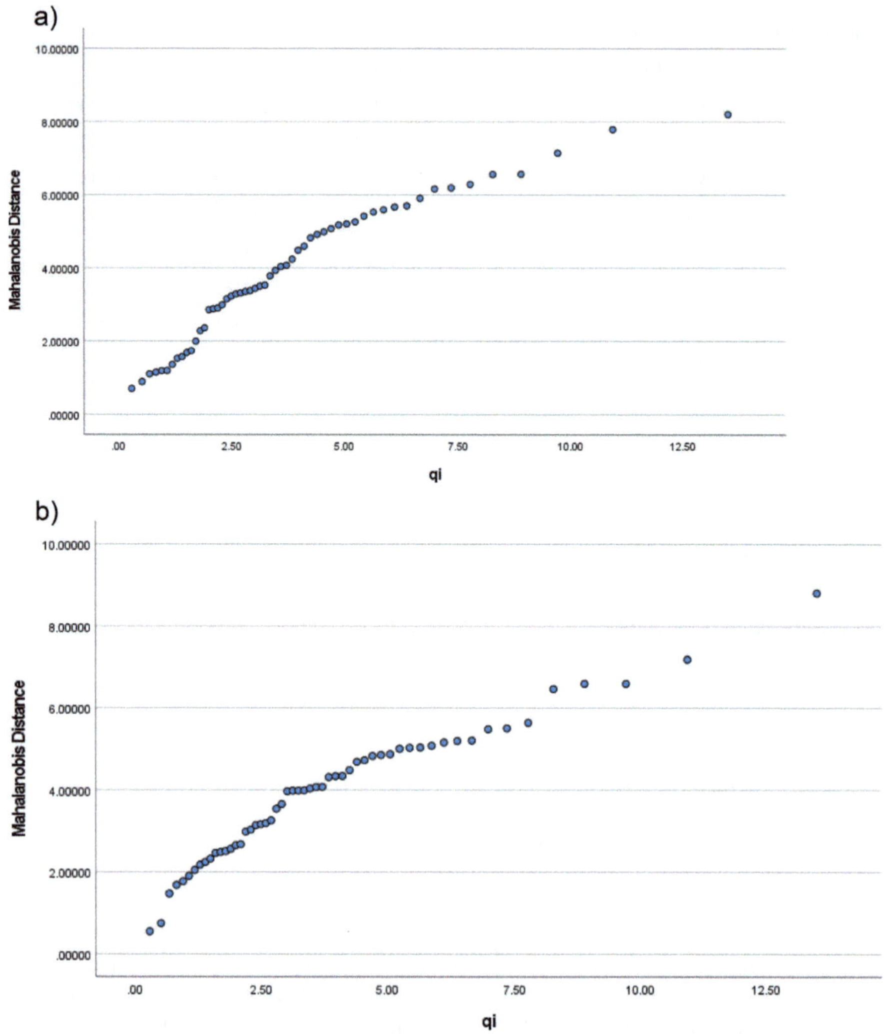

Fig. 2 QQ-plot **a** PT. PELNI **b** PT. DLU

Table 2 KMO result

Company	KMO MSA
PT. PELNI	0.514
PT. DLU	0.523

Table 3 VIF results

Company	Variable	VIF	Company	X_1	1.004
PT. PELNI	X_1	1.007	PT. DLU	X_2	1.040
	X_2	1.001		X_3	1.013
	X_3	1.007		X_4	1.047
	X_4	1.002		X_1	1.004

Table 4 Cluster validity index

Company	Number of cluster	Number of iteration	Time	*Silhouette*	*Xie Beni index*
PT. PELNI	2	283	0.18	0.377	2.555
PT. DLU		187	0.18	0.399	2.32

Table 5 Number of members of each cluster

Cluster	Number of member	Category
1	28	Not good
2	27	Good
1	25	Not good
2	30	Good

3.3 *Clustering Using the Fuzzy C-Means*

Table 4 shows how the fuzzy C-means approach groups companies into two clusters based on service quality: terrible and excellent. Table 4 shows that PT. DLU has a greater silhouette value and lower Xie Beni index than PT. PELNI with the same time, 0.18 s, and 2 clusters. Table 5 shows cluster memberships. It shows that most passengers that have used PT. PELNI, and PT. DLU believe that its service quality is superior since more passengers utilize it.

3.4 *Service Level Agreement*

Service level agreement (SLA) bridges the gap between service providers and consumers [16]. PT. PELNI offers ticket reservation among other services. The ticket reservation service works well since it answers quickly and is accessible 24/7 [17]. There is a cruise search service that provides ship itineraries and is accessible 24/7, and a cruise ship selection service that performs well [17]. PT. DLU offers ticket booking services similar to PT. PELNI, including ship schedule bookings and searches that can be accessible 24/7. PT DLU offers long or short ship routes and passenger age groups while looking for ticket timetables [18].

4 Conclusion

Descriptive statistics show that PT. PELNI's service is better than PT. DLU's, but the fuzzy C-means method shows that more people think PT. DLU's service is better in terms of cleanliness (X_1), security (X_2), comfort (X_3), and quick response (X_4). In fuzzy C-means clustering, PT. PELNI has 52.74% accuracy and PT. DLU 56.27%. To catch up with PT. DLU in the community's eyes, PT. PELNI may increase service quality by focusing on cleanliness (X_1), security (X_2), comfort (X_3), and response (X_4). In addition, PT. PELNI may innovate by digitizing the security system with CCTV or other technologies and introducing free Wi-Fi or a gaming area.

Acknowledgement Department of Marine Transportation Engineering, Institut Teknologi Sepuluh Nopember, Surabaya, Directorate General of Higher Education, Research, and Technology, Ministry of Education, Culture, Research, and Technology of the Republic of Indonesia are acknowledged for their support in this research.

References

1. Direktorat Jenderal Perhubungan Laut Republik Indonesia: Jumlah penumpang kapal menjelang hari raya. https://hubla.dephub.go.id/home/post/read/14211/hingga-h1-idul-fitri-2023-jumlah-penumpang-kapal-naik-5-persen-dibandingkan-tahun-2022-1 (2023). Accessed 8 Sept 2023
2. PT PELNI: Tentang kami. https://www.pelni.co.id/tentang-kami#:~:text=PT%20Pelayaran%20Nasional%20Indonesia%20atau,2%20tanggal%2028%20April%201952 (2023). Accessed 10 Sept 2023
3. PT PELNI: Visi dan Misi PT PELNI (Persero). https://www.pelni.co.id/visi-misi- (2023). Accessed 10 Sept 2023
4. Dharma Lautan Utama. https://dlu.co.id/ (2023). Accessed 12 Sept 2023
5. Dinas Kominfo Provinsi Jawa Timur: PT DLU jadi percontohan jasa pelayaran negeri-swasta. https://kominfo.jatimprov.go.id/berita/16420 (2023). Accessed 09 Sept 2023
6. Li, M., Ng, M.K., Cheung, Y., Huang, J.Z.: Agglomerative fuzzy k-means clustering algorithm with selection of number of clusters. IEEE Trans. Knowl. Data Eng. **20**(11), 1519–1534 (2008). https://doi.org/10.1109/TKDE.2008.88. Accessed 19 Sept 2023
7. Dewi, D.A., Pramita, D.A.: Analisis perbandingan metode elbow dan sillhouette pada algoritma clustering k-medoids dalam pengelompokan produksi kerajinan bali. J. Matrix **9**(3), 103 (2019). Accessed 18 Sept 2023
8. Sadaki, M., Ichihashi, H., Honda, K.: Algorithms for Fuzzy Clustering. Springer, New York (2008)
9. Kusumadewi, S., Purnomo, H.: Aplikasi logika fuzzy untuk pendukung keputusan. Graha Ilmu, Yogyakarta (2010)
10. Rahmatika, L., Suparti, S., Safitri, D.: Analisis kelompok dengan algoritma c-means dan gustafson kessel clustering pada indeks lq45. J. Gaussian **4**(3), 543–551 (2015). Accessed 01 Oct 2023
11. Johnson, R., Wichern, D.: Applied Multivariate Statistical Analysis, 6th edn. Prentice Hall International, New Jersey (2007). Accessed 08 Oct 2023
12. Hair, J., Black, W., Babin, B., Anderson, R.: Multivariate data analysis, 7th edn. Pearson, London (2009). Accessed 11 Oct 2023

13. Ghozali, I.: Aplikasi analisis multivariate dengan program spss. Badan Penerbit UNDIP, Semarang (2017). Accessed 02 Oct 2023
14. Saputra, D.: Analisis perbandingan metode fuzzy c-means dan substractive fuzzy c-means (SFCM). Undergradute Thesis, Universitas Brawijaya (2018). Accessed 10 Oct 2023
15. Azizah, A.N., Widiharih, T., Hakim, A.R.: Kernel k-means clustering untuk pengelompokkan sungai di kota semarang berdasarkan faktor pencemaran air. J. Gaussian **11**(2), 232 (2022). Accessed 12 Sept 2023
16. Trienekens, J.J., Bouman, J.J., Zwan, M.V.: Specification of service level agreements: problems, principles and practices. Softw. Qual. J. **12**(1), 43–57 (2004). https://doi.org/10.1023/B:SQJO.0000013358.61395.96. Accessed 11 Oct 2023
17. Khoiriyah, R.: Analisis peningkatan layanan reservasi tiket kapal laut pt.pelni menggunakan strategi it service management. Remik: Riset dan E-Jurnal Manajemen Informatika Komputer **6**(4), 817–826 (2022). https://doi.org/10.33395/remik.v6i4.11846. Accessed 12 Oct 2023
18. PT Dharma Lautan Utama. Jadwal Kapal DLU. https://tiket.dlu.co.id/ (2023). Accessed 01 Oct 2023

Optimization of Permits for Operation of Crossing Transport Vessels in Maluku Province Using Binary Logistics Regression

Murdhifin Zulmy Widya Kurniawan, Setyo Nugroho, and Muhammad Akbar

1 Introduction

Water transportation in Indonesia plays a crucial role in achieving national goals and improving the country's economy. According to Government Regulation of the Republic of Indonesia (PP) number 82 of 1999, passenger transportation in waters is essential for ensuring legal activities and passenger safety. A water service provider company must have a travel operation permit to ensure proper operations and minimize risks [1]. Business entities or individuals in Indonesia must have a business permit for vehicles in waters and business licenses for services related to transportation in waters. These permits are valid as long as sea transportation companies continue operating and are evaluated every 2 years by the minister, governor, or regent/mayor. The license validity period is long and requires completion immediately before the ship starts operating. Maluku province, one of Indonesia's provinces, pays close attention to the rules and standard operating procedures (SOPs) regarding passenger transport ships. Maluku Governor Regulation Number 18 of 2019 states that every ship operating to and from Maluku province must have a shipping permit issued by the government every 5 years and can be extended [1]. However, some companies providing sea transportation travel services manage and obtain permits for more than 5 days, making them less efficient in carrying out passenger transport trips. The time delay is influenced by several factors, including longer time

M. Z. W. Kurniawan (✉) · S. Nugroho · M. Akbar
Institut Teknologi Sepuluh Nopember, Kampus ITS Sukolilo, Surabaya, Jawa Timur, Indonesia
e-mail: murdhifin.zulmy@gmail.com

S. Nugroho
e-mail: snugroho@seatrans.its.ac.id

M. Akbar
e-mail: muhakbarabidin@gmail.com

R. S. Widjaja et al. (eds.), *Marine Technology 3*,
SpringerBriefs in Applied Sciences and Technology,
https://doi.org/10.1007/978-3-032-05613-9_9

to complete and validate documents, coordination time between the governor and the head of service, and the service system being fixed on initial rules that stipulate licensing services are 7 working days or more due to coordination from various parties or agencies other than the governor. The governor gives authority to heads of related agencies to issue permits to operate vessels, but the time required to issue these permits is considered inefficient and takes a long time. This research aims to optimize licensing time in Maluku province by analyzing factors influencing ship operating licensing service time using the binary logistic regression method. The study will focus on water transportation permit service time data in Maluku province to assess the effectiveness of obtaining permits within 5 days or less. The research will use binary logistic regression to examine the relationship between response variables (y) and predictor variables (x) [2]. The findings will help optimize the licensing process and improve community satisfaction.

2 Methodology

2.1 Binary Logistic Regression

Binary logistic regression is a regression analysis method in which the response variable is a categorical variable with values of 0 or 1 [3]. Binary logistic regression is a method for analyzing the relationship between one response variable made up of two kinds separated by zeros and one with numerous predictor variables whose scale is ratio or type [4]. Binary logistic regression has two kinds of response variables. One of the two types is used as a reference, while the other variables are utilized for comparison [5]. The binary logistic regression model takes the following general form:

$$\pi(x) = \frac{\exp(\beta_0 + \beta_1 x_1 + \cdots + \beta_p x_p)}{1 + \exp(\beta_0 + \beta_1 x_1 + \cdots + \beta_p x_p)} \tag{1}$$

The general form of a model, where p represents predictor variables and $\pi(x)$ represents the probability of a successful event, must be converted into a logit form due to its nonlinearity [2] (Fig. 1).

2.2 Independence Test

The chi-square test can determine whether a link or independence exists between two or more variables in a population. This is done by analyzing the contingency table [6] and applying the appropriate hypothesis. The chi-square approach can be used to perform an independence test by comparing the null hypothesis (H_0) of

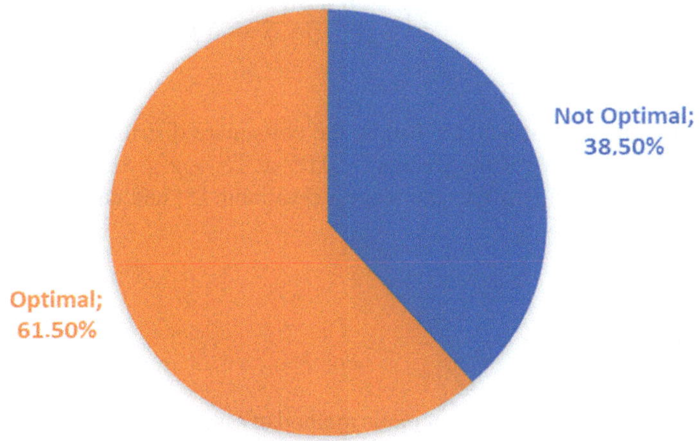

Fig. 1 Pie chart of response variable [22]

no correlation and the alternative hypothesis (H_1) of a correlation, which can be calculated by applying Eq. (2).

$$\chi^2 = \sum_{i=1}^{r}\sum_{j=1}^{c}\frac{\left(O_{ij} - E_{ij}\right)^2}{E_{ij}} \tag{2}$$

where O_{ij} represents the observed data in the ith row of the jth column, whereas E_{ij} represents the predicted value in the ith row of the jth column, given that the degrees of freedom are $df = (r-1)(c-1)$. The test findings indicate that H_0 should not be rejected if $\chi^2{}_{\text{count}} > \chi^2{}_{\text{table}}$ or p-value > α (0.05) which suggests that the assumption of independence has been met.

2.3 *Simultaneous Test*

A simultaneous test is a test of model parameters by involving all variables together to see the effect on the response variable. It is assumed that each hypothesis will be rejected or accepted as the appropriate test statistic exceeds or falls short of some critical point [7] with the hypothesis used:

$$H_0 : \beta_1 = \beta_2 = \cdots = \beta_p = 0; H_1 : \text{at least one } \beta_k \neq 0; k = 1, 2, \ldots, p$$

The simultaneous test can be mathematically calculated with the test statistics shown in Eq. (3) below.

$$G^2 = -2\ln\left(\frac{\left(\frac{n_1}{n}\right)^{n_1}\left(\frac{n_0}{n}\right)^{n_0}}{\prod_{i=1}^{n}\hat{\pi}_i^{y_i}(1-\hat{\pi}_i)^{(1-y_i)}}\right) \tag{3}$$

Tests using the G^2 test statistic follow the chi-square distribution, and it can be concluded that reject H_0 if the p-value < 0.05 or $G^2 > \chi^2{}_\alpha$ which means that all predictor variables jointly affect the response variable [8] and can be passed on to testing individually.

2.4 Individual Test

Individual or partial testing is a systematic method of assessing the impact of predictor variables on response variables. It involves presenting respondents with varying difficulty levels of items based on their answers [9], with the individual test employing the hypothesis in a specific manner. The individual test is done by using the hypothesis as follows:

$$H_0 : \beta_0 = 0;\ H_1 : \beta_k \neq 0;\ k = 1, 2, \ldots, p$$

To carry out individual tests, the method used is the Wald test with the test statistics being:

$$W = \frac{\hat{\beta}_k}{\mathrm{SE}(\hat{\beta}_k)} \tag{4}$$

The parameter estimates $\hat{\beta}_k$ and associated standard error estimate $\mathrm{SE}(\hat{\beta}_k)$ are utilized to draw a conclusion. The null hypothesis H_0 is rejected if the p-value is less than α (0.05), or if the absolute value of $|W|$ is larger than $Z_{\frac{\alpha}{2}}$ or if W^2 is bigger than the critical value $\chi^2{}_\alpha$ [3].

2.5 Model Fit Test

The Hosmer and Lemeshow test is a statistical assessment used to determine the adequacy of a binary logistic regression model. It involves comparing the observed data with the projected data [3]. The data are sorted based on $1-\hat{\pi}_{i0}$, which represents the estimated probability of the reference result [10]. The study aims to determine the appropriateness (H_0) or suitability of the resultant model. The study's hypothesis is whether the resulting model is appropriate (H_0) or not suitable (H_1).

The Hosmer and Lemeshow tests can be calculated through subsequent mathematical calculations,

$$\hat{C} = \sum_{h=1}^{g} \frac{(O_h - n_h\overline{\pi}_h)^2}{n_h\overline{\pi}_h(1 - n_h\overline{\pi}_h)} \quad (5)$$

The analysis uses variables g, O_h, $\bar{\pi}_h$, and n_h to represent the number of groups, response variable values, average estimated probability, and the number of subjects in the h group. The resulting decision is to reject the null hypothesis $\hat{C} > \chi^2{}_{\alpha}$ or if the p-value is greater than 0.05, indicating that the estimated results align with the actual observed data.

2.6 Classification Accuracy

Apparent error rate (APER) is a method used to calculate the classification accuracy of modeling results using binary logistic regression with the equation used [11]:

$$\begin{aligned} \text{Classification Accuracy} &= (1 - APER) \times 100\% \\ &= 1 - \frac{n_{12} + n_{21}}{n_{11} + n_{12} + n_{21} + n_{22}} \end{aligned} \quad (6)$$

In this context, n_{11} represents the number of subjects correctly classified in category Y_1, n_{12} represents the number of subjects classified in category Y_2 but belonging to Y_1, n_{21} represents the number of subjects classified in category Y_1 but belonging to Y_2, and n_{22} represents the number of subjects correctly classified in category Y_2.

2.7 Data Source

This study uses a survey of 39 private shipping service providers in Maluku province to analyze time to complete documents (X_1), time to validate documents (X_2), and coordination time between the governor and the agency (X_3). The research uses a minimum of 30 respondents and 30 samples for binary logistic regression, ensuring the results represent the entire population and the resulting model.

3 Results and Discussion

3.1 Descriptive Statistics

This study uses basic descriptive statistics to analyze research data, including mean, variance, minimum, and maximum values. The data is presented in Table 1, with an average value and small variance indicating a homogeneous or small range. A

Table 1 Descriptive statistics of research variables

Variable	Mean	Variance	Minimum	Maximum
X_1	3.513	1.520	2.000	7.000
X_2	2.667	0.912	1.000	5.000
X_3	3.513	1.151	2.000	5.000

Table 2 Independence test results

Variable	*P*-Value	Information
X_1	0.109	No relationship
X_2	0.414	No relationship
X_3	0.070	No relationship

categorical variable, response variable, is presented using a pie chart, with service operating permits marked as optimal and those not. Most shipping service providers believe the shipping licensing management system in Maluku province is optimal, with 61.05% of them stating so. However, 38.05% of providers consider it not optimal. The data is presented in a pie chart, with categories 1 and 0 indicating optimal and non-optimal systems.

3.2 Independence Test

The chi-square test is used in this study to examine the following hypotheses: The null hypothesis (H_0) posits that there is no correlation between the variables, whereas the alternative hypothesis (H_1) proposes that there is a correlation between the variables. The study's independence test results are presented in Table 2. If the *p*-value of all variables is 0.05, the research variables time to complete documents (X_1), time to validate documents (X_2), and coordination time between governors and agencies (X_3) have a *p*-value greater than 0.05, indicating no relationship between each variable, thus meeting the assumption of independence. This conclusion supports the study's findings.

3.3 Simultaneous Test

The simultaneous test is used after the assumption test to determine the influence of the predictor variable on the response variable. The hypothesis employed in this investigation is as follows:

$$H_0 : \beta_1 = \beta_2 = \beta_3 = 0; H_1 : \text{at least one } \beta_k \neq 0; k = 1, 2, 3$$

Table 3 Partial test results

Variable	*B*	Wald	*P*-Value
X_1	1.460	5.249	0.022
X_2	0.305	0.319	0.572
X_3	– 0.847	3.774	0.052
Constant	– 2.134	0.547	0.460

The simultaneous testing results in a G^2 value of 14.500 and a *p*-value of 0.002, which is less than the significance level of 0.05. Therefore, the null hypothesis (H_0) is rejected, indicating that at least one variable significantly influences the response variable, allowing for individual testing.

3.4 Individual Test

Individual or partial testing is conducted after simultaneous testing rejects H_0, identifying predictor variables significantly influencing response variables, with the hypothesis as follows:

$$H_0 : \beta_0 = 0; H_1 : \beta_k \neq 0; k = 1, 2, 3$$

Table 3 indicates that the most influential factor in optimizing shipping operation permits is the time taken to complete documents (X_1). Thus, the resulting regression model is:

$$\hat{g}(x) = -\ 2.134 + 1.460X_1.$$

3.5 Model Fit Test

The study examines the suitability of a model for predicting and explaining the response variable related to optimizing shipping operation permits using a hypothesis. The null hypothesis (H_0) suggests that the model produced is suitable, while the alternative hypothesis (H_1) indicates that the model produced is not suitable. The Hosmer and Lemeshow test method was used to assess the model's suitability, with a test statistic value of 7.299 and a *p*-value of 0.398, indicating that the model is suitable, rejecting the hypothesis H_0.

Table 4 Classification accuracy

		Predicted		
		Y		Percentage correct
	Observed	0	1	
Y	0	9	6	60.0
	1	2	22	91.7
	Overall percentage			79.5

Table 5 Classification accuracy

Variable	Odds ratio
X_1	4.305

3.6 Classification Accuracy

The study's classification accuracy, as demonstrated in Table 4, is 79.5%, indicating a high level of precision in estimating the observational data used in the analysis, indicating the binary logistic regression method's effectiveness in achieving this high accuracy.

3.7 Model Interpretation

The odds ratio value interprets the model by assessing the impact of document completion time (X_1) on response variables. Table 5 demonstrates that increasing X_1 by 1 unit increases the likelihood of an optimal time of 4305 times.

4 Conclusion

The shipping licensing management system in Maluku province is considered optimal by most shipping service providers, with 61.05% believing it is. However, 38.05% still believe it is not optimal. The most influential variable is the time to complete documents, which should be optimized. To comply with government instructions, minimal requirements and procedures should be provided, or an online recording system should be established. The study has a classification accuracy of 79.5%, but it is hoped that future studies will increase the accuracy and consider other variables as well. The study's findings highlight the need for further improvement in the shipping licensing management system.

Acknowledgements Thank you for the support for this research, Department of Marine Transportation Engineering, Institut Teknologi Sepuluh Nopember, Surabaya, Directorate General of Higher Education, Research, and Technology, Ministry of Education, Culture, Research, and Technology of the Republic of Indonesia.

References

1. Kementerian Perhubungan: Persetujuan Pengoperasian Kapal Angkutan Penyeberangan Kementerian Perhubungan Republik Indonesia. https://dephub.go.id/post/read/persetujuan-pengoperasian-kapal-angkutan-penyeberangan (2014). Accessed 12 Aug 2023
2. Rizki, F., Widodo, D.A.A., Hakim, J.A.R.: Faktor Risiko Penyakit Anemia Gizi Besi pada Ibu Hamil di Jawa Timur Menggunakan Analisis Regresi Logistik. J. Sains dan Seni **4**(2), 305–310 (2015)
3. Hosmer, D.W., Lemeshow, S., Sturdivant, R.X.: Applied Logistic Regression. Wiley, Canada (2013)
4. Sepang, F., Komalig, H., Hatidja, D.: Penerapan Regresi Logistik untuk Menentukan Faktor-Faktor yang Mempengaruhi Pemilihan Jenis Alat Kontrasepsi di Kecamatan Modayag Barat. Universitas Sam Ratulangi. Manado. J. MIPA Unsrat Online **1**, 1–5 (2012)
5. Pratama, R.N.: Regresi Logistik Biner Untuk Mengetahui Faktorfaktor Yang Mempengaruhi Penerimaan Mahasiswa Melalui Jalur Masuk Perguruan Tinggi Snmptn Fmipa. Undergraduate Thesis, Universitas Brawijaya (2018)
6. Daniel, W.W.: Statistika Nonparametrik Terapan. Gramedia, Jakarta (1989)
7. Jansen, T., Gislason, H.: Temperature affects the timing of spawning and migration of North Sea mackerel. Cont. Shelf Res. **31**, 64–72 (2011). https://doi.org/10.1016/j.csr.2010.11.003
8. Varamita, A.: Analisis Regresi Logistik dan Aplikasinya pada Penyakit Anemia untuk Ibu Hamil di Rskd Ibu dan Anak Siti Fatimah. Undergraduate Thesis, Universitas Makassar (2017)
9. Yaşar, M.: A study on individualized tests. J. Educ. Sci. Res. **7**, 269–284 (2017)
10. Fagerland, M., Hosmer, D.: A generalized Hosmer Lemeshow goodness of fit test for multinomial logistic regression models. Stata J. **12**, 447–453 (2012). https://doi.org/10.1177/1536867X1201200307
11. Johnson, R.A., Wichern, D.W.: Applied Multivariate Statistical Analysis. Pearson Prentice Hall, Upper Saddle River, New Jersey (2007)

Grouping Policy Ticket Price for Passenger Ship Services PT. Dharma Lautan Utama Using the *K*-Means Clustering Method

Fajar Gumelar, Filemon, and Tri Achmadi

1 Introduction

Indonesia, with a vast sea area of 5.8 million km^2, constituting 75.7% of its territory, holds a unique maritime advantage [1]. This extensive coastline significantly impacts local and national economies, evident in various economic activities such as fishing, business transactions requiring sea connectivity, and cost-effective maritime transportation. Sea transport, remaining popular despite air travel availability, made a substantial contribution to the economy in 2022, with the Directorate General of Sea Transportation realizing revenues of 4.62 trillion, exceeding the set target [2]. Entities like PT. Dharma Lautan Utama (PT. DLU) actively harness Indonesia's maritime resources, aiming for high competitiveness by offering affordable inter-island crossing and shipping services [3]. To balance social, economic, and business missions, PT. DLU charges reasonable fees while adhering to government regulations [3]. The pricing approach has led to varying customer perceptions, prompting this study's proposal to use *K*-means clustering for tariff policy evaluation and optimization. The results are expected to provide valuable insights for both PT. DLU and the government in formulating a new tariff policy to enhance customer satisfaction.

F. Gumelar (✉) · Filemon · T. Achmadi
Institut Teknologi Sepuluh Nopember, Kampus ITS Sukolilo, Surabaya, Indonesia
e-mail: gumelarfajar51@gmail.com

Filemon
e-mail: filemontarigan@gmail.com

T. Achmadi
e-mail: triachmadi@seatrans.its.ac.id

R. S. Widjaja et al. (eds.), *Marine Technology 3*,
SpringerBriefs in Applied Sciences and Technology,
https://doi.org/10.1007/978-3-032-05613-9_10

2 Methodology

2.1 K-*Means Clustering*

Clustering is a method that group's data based on predetermined similarities [4]. The goal is to classify data into groups with high internal similarity and low between-group similarity [5]. The K-means algorithm partitions data into k sub-regions, with each data belonging to a specific cluster [6]. To perform K-means clustering, specify the desired number of clusters (K), and the algorithm assigns each observation to one of the K clusters [7]. Known for its simplicity and interpretability, K-means is a popular machine learning algorithm [8]. Effective cluster analysis captures statistically important clusters using as few clusters as possible [9]. This study uses the icdrate value to indicate variation between clusters, where a smaller icdrate signifies better-formed clusters [10].

2.2 *Clustering Assumption Testing*

The basic assumptions in clustering are useful in order to prevent errors that occur when clustering. The assumptions in clustering are:

1. Multivariate Normal Assumptions

   Testing the multivariate normal assumptions can be done by looking at the Q-Q plot formed between square distances $\left(d_j^2\right)$ with the quartile values of the chi-square distribution $\left(\frac{j-0.5}{n}\right)$ with a hypothesis,

   H_0: data normally distributed multivariate.

   H_1: data not normally distributed multivariate.

   If the resulting Q-Q plot is a straight line and p-value > 0.05 or $d_j^2 \le \chi^2_{p;0.05}$, then it fails to reject H_0, which means that the data meets the multivariate normal assumptions [11].
2. Data Sufficient Assumptions

   The method used to measure the adequacy of the data used in the research is the Kaiser-Meyer-Olkin (KMO) method which aims to determine whether all the data to be used in the research is sufficient to be analyzed with a hypothesis [12].

   H_0: the amount of data is sufficient to be analyzed.

   H_1: the amount of data is not enough to be analyzed.

   The mathematical calculation of KMO is as follows:

$$\text{KMO} = \frac{\sum_{j=1}^{p}\sum_{k=1}^{p} m_{jk}^2}{\sum_{j=1}^{p}\sum_{k=1}^{p} m_{jk}^2 + \sum_{j=1}^{p}\sum_{k=1}^{p} n_{jk}^2} \tag{1}$$

with

j: 1,2,3, ..., p and $k = 1,2,3, \ldots, p$.

m_{jk}: Correlation coefficient between variable j and variable k.

n_{jk}: Partial coefficient between variable j and variable k.

The amount of data is not enough to be analyzed if the resulting KMO value is less than 0.5.

3. Multicollinearity Assumptions

Multicollinearity, in essence, artificially enlarges the standard errors associated with coefficients in a statistical model. This escalation in standard errors implies that the coefficients of certain independent variables might appear to be not significantly different from 0 [13]. The purpose of the multicollinearity test is to determine if there is a strong correlation among the independent variables in a dataset [12].

1. The value of the variance inflation factor (VIF) is more than 10.
2. If the correlation between independent variables exceeds 0.90, it indicates a very strong relationship.
3. When the coefficient of determination R^2 is high but there are no significant independent variables.

2.3 Data Source

In this study, survey data is used on 65 people who have used the services of PT. DLU using the variable price compatibility with the service (X_1) and the percentage of price compatibility with the infrastructure (X_2) and there is also a price or tariff variable charged to the community (X_3) which is divided into 2 categories, namely cheap is denoted by 1 and expensive is denoted by 2.

3 Results and Discussion

3.1 Data Characteristics

This study presents basic descriptive statistics to highlight characteristics of research data. Table 1 reveals that the data variance of the percentage of price compatibility with infrastructure (X_2) is greater than that of price compatibility with service (X_1), indicating greater diversity in X_2. Figure 1 illustrates the origin of people using

Table 1 Descriptive statistics

Variable	Mean	Variance	Min	Max
X_1	72.82	93.90	55.000	89.000
X_2	74.45	202.41	51.000	95.000

Fig. 1 Map of the distribution of PT. DLU

shipping facilities at PT. DLU, with East Nusa Tenggara province having the majority followed by East Java province.

3.2 Clustering Assumption Testing

There are three clustering assumptions that must be met before carrying out the classification process so that there are no basic problems that can interfere with the classification results:

1. Multivariate Normal Assumptions
   The multivariate normal assumption is visually confirmed by examining the Mahalanobis distance data and cluster member distances. Figure 2's Q-Q plot illustrates the normal distribution of the study data, forming a straight line around the distribution line. Additionally, the correlation coefficient between Mahalanobis distance and inter-cluster distance at PT. PELNI is 0.958, with a p-value < 0.001, indicating high correlation and multivariate normal distribution.
2. Data Adequacy Assumptions
   The KMO coefficient value generated in this study using the data variable percentage of price compatibility with services (X_1) and the percentage of price compatibility with infrastructure (X_2) is 0.5, where this value is not smaller than 0.5 so the data used is sufficient for analysis or factored by the method to be used [14].
3. Multicollinearity Assumptions

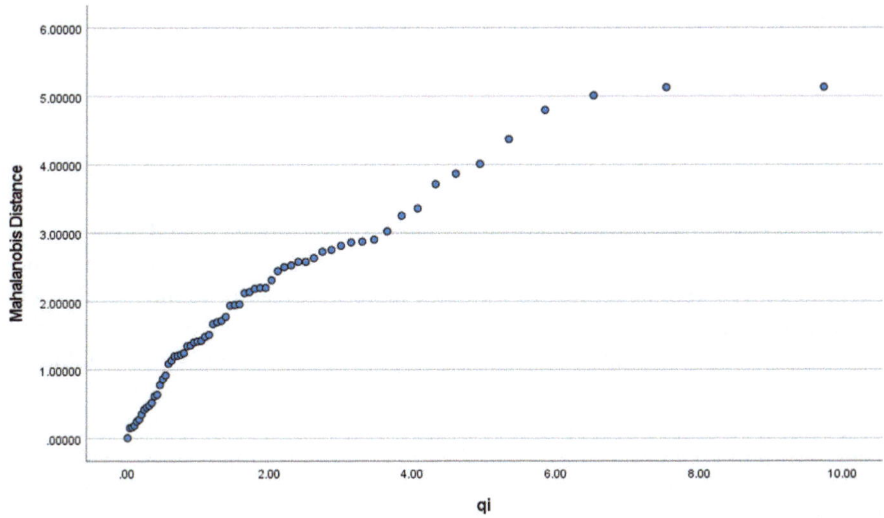

Fig. 2 QQ-plot PT. DLU

Table 2 VIF results

Variable	VIF
X_1	1.000
X_4	1.000

Extreme multicollinearity can lead to high variability in the regression coefficients, resulting in coefficients that are difficult to interpretation and unreliable. The variance inflation factor (VIF) is used to detect multicollinearity. Table 2 shows that both the service price-matching percentage variable (X_1) and the infrastructure price-fit percentage variable (X_2) have VIF values < 10, indicating the fulfillment of the multicollinearity assumption.

3.3 Clustering Using the K-Means Clustering

The K-means algorithm, a widely utilized partitioning-based clustering method, is valued for its theoretical reliability, simplicity, quick convergence, and efficiency in handling small–to medium-sized datasets [15]. The simplicity of its implementation, along with its computational effectiveness and minimal memory usage, continues to make it a preferred choice over other clustering methods [9]. In this study, clustering with a predetermined number of clusters (2) is conducted, and initial clusters are obtained, as seen in Table 3. The initial cluster centers table reveals low values for the centers of each variable in each cluster, prompting further iterations to identify changes in group members (Table 3). After five iterations, a new cluster center is

Table 3 Output initial cluster centers

	Cluster	
	1	2
X_1	1.257	− 1.838
X_2	− 1.578	1.304

formed, and the final cluster centers table (Table 3) indicates increased homogeneity within each cluster. Examination of variable differences is performed using F values and probability values (sig) for each variable (Table 3), revealing minimal differences between variables in the two clusters. The subsequent step involves determining the number of members in each cluster, which is detailed in the output (Table 3). Cluster 1 comprises individuals perceiving tariffs as cheap (red color), while cluster 2 represents those considering tariffs as expensive (blue color), with nearly equal membership in both clusters. The distribution of classification results is visualized in Fig. 3. The Icdrate value, generated using the K-means method with 2 clusters, is 0.470, and the classification accuracy is 53.00%. Despite the relatively low Icdrate value, the classification accuracy remains high, indicating satisfactory results in the obtained classification (Tables 4, 5, 6, and 7).

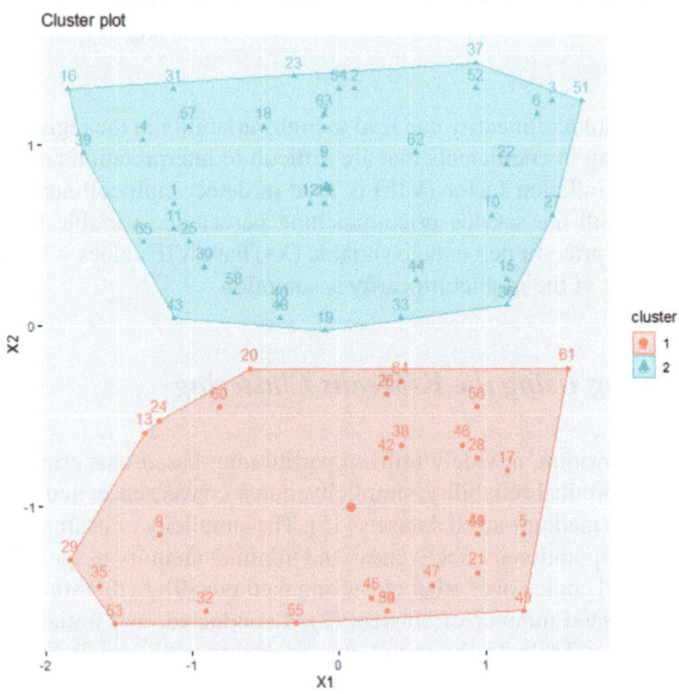

Fig. 3 Clustering result

Table 4 Output iteration history

Iteration	Change in cluster centers	
	1	2
1	1.221	1.402
2	0.164	0.129
3	0.168	0.154
4	0.041	0.045
5	0.000	0.000

Table 5 Final cluster center

	Cluster	
	1	2
X_1	0.858	− 0.832
X_2	− 0.284	0.276

Table 6 Output ANOVA

	Cluster mean square	Error mean square	*F*	Sig.
X_1	46.348	0.280	165.418	< 0.001
X_2	5.082	0.935	5.435	0.023

Table 7 Output number of cases in each cluster

Cluster	1	32.000
	2	33.000
Valid		65.000
Missing		0.000

4 Conclusion

This study employs a classification analysis adhering to normal multivariate, data adequacy, and multicollinearity assumptions. Findings reveal that PT. Dharma Lautan Utama (DLU) charges relatively high tariffs, with a classification accuracy of 53.00%. A significant portion of DLU's clientele is from East Nusa Tenggara, followed by East Java. Considering the perceived expensiveness, an evaluation is recommended to align services and infrastructure value with prices, mitigating customer dissatisfaction despite higher costs.

Acknowledgments Support for this research by the Ministry Transportation of Republic Indonesia, Institut Teknologi Sepuluh Nopember, and Rotterdam University of Applied Sciences is acknowledged.

References

1. Nikawanti, G., Aca, R.: Ecoliteracy: membangun ketahanan pangan dari kekayaan maritim Indonesia. J. Kemaritiman: Indonesian J. Maritime **150** (2021)
2. Direktorat Jenderal Perhubungan Laut: Target realisasi penerimaan pnbp tahun 2023. https://hubla.dephub.go.id/home/post/read/13147/capai-rp-4-62-triliun-di-tahun-2022-ini-langkah-kemenhub-capai-target-realisasi-penerimaan-pnbp-tahun-2023. Accessed 19 Oct 2023
3. Dharma Lautan Utama: https://dlu.co.id/#1482113419916-693aad9e-f2be. Accessed 15 Oct 2023
4. Widodo: Psikologi belajar. Jakarta: Rineka Cipta (2004)
5. Hastie, T., Tibshirani, R., Friedman, J.: The Elements of Statistical Learning: Data Mining, Inference and Prediction. Springer, Verlag (2000)
6. Metisen, B.M., Sari, H.L.: Analisis clustering menggunakan metode K-means dalam pengelompokkan penjualan produk pada swalayan fadhila. J Media Infotama. **113** (2015)
7. Steorts, R.C.: K-Means Clustering. Duke University, California (2022)
8. Adams, R.P.: K-Means Clustering and Related Algorithms. Princeton University, New Jersey (2020)
9. Morissette, L., Chartier, S.: The K-means clustering technique: general considerations and implementation in mathematica. Tutorials Quant. Methods Psychol. **15** (2013)
10. Putri, M.M.: Pengelompokan kabupaten/kota di provinsi jawa timur berdasarkan indikator kesehatan masyarakat menggunakan metode kohonen som dan K-means. Surabaya: Institut Teknologi Sepuluh Nopember (2015)
11. Johnson, R., Wichern, D.: Applied Multivariate Statistical Analysis, 6th edn. Prentice Hall International, New Jersey (2007)
12. Hair, J., Black, W., Babin, B., Anderson, R.: Multivariate Data Analysis, 7th edn. Pearson (2009)
13. Akinwande, M.O., Dikko, H.G., Samson, A.: Variance inflation factor: as a condition for the inclusion of suppressor variable(s) in regression analysis 755 (2015)
14. Malholtra, N.K., Birks, D.: Marketing Research: An Applied Approach, 3rd edn. Prentice Hall, Upper Saddle River (2006)
15. Zhao, Y., Zhou, X.: K-means clustering algorithm and its improvement research. J. Phys.: Conf. Ser. **1** (2021)

The Impact of Infrastructure Development on Improving Indonesia's Logistics Performance

Agung Yudhantara and Tri Achmadi

1 Introduction

The logistics performance index (LPI), created by the World Bank, is used to assess various parameters of logistics performance. The LPI is utilized globally to assist nations in identifying the obstacles in trade logistics and suggesting solutions and opportunities for enhanced efficiency [1]. LPI assesses six specific aspects of logistics performance, which are the ability to monitor and trace shipments, the capabilities and quality of logistics services, the ease of arranging cost-effective shipments, the effectiveness of customs clearance procedures, the frequency of punctual shipments, and the quality of infrastructure related to trade and transportation. The six LPI indicators are categorized into two primary classifications [1]:

- Policy regulation areas that influence the supply chain include customs, infrastructure, and services.
- Supply chain performance outcomes are measured by LPI indicators such as time, cost, and reliability—timeliness, international shipments, and tracking and tracing.

The most important factor contributing to the LPI is the standard of trade and transport-related infrastructure [2]. Moreover, infrastructure is regarded as the primary element in determining logistics performance, carrying a weight of 0.24, which is more than double the significance of tracking and tracing, the least influential factor, with a weight of 0.10. However, several aspects that could significantly

A. Yudhantara (✉) · T. Achmadi
Department of Marine Transportation Engineering, Institut Teknologi Sepuluh Nopember, Surabaya, Jawa Timur, Indonesia
e-mail: agung07yudhantara@gmail.com

T. Achmadi
e-mail: triachmadi@seatrans.its.ac.id

R. S. Widjaja et al. (eds.), *Marine Technology 3*,
SpringerBriefs in Applied Sciences and Technology,
https://doi.org/10.1007/978-3-032-05613-9_11

impact logistics performance have not yet been incorporated into the LPI [3]. By investing in modern and efficient port infrastructure, Indonesia aims to address logistical challenges, reduce transportation costs, and enhance its competitiveness on the global stage [4]. Additionally, efficient port infrastructure supports the development of logistics and transportation sectors [5]. It encourages the growth of logistics service providers, freight forwarders, and other related industries. This creates a multiplier effect on the economy, generating employment and fostering innovation and technological advancements in the logistics sector [6]. Nevertheless, LPI reports a decline of 0.15 in Indonesia's score in 2023, as compared to the previous period in 2018. In order to enhance LPI, it is important to invest in port infrastructure development and increase port capacity [7].

The objective of this research project is to determine the correlation between infrastructure and logistic performance in Indonesia. This can assess the economic development in relation to competitive factors for the Indonesian economy. The secondary data is obtained from the database of the World Bank and the development infrastructure in Indonesia from 2014 to 2023. Moreover, Indonesia LPI score will compare with major countries in Southeast Asia to give insight toward improvement for infrastructure that can contribute for logistics performance in Indonesia.

2 Methodology

The research methodology involves collecting data (such as the LPI) and analyzing its correlation with infrastructure development. Descriptive statistics is utilized to examine comparative data and derive conclusive findings. By employing this approach, the study seeks to enhance our understanding of the relationship between logistics performance and infrastructure development. Figure 1 illustrates the framework of the methodology.

The research utilizes information obtained from eight countries in Southeast Asia covering 2014, 2016, 2018, and 2023. The selection of countries was predicated upon two fundamental factors. Firstly, it is imperative that the data pertaining to the chosen variables be accessible for every country and year. Secondly, in order to prevent erroneous estimations caused by changes in development phases, the country must maintain the same region and development classification throughout the research period. This study employs infrastructure as the primary variable in LPI, and the data analysis and processing used in this research were conducted using the Minitab program.

Additionally, it is analyzed the influence of the trade and transport infrastructure's quality on the logistics performance in Indonesia and compared to Southeast Asian countries (Singapore, Malaysia, Thailand, Philippines, Vietnam, Cambodia, and Laos). The input variables for LPI consist of six indicators: service quality and competence (LPCQ), tracking and tracing (LPTT), infrastructure (LPQI), customs (LPC), international shipments (LPIS), and timeliness (LPT). LPI rating is from 1

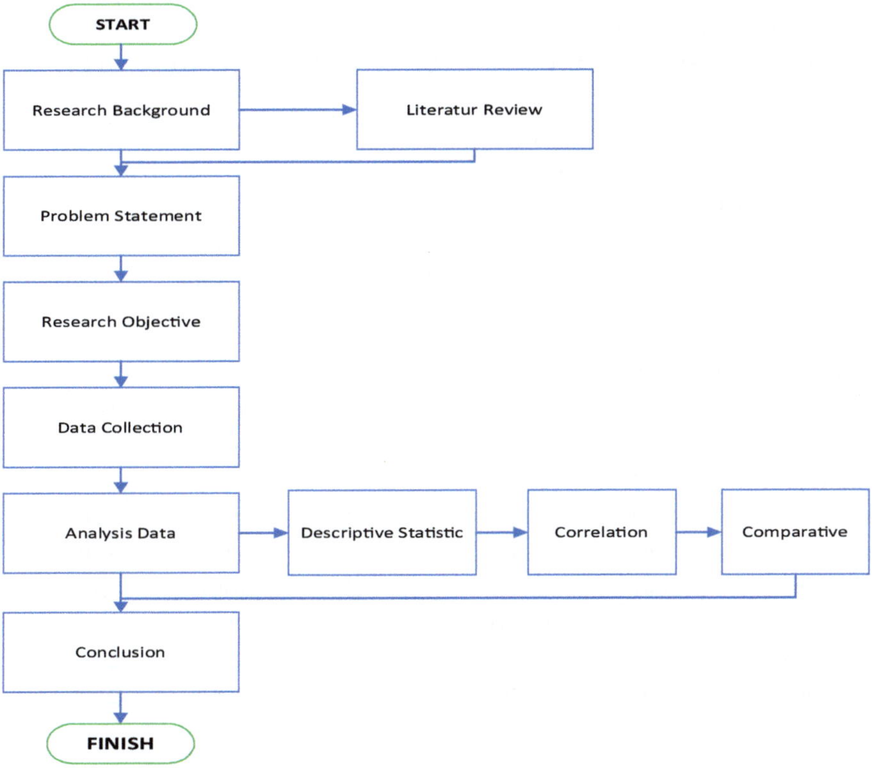

Fig. 1 Methodology framework

to 5, with 1 indicating inadequate logistics performance and 5 representing strong logistics performance.

The findings of the correlation matrix analysis shown in Table 1 indicate that all the criteria examined in this research make a significant impact to the LPI score. Among these criteria, the logistics performance services quality and competence (LPCQ) demonstrate the highest level of contribution, suggesting that improvements in customs procedures and infrastructure quality have a particularly strong impact on enhancing the overall effectiveness of logistics. However, the logistics performance international shipments (LPIS) criterion exhibits the lowest level of contribution, implying that factors related to international shipments have a relatively lesser influence on the LPI score. These results highlight the varying degrees of importance and impact that different criteria have on the overall logistics performance, providing valuable insights for policymakers and stakeholders in identifying areas of focus for improving logistics efficiency and effectiveness; see Table 2. The results indicate that all variables have positive mean values, illustrating that they follow a normal distribution with a tendency toward higher values. Specifically, Singapore stands out with the highest average LPI score of 4.1, indicating a relatively superior logistics

Table 1 Correlation indicator LPI

Criteria	LPI Score	LPC	LPQI	LPIS	LPCQ	LPT
LPC	0.971					
LPQI	0.983	0.964				
LPIS	0.938	0.898	0.899			
LPCQ	0.988	0.955	0.976	0.899		
LPT	0.938	0.871	0.899	0.863	0.927	
LPTT	0.967	0.921	0.941	0.906	0.961	0.885

Table 2 Descriptive statistics

Countries	Mean	StDev	Min	Me	Max
Singapore	4.111	0.143	3.9961	4.074	4.300
Malaysia	3.4594	0.1779	3.2209	3.5084	3.600
Thailand	3.3989	0.1033	3.2551	3.4203	3.500
Vietnam	3.1763	0.1474	2.9766	3.2144	3.300
Indonesia	3.0541	0.0769	2.9845	3.0409	3.150
Philippines	3.0159	0.1991	2.8563	2.9538	3.300
Cambodia	2.6302	0.1799	2.400	2.660	2.801
Lao PDR	2.389	0.259	2.067	2.395	2.700

performance during the observed period. On the other hand, Indonesia is in fifth place with a 3.015 average and a max of 3.15 during the same period.

3 Results and Discussion

3.1 *Result*

According to the analysis of descriptive statistics and correlation data for the LPI between the years 2014 and 2023, it is possible to assess the impact and contribution of logistics performance quality infrastructure (LPQI) in enhancing the overall LPI score. Figure 2 shows the scatterplot graph, along with the regression line, and provides insights into the performance of different countries during this period. Notably, Laos exhibited the lowest performance in terms of LPI, while Singapore emerged as the most stable country in terms of logistics performance.

The comparative analysis of LPQI (see Fig. 3) across Southeast Asian countries reveals varying outcomes. Singapore stands out with an impressive average LPQI score of 18%, indicating a significant contribution of LPQI to their logistics performance. Conversely, both Indonesia and Vietnam exhibit a lower average LPQI score

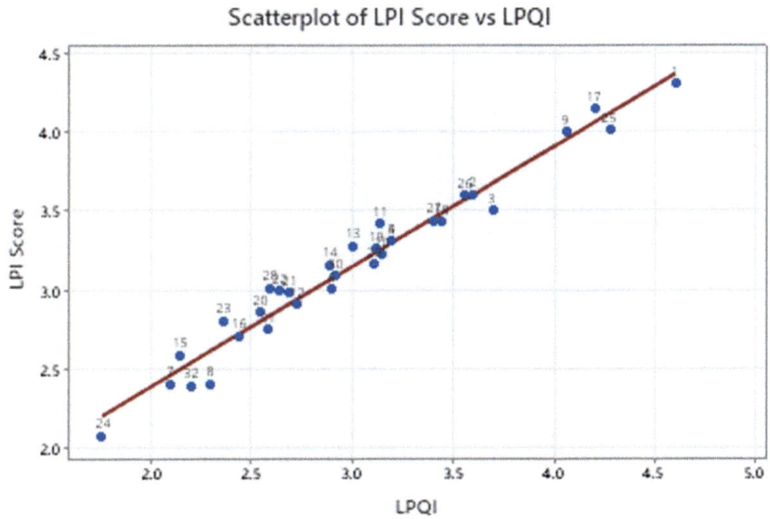

Fig. 2 Scatterplot with regression line

of 12%. However, it is noteworthy that Vietnam surpasses Indonesia in terms of average LPI score. Among the Southeast Asian countries, Laos records the lowest average LPQI score of 9% (Fig. 4).

The comparison between infrastructure and logistic performance can be described in the data correlation for each country. The specific measure that quantifies the intensity of the linear relationship between two variables in a correlation analysis is also presented in Table 3. This data can be used to figure out how infrastructure affects the performance of logistics, using the following correlation formula:

$$r = \frac{n\sum XY - \sum X \sum Y}{\sqrt{\left(n\sum X^2 - \left(\sum X\right)^2\right)\left(n\sum Y^2 - \left(\sum Y\right)^2\right.}}$$

$$r = \frac{4 \cdot 34.7 - 138.76}{\sqrt{(4 \cdot 37.33 - (149.24)((4 \cdot 32.31 - 129.02)}}$$

$$r = 0.592$$

In Indonesia, the correlation between LPI and LPQI is 0.592, as indicated in Table 3; this corresponds to a medium correlation. Consequently, the advancement of infrastructure in Indonesia is unable to increase the logistic performance. On the other hand, there are robust correlations in other countries, with infrastructure development serving as the primary indicator of logistic performance.

The examination of additional elements that may enhance logistics performance in Indonesia demonstrates that LPT makes a substantial contribution of around 18% throughout each period.

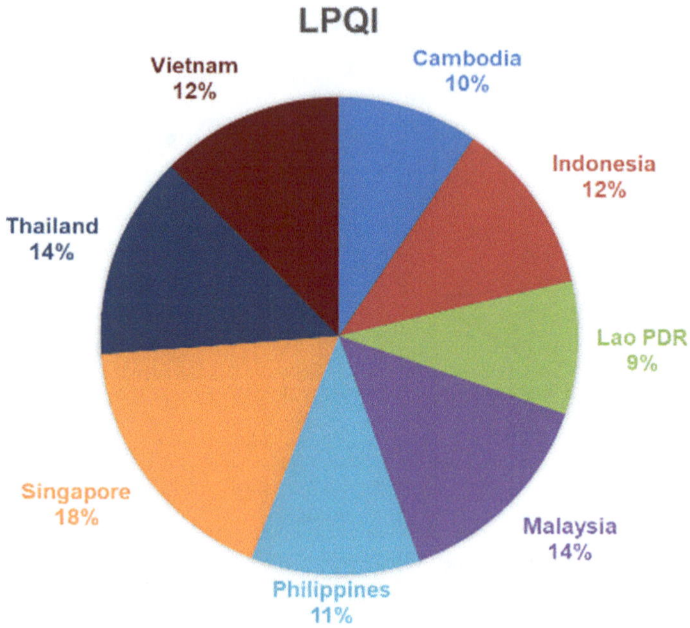

Fig. 3 Pie chart LPQI

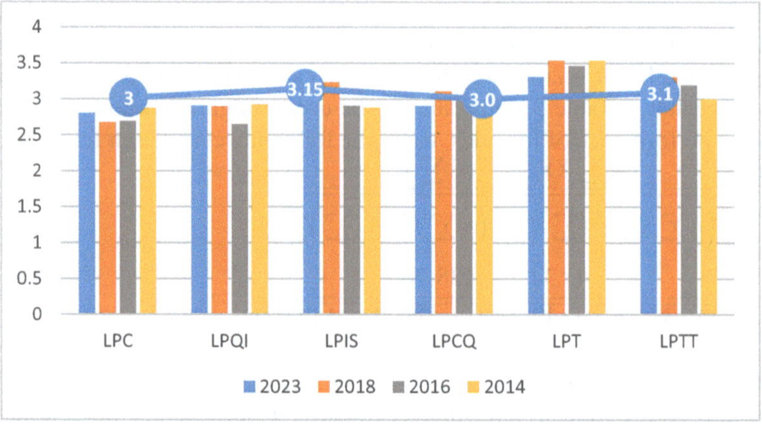

Fig. 4 LPI Indonesia

3.2 Discussion

Existing research [2, 3, 6, 8] states that the infrastructure sector assumes a critical role in facilitating logistics performance and propelling the economic advancement of a nation. Moreover, a study by [9] indicated that the regional economic growth

Table 3 Correlation data LPI versus LPQI

Correlations (r)	Singapore		Malaysia		Thailand		Philippine		Vietnam		Indonesia		Cambodia		Lao	
	LPI	LPQI	LPI	LPQI	LPI	LPQI	LPI	LPQI	LPI	LPQI	LPI	LPQI	LPI	LPQI	LPI	LPQI
2023	4.3	4.6	3.6	3.6	3.5	3.7	3.3	3.2	3.3	3.2	3	2.9	2.4	2.1	2.4	2.3
2018	4.0	4.1	3.2	3.1	3.4	3.1	2.9	2.7	3.3	3.0	3.15	2.9	2.6	2.1	2.7	2.4
2016	4.1	4.2	3.4	3.4	3.3	3.1	2.9	2.6	3.0	2.7	3.0	2.6	2.8	2.4	2.1	1.8
2014	4.0	4.3	3.6	3.6	3.4	3.4	3.0	2.6	3.2	3.1	3.1	2.9	2.7	2.6	2.4	2.2
Correlations (r)	0.852		0.986		0.797		0.927		0.859		0.592		0.802		0.955	

is substantially and relatively consistently influenced by infrastructure connectivity. Conversely, despite the correlations between infrastructure and logistics performance in Indonesia, the development of infrastructure has not been successful in enhancing the logistics performance index.

Improving performance in countries with low LPI scores is primarily dependent on enhancing infrastructure. However, achieving consistently good logistics performance requires a more comprehensive range of actions that encompass policy and the development of the private sector [10].

4 Conclusion

The literature evaluations indicate that infrastructure is regarded as the most fundamental for improving the logistics performance. In other countries, it has been seen that a higher LPQI and a strong correlation may serve as drivers to improve logistic performance. However, the findings of this research have revealed a lack of effectiveness in enhancing Indonesia's logistics performance through infrastructure development. This is evident in the correlation between LPI and LPQI in Indonesia, which stands at a mere 0.592, the lowest of all Southeast Asian countries. It was suggested that Indonesia's logistic performance cannot be effectively enhanced through infrastructure. Despite recognizing the importance of infrastructure development, the research has identified several challenges and limitations that hinder its effectiveness in enhancing logistics performance in the country.

Finally, this study advocates for additional research aimed at analyzing and overcoming the obstacles hindering the realization of the whole potential of infrastructure development and the improvement of logistics performance in Indonesia. In addition, identifying best practices from other countries that have successfully improved their logistics performance through infrastructure development.

Acknowledgement The author is deeply grateful and appreciative of the valuable perspective and insightful recommendations offered by Dr. Ing. Ir. Setyo Nugroho, Dr. Eng. IG.N. Sumanta Buana S.T., M.Eng, Dr. Dra. Agnes Tuti Rumiati, M.Sc., participants of this research for their willingness to share their knowledge and expertise, Institut Teknologi Sepuluh November for providing chance and resources to create this paper, and lastly the broader academic community for the extensive knowledge and research that inspired this work.

References

1. Arvis, J.-F., Alina Mustra, M., Panzer, J., Ojala, L., Naula, T.: Connecting to Compete 2007, Washington DC (2007). https://doi.org/10.1596/24600
2. Sharapiyeva, M.D., Antoni, A., Yessenzhigitova, R.: The impact of port transport-logistics infrastructure and LPI for economic growth: on the example of landlocked countries. Pomorstvo **33**(1), 63–75 (2019). https://doi.org/10.31217/p.33.1.7

3. Rezaei, J., van Roekel, W.S., Tavasszy, L.: Measuring the relative importance of the logistics performance index indicators using best worst method. Transp Policy (Oxf) **68**, 158–169 (2018). https://doi.org/10.1016/j.tranpol.2018.05.007
4. Saini, M., Hrušecká, D.: Influence of logistics competitiveness and logistics cost on economic development: an fsQCA qualitative approach. Ekon. Manag. **24**(2), 51–65 (2021). https://doi.org/10.15240/tul/001/2021-2-004
5. Sénquiz-Díaz, C.: Transport infrastructure quality and logistics performance in exports. ECONOMICS-Innov. Econ. Res. J. **9**(1), 107–124 (2021). https://doi.org/10.2478/eoik-2021-0008
6. Ben Haj Ahmed, R., Bouzir, A., Benhadj Mbarek, M.H., Benammou, S.: The impact of logistics performance index on port infrastructure quality: comparative study tunisia morocco. Arch. Transp. **67**(3), 71–89 (2023). https://doi.org/10.5604/01.3001.0053.7212
7. Mahpour, A., Farzin, I., Baghestani, A., Ashouri, S., Javadi, Z., Asgari, L.: Modeling the impact of logistic performance, economic features, and demographic factors of countries on the seaborne trade. Asian J. Shipp. Logist. **39**(2), 60–66 (2023). https://doi.org/10.1016/j.ajsl.2023.02.003
8. Meersman, H., Nazemzadeh, M.: The contribution of transport infrastructure to economic activity: the case of Belgium. Case Stud Transp Policy **5**(2), 316–324 (2017). https://doi.org/10.1016/j.cstp.2017.03.009
9. Li, W., Bai, X., Yang, D., Hou, Y.: Maritime connectivity, transport infrastructure expansion and economic growth: a global perspective. Transp. Res. Part A: Policy Pract. **170** (2023). https://doi.org/10.1016/j.tra.2023.103609
10. Arvis, J.-F., Ojala, L., Shepherd, B., Ulybina, D., Wiederer, C.: Connecting to Compete 2023: Trade Logistics in an Uncertain Global Economy—The Logistics Performance Index and Its Indicators, Washington DC (2023). https://doi.org/10.1596/39760

Development of Smart Ports in Indonesia Using Blockchain Technology: A Literature Review

Shantika Dita Putri, Gilang Prakasa, and Setyo Nugroho

1 Introduction

In recent years, industries worldwide have adopted digital technologies to optimize their operations, streamline processes, and enhance competitiveness. Notteboom et al. [1] said digitalization is about integrating information technology into business processes, changing how they work, and delivering valuable customer services. The International Energy Agency [2] describes digitalization as transforming business operations and processes using digital technologies. Another perspective by Parvianen et al. [3] stated that business environments, as well as the way business is done in general, are being affected by digitalization, and lack of digitalization in highly competitive markets may lead to a risk that the game is lost. Digitalization is linked to Industry 4.0, which integrates digital technologies into manufacturing. Industry 4.0 drives digitalization in sectors, and digitalization is necessary for its implementation [4, 5]. Both Industry 4.0 and digitalization offer numerous benefits to industries, including improved efficiency, cost reduction, enhanced customer experience, increased competitiveness, improved data management, and sustainability [5–7].

Digitalization is crucial for modern business operations; even ports are adopting Industry 4.0 technologies. These technologies have gained momentum and are predicted to impact ports, leading to the acceptance of new technology and business models [8]. The digitalization of ports offers significant opportunities, especially in improving efficiency, productivity, security, and sustainability [9]. Ports are essential

S. D. Putri (✉) · G. Prakasa · S. Nugroho
Department of Marine Transportation Engineering, Sepuluh Nopember Institute of Technology, Kampus ITS Sukolilo, Surabaya, Indonesia
e-mail: shantikadp@gmail.com

S. Nugroho
e-mail: snugroho@seatrans.its.ac.id

R. S. Widjaja et al. (eds.), *Marine Technology 3*,
SpringerBriefs in Applied Sciences and Technology,
https://doi.org/10.1007/978-3-032-05613-9_12

for the safe and efficient transportation of goods worldwide [10], achieving global supply chain performance and national economic growth. At the same time, their problems, such as rapid developments and market liberalization, resulted in port competition and awareness of environmental challenges. As a result, ports need to modernize and become smart ports [11].

The concept of smart ports utilizes advanced technologies and data-driven solutions to enhance port operations, management, safety, and logistics; smart infrastructure, efficient traffic flows, and practical logistics are the core components of smart ports [12]. The smart port project aims to contribute sustainably by setting energy savings with new management based on low environmental impact and triggering technological innovation [13]. Some of the most commonly used digital technologies in smart ports include the Internet of Things (IoT), artificial intelligence (AI), blockchain, digital twin technology, and big data [14].

Blockchain technology has been identified as a potential solution to improve efficiency, transparency, and security in port operations and logistics management [15]. Blockchain is a decentralized, encrypted, and distributed database with a wide range of applications in manufacturing. It enables the creation of tamper-proof real-time records and is closely linked to intelligent manufacturing and Industry 4.0 [16].

This research will explore the use of blockchain technology in developing smart ports. It will discuss the implementation of this technology in ports, particularly in Indonesia, along with recommendations and future research.

2 Methodology

This research starts by searching for the keywords "digitalization", "smart port", and "blockchain" in academic databases, specific journals, conference proceedings, and government or company reports related to digitalization in port and blockchain technology.

The research is divided into four sections. The first explains smart ports, the second discusses blockchain technology, the third covers Indonesian ports and blockchain possibilities, and the last section covers challenges and recommendations.

3 Findings

3.1 Smart Ports Concept

"Smart Port" is an upgraded version of traditional ports that uses advanced technology and infrastructure to optimize operations, increase flexibility, promote sustainability, and enhance safety and security [17]. Another perspective from Sari et al. [13] about the aims of smart port implementation is to make a sustainable contribution by establishing energy savings with low environmental impact and technological innovation. In general, smart ports are an evolution of port development that uses modern technology to support sustainably growing economies [14].

Based on Min [12], the essential elements of a smart port architecture in the era of Industry 4.0 include:

1. **Smart infrastructure**: using advanced technologies like IoT, sensors, cloud computing, and automation to create a connected and intelligent port infrastructure, including smart buildings, lighting, and energy management systems.
2. **Smart traffic flows**: using data analytics and predictive modeling to optimize port traffic. It involves real-time monitoring of traffic, automated guided vehicle (AGV) and intelligent traffic system (ITS) implementation, and dynamic routing to reduce congestion.
3. **Smart logistics**: advanced technologies like blockchain, AI, and robotics can optimize logistics in ports by enabling real-time cargo tracking, automated handling systems, and predictive maintenance.

According to Heikkilä et al., Belmoukari et al., and Min [8, 11, 12], there are some benefits to smart port development, such as:

1. Increase port operations' efficiency, safety, and reliability, such as cargo handling, traffic management and security, using automation, IoT, artificial intelligence, and simulation technologies.
2. Reduce port activities' environmental and social impacts, such as emissions, energy consumption, waste, and noise, using renewable energy sources, smart grids, waste management systems, and blockchain technologies.
3. Enhance the visibility, transparency and flexibility of port operations and supply chains by using cloud computing, big data analytics, blockchain, and digital platforms to facilitate collaboration among port actors, such as port authorities, terminal operators, shipping lines, logistics providers and customers.
4. Create value and competitive advantage for ports by offering innovative solutions and services that meet the changing needs and expectations of customers and stakeholders in the era of Industry 4.0.

Brunila et al., Belmoukari et al., and Min [9, 11, 12] also describe the challenges and the solutions for the port becoming a smart port; see Table 1.

3.2 Blockchain Technology in Ports

There are some discussions among researchers about what blockchain is. Blockchain is a decentralized database storing and executing transactions on a peer-to-peer network. Each participant maintains a copy of the ledger, and the data history of finished transactions is linked and impossible to modify. All parties with access to

Table 1 Challenges and the solutions for port becoming a smart port

	Challenges	Solutions
1	High investment costs for implementing smart technologies and infrastructures	Looking for public-private partnerships, subsidies, grants, or loans to finance smart port projects
2	Cybersecurity risks and vulnerabilities due to increased digitalization and data sharing	Adopt robust security measures, standards, and protocols to protect data and systems from cyberattacks
3	Workforce adaptation and skill development for smart port operations	Provide training, education, and incentives for workers to acquire new skills and competencies for smart port tasks
4	Collaboration and coordination among port stakeholders and external partners	Establish clear governance structures, roles, and responsibilities for smart port initiatives and foster trust and communication among port actors

the data are considered owners, and no single entity owns the data within a blockchain system [15, 18, 19].

The characteristics of the blockchain which distinguish it from today's port digitalization, include [15]:

1. **Transparency**, since blockchain is a decentralized storage system, ensuring that all network nodes have duplicate records of transactions and data.
2. **Availability** of data will facilitate stakeholders to make decisions more efficiently because the data is always accessible to authorized users using a distributed and shared ledger.
3. **The integrity** of data and transactions in the blockchain is maintained using asymmetric key cryptography and timestamps.
4. **Audit** and data provenance since transactions on a blockchain are immutable.
5. **Authorization** as blockchain assures only authorized and certified parties are allowed to perform transactions. A level of privacy is ensured through private channels.
6. **Privacy** is guaranteed through private channels.

Several blockchain applications, including smart contracts and the Internet of Things, are built on the blockchain infrastructure and protocols [15, 20].

1. Smart contract

   Smart contracts are self-executing codes that enable the transfer of funds or assets between parties in a transparent and auditable manner. They reduce the risk of errors and fraud while also increasing the efficiency and credibility of transactions.
2. Internet of Things (IoT)

   IoT links smart electronic devices to the Internet for data exchange. Blockchain could create a continuously updated database verified by multiple parties. Records validated at inspection points can be sent to the blockchain as timestamps.

4 Discussions, Implications, and Recommendations

4.1 Potential Blockchain Applications in Indonesian Ports

Indonesia's ports have recognized the importance of technology and digitalization to enhance their efficiency and competitiveness. The government supports initiatives to improve digitalization in the logistics and port sectors. However, technology adoption varies at different ports, with major ports like Tanjung Priok, Tanjung Perak, and Belawan leading in technology integration.

Amidst the rise of digitalization in Indonesian ports, there are some potential blockchain applications in Indonesian ports, such as [15, 18, 19]:

1. **Port terminal operations and management**: Blockchain can be used in port automation operations to share data, coordinate logistics parties, automate agreements, support autonomous vehicles, and integrate cloud computing and IoT technologies.
2. **Real-time cargo shipment tracing**: Blockchain allows stakeholders to track the location, condition accurately, and other essential parameters of cargo, such as containers, enhancing the monitoring of goods in transit.
3. **Document workflow tracing and protection**: Blockchain technology can digitalize this workflow with digitization of documents, provide smart contracts, enable real-time data sharing, and implement security measures to enhance transparency and trust between partners and reduce fraud.
4. **Financial process**: Blockchain can be implemented in the financial process of port logistics in possible ways, such as automated payment orders using smart contracts, providing real-time status orders, establishing a secure and trusted information flow, and integrating with existing systems to enhance the efficiency of the financial process.

4.2 The Challenges of Blockchain Implementations in Indonesian Ports

Applying blockchain technology in Indonesian ports has a significant potential to transform them into smart ports. However, the stakeholders involved will face several challenges in implementing it, some of which are as follows:

1. **Financial cost**: Implementing blockchain is expensive due to high initial investment, maintenance, and operation costs. Modifying existing systems and data is risky and costly for IT and human resources departments [19, 21]. Especially in Indonesia, where the coverage distribution of Internet connectivity is still not even.
2. **Lack of trust**: Blockchain requires trust among stakeholders, which is challenging in ports where data sharing is a concern [18].

3. **Lack of awareness and skills**: There is still a lack of awareness of the potential of blockchain technology and the expertise for developing the technology, especially in port and logistics [21].
4. **Regulations**: The lack of regulations and legal issues hinder the adoption of blockchain, causing ambiguity and distrust among stakeholders [15].
5. **Interoperability and scalability**: Ports have multiple stakeholders with different needs and interests. The complexity of port operations requires a lot of data, which can impact the scalability and interoperability of blockchain due to its diversity and the increasing size and capacity of platforms [22].

4.3 Recommendations

The development of smart ports using blockchain in Indonesia can be approached with the following recommendations:

1. Invest in infrastructure such as servers, networks, and blockchain platforms.
2. Provide staff training on blockchain technology and its applications in port operations.
3. Implement pilot projects in specific areas of port operations.
4. Collaborate with relevant stakeholders to develop and implement blockchain solutions.
5. Work with government agencies and industry associations to develop regulations and standards.
6. Ensure data security and privacy.
7. Integrate blockchain with other smart port technologies such as IoT, AI, and big data analytics.

5 Conclusion and Future Works

This research presents a recent literature review about the smart port concept and development with blockchain technology in Indonesian ports. The research can conclude that there are some challenges to implement this concept and technology in Indonesia due to the limitation of infrastructure development, regulations, and technology improvements. However, there is still potential and opportunities to make it happen because of the support from the Indonesian government, the implementation of the national logistics ecosystem (NLE), and the development of infrastructure and digitalization in Indonesian ports that are still running.

For future research, two possibilities can be presented. The first is about the concept of smart port implementation in Indonesia due to the lack of resources that discuss it. The second is to explore more about implementing some of the future trends in technology, such as AI, big data, and blockchain in the port industry in Indonesia, to develop Indonesian ports and make them more competitive with other ports worldwide.

References

1. Notteboom, T., Pallis, A., Rodrigue, J.-P.: Port Economics, Management and Policy, 1st edn. Routledge, London (2022). https://doi.org/10.4324/9780429318184
2. IEA: Digitalisation and Energy. IEA, Paris. https://www.iea.org/reports/digitalisation-and-energy (2017).. Accessed 9 Sept 2023
3. Parviainen, P., Tihinen, M., Kääriäinen, J., Teppola, S.: Tackling the digitalization challenge: how to benefit from digitalization in practice. Int. J. Inf. Syst. Proj. Manag. **5**, 63–77 (2017) https://doi.org/10.12821/ijispm050104
4. Ghobakhloo, M.: Industry 4.0, digitization, and opportunities for sustainability. J. Clean. Prod. **252**, 119869 (2020). https://doi.org/10.1016/j.jclepro.2019.119869
5. Calderon-Monge, E., Ribeiro-Soriano, D.: The role of digitalization in business and management: a systematic literature review. Rev. Manag. Sci. (2023). https://doi.org/10.1007/s11846-023-00647-8
6. Kraus, S., Jones, P., Kailer, N., Weinmann, A., Chaparro-Banegas, N., Roig-Tierno, N.: Digital transformation: an overview of the current state of the art of research. SAGE Open **11**(3), 21582440211047576 (2021). https://doi.org/10.1177/21582440211047576
7. Trittin-Ulbrich, H., Scherer, A.G., Munro, I., Whelan, G.: Exploring the dark and unexpected sides of digitalization: toward a critical agenda. Organization **28**(1), 8–25 (2021). https://doi.org/10.1177/1350508420968184
8. Heikkilä, M., Saarni, J., Saurama, A.: Innovation in smart ports: future directions of digitalization in container ports. J. Mar. Sci. Eng. **10**(12) (2022). https://doi.org/10.3390/jmse10121925
9. Brunila, O.-P., Kunnaala-Hyrkki, V., Inkinen, T.: Hindrances in port digitalization? Identifying problems in adoption and implementation. Eur. Transp. Res. Rev. **13**(1), 62 (2021). https://doi.org/10.1186/s12544-021-00523-0
10. Heilig, L., Lalla-Ruiz, E., Voß, S.: Digital transformation in maritime ports: analysis and a game theoretic framework. NETNOMICS **18**(2), 227–254 (2017). https://doi.org/10.1007/s11066-017-9122-x
11. Belmoukari, B., Audy, J.-F., Forget, P.: Smart port: a systematic literature review. Eur. Transp. Res. Rev. **15**(1), 4 (2023). https://doi.org/10.1186/s12544-023-00581-6
12. Min, H.: Developing a smart port architecture and essential elements in the era of Industry 4.0. Marit. Econ. Logist. **24**(2), 189–207 (2022). https://doi.org/10.1057/s41278-022-00211-3
13. Sari, Y.A., Pamadi, M.: The smart port concept of Batu Ampar port in Batam. IOP Conf. Ser.: Earth Environ. Sci. **343**(1), 012095 (2019). https://doi.org/10.1088/1755-1315/343/1/012095
14. Pham, T.Y.: A smart port development: systematic literature and bibliometric analysis. Asian J. Shipping Logistics **39**(3), 57–62 (2023). https://doi.org/10.1016/j.ajsl.2023.06.005
15. Ahmad, R.W., Hasan, H., Jayaraman, R., Salah, K., Omar, M.: Blockchain applications and architectures for port operations and logistics management. Res. Transp. Bus. Manag. **41**, 100620 (2021). https://doi.org/10.1016/j.rtbm.2021.100620
16. Javaid, M., Haleem, A., Pratap Singh, R., Khan, S., Suman, R.: Blockchain technology applications for industry 4.0: a literature-based review. Blockchain: Res. Appl. **2** (4), 100027 (2021). https://doi.org/10.1016/j.bcra.2021.100027
17. Yen, B.T.H., Huang, M.-J., Lai, H.-J., Cho, H.-H., Huang, Y.-L.: How smart port design influences port efficiency—a DEA-Tobit approach. Res. Transp. Bus. Manag. **46**, 100862 (2023). https://doi.org/10.1016/j.rtbm.2022.100862
18. Weernink, M. O., van den Engh, W., Francisconi, M., Thorborg, F.: The Blockchain Potential for Port Logistics. https://smartport.nl/wp-content/uploads/2017/10/White-Paper-Blockchain.pdf (2017)
19. Tsiulin, S., Reinau, K.H., Hilmola, O.-P., Goryaev, N., Karam, A.: Blockchain-based applications in shipping and port management: a literature review towards defining key conceptual frameworks. Rev. Int. Bus. Strategy **30**(2), 201–224 (2020). https://doi.org/10.1108/RIBS-04-2019-0051

20. Francisconi, M.: An explorative study on blockchain technology in application to port logistics.. https://repository.tudelft.nl/islandora/object/uuid%3Ab1de98dd-f1a7-456a-b8cb-4d87dc2c4f9f (2017). Accessed 9 Sept 2023
21. Alimohammadlou, M., Alinejad, S.: Challenges of blockchain implementation in SMEs' supply chains: an integrated IT2F-BWM and IT2F-DEMATEL method. Electron. Commer. Res. (2023). https://doi.org/10.1007/s10660-023-09696-3
22. Dutta, P., Choi, T.-M., Somani, S., Butala, R.: Blockchain technology in supply chain operations: applications, challenges and research opportunities. Transp. Res. Part E: Logistics Transp. Rev. **142**, 102067 (2020). https://doi.org/10.1016/j.tre.2020.102067

Establishment of Sustainable Green Fishing Port Principles Towards Indonesian Blue Economy and Sustainable Maritime Development

Sunarsih, Sukandar, and Mukhammad Risfan Baihaqi

1 Introduction

Indonesia is a maritime country blessed with strategic location and plentiful maritime resources. The Pacific and Indian oceans where the country lies are one of the most popular shipping areas in the world. This provides opportunities for the country to develop and enhance the maritime-based industries and become the world maritime axis state. Integrated sea transportation and other maritime management systems are needed to support optimum supply chains and reach the intended goal. Fishing ports roles as the front layers in administering the resources are obvious and unexceptional.

Conventional fishing ports are generally less efficient and do not align with the sustainable development goals (SDGs) development set by the government through the Ministry of Maritime Affairs and Fisheries (Kementrian Kelautan dan Perikanan, KKP). In KKP's 2020–2024 strategic plan [1], development of green and sustainable fishing ports has been included as one of its priorities to strongly support the ministry and government efforts in implementing SDG 14 to preserve marine fisheries and resources for sustainable development. In accordance with the objective, fishing port development which accommodates fisheries sector has to take into account four basic activities entailing in the management and utilization of marine fisheries and resources starting from pre-production, production, post-production and processing to marketing [2]. However, there is no clear line, guidance or standard for developing a new or upgrading the existing ones into green and sustainable fishing ports.

Sunarsih (✉) · M. R. Baihaqi
Department of Marine Engineering, Faculty of Marine Technology, Institut Teknologi Sepuluh Nopember, Surabaya, Indonesia
e-mail: sunarsihits@gmail.com

Sukandar
Faculty of Fisheries and Marine Science, Universitas Brawijaya, Malang, Indonesia
e-mail: sukandar@ub.ac.id

R. S. Widjaja et al. (eds.), *Marine Technology 3*,
SpringerBriefs in Applied Sciences and Technology,
https://doi.org/10.1007/978-3-032-05613-9_13

In shipping industry and related areas, a term 'green port' has been proposed and developed from research activities related to sustainability [3, 4]. Similar terms of 'eco-port' which emphasize more on the environmental protection (eco/environmentally friendly) and 'sustainable port' which in line with the SDG concept are also emerged [5]. While ecocentrism is indispensable for the sustainability (long term) goal, the scope and definition amongst the three are often overlap, interchangeable or even inclusive.

This paper elaborates a baseline study of a SGFP conceptual framework for enhancement of Indonesian fishing ports. To begin with, the work referred to the concept of ecofishing port (EFP) introduced previously [6] and widely addressed by the KKP. To develop the SGFP principles, the work performed a desktop study to identify main aspects and criteria representing the concept. Afterwards, the work carried out a survey to related stakeholders comprising fishing port regulator, managerial and user to get their opinions on the aspects and criteria set. Finally, an analytical hierarchy process (AHP) was conducted to determine the consistency and scale of each aspect and criteria defined previously.

2 Material and Methodology

2.1 Identification Aspect and Criteria of an SGFP

The conceptual framework of the SGFP adopted the EFP concept where a balance between economic (commercial) benefits and environmental aspects is maintained to support sustainable fisheries management. To identify the criteria of SGFP, the research analysed various scientific publications [7–21]. The criteria were grouped into five main categories representing generic aspects mostly referred by fishing port studies namely port facilities (A1), policy and regulation (A2), community empowerment (A3), energy efficiency (A4), and port management and monitoring (A5) defined as follows:

1. *Port facilities* comprise facilities and infrastructure to support pre-production, processing, production and marketing activities at the port. Amongst facilities that must be available include waste treatment, clean water supply and other similar eco-friendly facilities and support systems.
2. *Port operation and management* depict port handling and control of other aspects including implementation of the applied policy and regulations. This aspect governs and maintains port processes and activities to take place in a sustainable, efficient and safe manner.
3. *Energy efficiency* is specifically added to SGFP aspect to encourage maritime decarbonization and use of renewable energy in various port processes and activities hence contributes to SDG achievement.

4. *Policy and regulations* refer to a set of rules, policies and regulations to govern various aspects and activities in the port. This aspect aims at directing the control of SGFP and optimizing the technical results obtained by the port.
5. *Community empowerment* contains any approaches taken by the port to ensure the community to get benefited from the SGFP existence while participating in maintaining the port sustainability and balance of the marine ecosystem.

2.2 Questionnaire Survey

Sustainability is a long-term goal and a product of collective performance. All parties individually and/or mutually involved have various responsibilities for its achievement. The research carried out a questionnaire survey addressed to relevant fishing port stakeholders comprising regulator, managerial and user to get their response on the stated aspects and criteria obtained from the desk study. Considering the roles and responsibilities, port regulators and managers were asked to convey their views on all SGFP aspects and criteria while port users were asked to respond only on port facilities, energy efficiency and community empowerment. However, aside from the defined ones, respondents were encouraged to propose other criteria within the context based on their role, expertise and experience.

2.3 Analytical Hierarchy Process

Analytical hierarchy (AHP) is a comprehensive decision support system (DSP) technique that takes into account both quantitative and qualitative matters and allows quantification of the weights of the decision criteria. For a set of alternatives (options to be screened, selected or ranked) and attributes (criteria which characterize the alternatives and enable evaluation of their levels) of a multicriteria decision-making (MCDM) problem, the technique employs pairwise comparison questions to produce pairwise comparisons matrices (PCM). The PCM contains judgments of the relative preference from each pair of alternatives against each attribute and judgments of the relative importance of each pair of the attributes to compute the priorities. As such, the technique enables a structure of a system and environment into interacting components before joining them by measuring and managing the impact of system error components [21, 22].

3 Results and Discussion

3.1 SGFP Aspects and Criteria

Development of aspects and criteria is important to comprehend the overall concept of SGFP. Both variables can be used as a platform to standardize fishing ports in a green and sustainable manner. Table 1 outlines the SGFP criteria and the definition or scopes bear within each criterion.

3.2 Stakeholders Opinion on SGFP Concept

Stakeholders opinions are critical in developing a comprehensive SGFP framework to confirm the existing aspects and criteria from their technical and operational point of views, expertise and experiences. Figure 1 summarizes each stakeholder priority on the five SGFP aspects and the global weightage for each aspect. Observed from the figure that port regulators, managerial and users have different views of importance on each aspect. Port regulators consider port operation and management as the most important aspect by 0.40 while port managerial prioritize port facilities and energy efficiency the most and equally by 0.29. On the other hand, port facilities became the most important aspect to port users by a significant weight of 0.72 followed by community empowerment which weighs only 0.19. The evaluations made port facilities as the most important aspect globally by 0.37 while port operation and management became secondary by 0.26.

Figure 2 details out the global weight of SGFP criteria grouped by each aspect. Noticed from the figure that within port facilities which is considered as the most important aspect, facilities to supply clean water for port operation gained much attention and recorded the highest weightage globally by 0.13. This level of importance is similar to the application of renewable energy which for the time being widely used solar cells to save energy usage in the port.

3.3 Stakeholders Feedback on SGFP Aspects and Criteria

Aside from the defined criteria and outlined in the questionnaire survey, stakeholders have proposed some other criteria regarded as important to be included in SGFPs based on their views or experience in the field. The criteria include facilities for the disabled, equipment for ship safety, fish tracking for export marketing, docking facilities/piers, post-training monitoring and hygiene fish market. Table 2 summarizes the criteria grouped into existing SGFP aspects.

Table 1 SGFP aspects and criteria

Aspect/Criteria	Definitions/Scopes
A1 Port facilities	
C1. Clean water supply	Facilities to supply fresh water for port activities
C2. Drainage and sanitary systems	Facilities for waste water runoff and removal
C3. Local marketing	Facilities for fish cleaning, auction and direct selling
C4. Waste management	Facilities for waste 3R (reduce, reuse and recycle) and hazardous waste treatment
C5. Green space	Facilities for mitigating air pollution and reducing heat effects
C6. Global marketing	Facilities to support export activities and worldwide marketing
A2 Port operation and management	
C1. Pollution prevention	Monitoring of air, land and water pollution within port and adjunct areas
C2. Good governance	Fulfilment of available, accessible, acceptable and affordable port governance
C3. Marine products	Reporting and monitoring of catches
C4. Clean water and sanitation	Monitoring of clean water supply and sanitary systems
C5. Port regulation	Monitoring of regulations on port operation and management
A3 Energy efficiency	
C1. Use of renewable energy	Exploration and exploitation of renewable energy for port infrastructures and activities
A4 Policy and regulations	
C1. Energy and resource processing	Regulations on the use of clean energy and exploitation of marine resources
C2. Environmental control	Regulations on port eco-friendliness and environmental protection
C3. Bureaucratic reformation	Guidelines to promote good port and governance systems
C4. Industry 4.0	Guidelines to promote smart port and operation systems
C5. Fishing gears	Regulations on fishing gears to support marine ecosystem and sustainability
A5 Community empowerment	
C1. Community wealth building	Training/upgrading to improve fishermen's skills and product value
C2. Sustainability campaign	Campaign on marine ecosystem conservation for sustainable marine fisheries
C3. Gender equality	Empowerment programmes for women fishers and related fisheries sectors

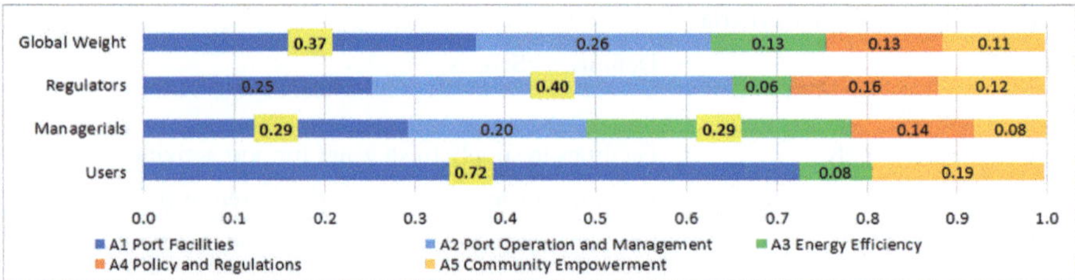

Fig. 1 Stakeholder priority and global weight of SGFP aspects

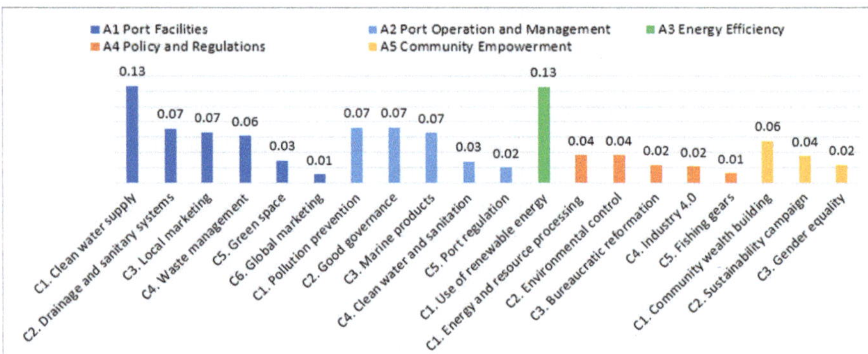

Fig. 2 Global weight of SGFP criteria

Table 2 Appended SGFP criteria

Aspect/Criteria	Definitions/Scopes
A1 Port facilities	
Disabled facility	Facilities to support the health, safety or comfort of people with disabilities
Early warning system	Facilities to early detect natural disaster or other harmful conditions
Piers/Docking system	Facilities to land or moor boats
Hygiene market	Facilities to ensure the health and cleanliness of port and the products
A2 Port operation and management	
Training evaluation	Monitoring of conducted training for post-mortem evaluation
A4 Policy and regulations	
Ship safety system	Regulations for onboard safety equipment
Fish tracking system	Regulations for product identification (i.e. fish barcoding)

4 Conclusion

Sustainable green fishing ports reflect long-term approaches, strategies and efforts to create fishing port infrastructure that is not only environmentally friendly but also human and economic friendly for the survival of current and future generations. Indonesian government has highlighted the significance of SGFP and initiated its development despite no clear standard for further reference. As such, the aspects and criteria identified by the study can be used as initial conceptual framework for SGFP operable model by the KKP. Pilot studies targeting potential fishing ports across the country can be carried out as a first step towards realizing Indonesian blue economy and sustainable maritime development.

Acknowledgement The authors gratefully acknowledged financial support from the Institut Teknologi Sepuluh Nopember for this work, under project scheme of the Publication Writing and IPR Incentive Program (PPHKI) 2023.

References

1. Database Peraturan: Jaringan Dokumentasi dan Informasi Hukum BPK Republik Indonesia. https://peraturan.bpk.go.id/Details/159375/permen-kkp-no-17permen-kp2020-tahun-2020 (2020). Accessed 20 Jan 2023
2. Database Peraturan: Jaringan Dokumentasi dan Informasi Hukum BPK Republik Indonesia. https://jdih.maritim.go.id/id/permen-kkp-no-59-tahun-2021 (2021). Accessed 20 Jan 2023
3. Wang, X., Yuen, K.F., Wong, Y.D., Li, K.X.: How can the maritime industry meet sustainable development goals? An analysis of sustainability reports from the social entrepreneurship perspective. Transp. Res. D Transp. Environ. **78** (2020). https://doi.org/10.1016/j.trd.2019.11.002
4. Anastasopoulos, D., Kolios, S., Stylios, C.: How will Greek ports become green ports. Geo-Eco-Marina **17** (2011)
5. Wu, X., Zhang, L., Yang, H.C.: Integration of eco-centric views of sustainability in port planning. Sustainability **12** (2020)
6. Lubis, E.: Pelabuhan perikanan. IPB Press (2019)
7. Kementerian Kelautan dan Perikanan Republik Indonesia: Rencana Strategis KKP Tahun 2020–2024. https://kkp.go.id/artikel/21471-rencana-strategis-kkp-tahun-2020-2024 (2020). Accessed 20 Jan 2023
8. Scheffczyk, R.B.: Fishing port management, the forgotten subject. Fish. Aquaculture. **1** (2009)
9. Satır, T., Doğan-Sağlamtimur, N.: The protection of Marine Aquatic Life: Green Port (ecoport) model inspired by Green Port concept in selected ports from Turkey, Europe and the USA. Period. Eng. Nat. Sci. **6** (2018). https://doi.org/10.21533/pen.v6i1.149
10. Wicaksono, A., Yanuwiadi, B., Dwiyanto, A.: Eco-fishing port assessment model as an environmental management tool on coastal fishing port "PONDOKDADAP"—Indonesia. In: 11th Asia Pacific Transportation and the Environment Conference (APTE 2018) (2019). https://doi.org/10.2991/apte-18.2019.22
11. Permana, A.R., Boesono, H., Jayanto, B.B.: Analisis Pengembangan Fasilitas Pelabuhan yang Berwawasan Lingkungan (Ecoport) di Pelabuhan Perikanan Nusantara (PPN) Kejawanan, Kota Cirebon, Jawa Barat. J. Fish. Resour. Util. Manag. Technol. **7** (2018)
12. Rodrigues, J., Kaming, P.F., Koesmargono, A.: Evaluasi Pelabuhan Tibar Timor-Leste dengan Indikator Hijau. J. Teknik Sipil **16** (2021). https://doi.org/10.24002/jts.v16i2.4779

13. Supriyanto, S.: Analisis Pengelolaan Pelabuhan Perikanan Berwawasan Lingkungan di Pelabuhan Perikanan Samudera Nizam Zachman Jakarta. J. Ilmu Lingkungan **7** (2013)
14. Wahyuni, D.M., Mustaruddin, M., R.: Penilaian Pengelolaan Lingkungan Pelabuhan Perikanan Samudera Kutaraja Berdasarkan Parameter Eco-Fishingport. ALBACORE **6** (2022) https://doi.org/10.29244/core.6.2.123-137
15. Satari, F., Rosyid, A., Wibowo, B.A.: Analisis kesesuaian fasilitas fungsional dan fasilitas penunjang pelabuhan perikanan berbasis ecoport di Pelabuhan Perikanan Pantai Tegalsari, Tegal. J. Fish. Resour. Util. Manag. Technol. **4** (2015)
16. Muninggar, R., Lubis, E., Iskandar, B.H., Haluan, J.: Simulasi Pengelolaan Ecofishingport Di Pelabuhan Perikanan Samudera Nizam Zachman Jakarta Dengan Pendekatan Sistem Dinamik. ALBACORE **3** (2019)
17. Mbay, L.O., Nugraha, R.B.A., Kusyanto, D.: Kajian Konsep Fishing Ecoport untuk pengembangan Pelabuhan Perikanan di Indonesia. J. Kelautan Nasional **9** (2014)
18. Hamzah, A., Rahmawati, A.: Penerapan Eco-Fishing Port di Pelabuhan Perikanan Nusantara Karangantu. Provinsi Banten. Akuatika Indonesia **6** (2021) https://doi.org/10.24198/jaki.v6i2.35137
19. The United Nations: The 17 Sustainable Development Goals (SDGs). https://sdgs.un.org/goals. Accessed 23 Jan 2023
20. European Sea Ports Organisation: Sustainability Report 2017. https://www.espo.be/publications/sustainability-report-2017. Accessed 25 Jan 2023
21. Saaty, T.L.: The Analytic Hierarchy Process: Planning, Priority Setting, Resource Allocation. McGraw-Hill (1980)
22. Saaty, T.L., Vargas, L.G.: Models, Methods, Concepts & Applications of the Analytic Hierarchy Process, p. 175. Springer Science & Business Media (2012)

Green Port Priority Development Strategy with the Fuzzy AHP Approach: A Case Study at the Port of Tanjung Emas Semarang

I Made Wahyu Sandika Putra, Rufiajid Navy Abritia, and I Gusti Ngurah Sumanta Buana

1 Introduction

Ports are a part of multimodal transportation systems, which are fundamentally essential to global supply chains and international logistics, where ports contribute about 70% by value and 90% of trade by volume worldwide [1, 2]. Ports significantly contribute to global air pollution and are major consumers of energy, thus playing a considerable role in environmental pollution [3]. Over the past years, scholars and policymakers have increasingly focused their attention to the impact of the port business, technological change, and environmental policy, which results in a decrease in environmental quality [4, 5]. Therefore, the business strategy formulated by ports should align with environmental management systems, aim to reduce waste, and incorporate environmental considerations like voluntary green programs, mandatory regulations, and corporate eco-innovation [6].

In the port of Tanjung Emas, overall office and operational activities at the port are still running not in a sustainable way. In the last two years, the recorded greenhouse gases (GHG) emissions consisting of CO_2, CH_4, and N_2O emissions were 1161.09 tons CO_2eq in 2020 and it is increased fivefold the following year to 5889.55 tons CO_2eq in 2021 [7]. The Tanjung Emas port has not yet fully implemented the green concept for sustainable business and operation. Hence, it is necessary to prioritize

I. M. W. Sandika Putra (✉) · R. N. Abritia
Politeknik Ilmu Pelayaran Semarang, Semarang, Jawa Tengah, Indonesia
e-mail: i_made_wahyu@pip-semarang.ac.id

R. N. Abritia
e-mail: rufiajid_navy@pip-semarang.ac.id

I. G. N. Sumanta Buana
Marine Engineering, Institut Teknologi Sepuluh Nopember, Surabaya, Indonesia
e-mail: buana@seatrans.its.ac.id

R. S. Widjaja et al. (eds.), *Marine Technology 3*,
SpringerBriefs in Applied Sciences and Technology,
https://doi.org/10.1007/978-3-032-05613-9_14

several criteria and indicators into strategies for developing a green port at the port of Tanjung Emas Semarang.

Since the sustainability in port operations is an essential issue, several studies have been made to develop environmentally friendly ports (green ports). However, while much of this work has been carried out for ports in Europe and North America, there has been limited research focused on ports in Asia [1]. Some previous research related to sustainable and green ports has been carried out by several researchers, such as [3, 4, 8, 9], which were studies of factors, performances, and evaluations on green port indicators in the implementation of green ports practice in several ports in the world. However, very limited research is available to analyze and prioritize green port development strategies and there is no study particularly conducted in terms of green port development at the port of Tanjung Emas Semarang. Thus, this research aims to analyze the green port strategy suitable for short-term, middle-term, and long-term at the port of Tanjung Emas Semarang.

2 Research Methodology

This quantitative research design uses the fuzzy analytical hierarchy process (FAHP). The FAHP model is used to analyze the development priority of each criterion and alternatives. The primary data was obtained from the questionnaires which were given to experts as respondents. Then, the secondary data was collected from books, national and international journal publications, conference papers, government regulations, company internal documents, and websites related to the green port and sustainable port operations.

Implementing FAHP has minimal downsides due to its use in a controlled setting, the instability of the measuring scale, and the subjective nature of the method [8]. FAHP can diminish or completely remove fuzziness; in numerous decision-making scenarios, fuzziness can result in vague evaluations by decision-makers who rely on traditional AHP methods [10]. Thus, the fundamental aspect of the fuzzy AHP approach lies in conducting pairwise comparisons using a ratio scale that is linked to a fuzzy scale value.

The evaluations made by experts are articulated as triangular fuzzy values, and these are employed to build fuzzy pairwise comparison matrices [11, 12]. Within the framework of fuzzy AHP, the adoption of triangular fuzzy numbers enhances the scaling mechanism in judgment matrices, and interval arithmetic is employed to ascertain the fuzzy eigenvector [11]. Then, a triangular fuzzy number is represented as $\tilde{M} = (a, b, c)$ where $a \leq b \leq c$, [13]. This representation is interpreted as the following membership functions:

$$\mu_{(A)}(x) = \begin{cases} 0, & x < a_1 \\ \frac{x-a_1}{a_2-a_1}, & a_1 \le x \le a_2 \\ \frac{a_3-x}{a_3-a_2}, & a_2 \le x \le a_3 \\ 0, & x > a_3 \end{cases}$$

Chang [14] suggested a method that utilizes triangular fuzzy numbers for handling scales in pairwise comparisons. The initial step of this technique is to use triangular fuzzy numbers for pairwise comparison using the FAHP scale. According to [14] if two fuzzy triangular numbers M_1 and M_2 where $\tilde{M}_1 = (l_1, m_1, u_1)$and $\tilde{M}_2 = (l_2, m_2, u_2)$, then the mathematical operation rules for fuzzy triangular numbers are:

$$(l_1, m_1, u_1) \otimes (l_2, m_2, u_2) = (l_1 l_2, m_1 m_2, u_1 u_2). \tag{2}$$

$$(\lambda, \lambda, \lambda) \otimes (l_2, m_2, u_2) = (\lambda l_2, \lambda m_2, \lambda u_2), \text{ where } \lambda > 0, \lambda \in R. \tag{3}$$

$$(l_1, m_1, u_1)^{-1} = \left(\frac{1}{u_1}, \frac{1}{m_1}, \frac{1}{l_1}\right). \tag{4}$$

Thus, the steps in FAHP methodology in the context of ranking sustainability tools are listed below [8]:

1. *First*: Selection of assessing criteria for the selected tools.
2. *Second*: Constitution of an expert team comprised of "sustainability experts" with experience in applying sustainable practices across multiple enterprises.
3. *Third*: Identifying the case process and developing a process model to comprehend the numerous inputs and outputs.
4. *Fourth*: Selection of suitable tools for a candidate process from a triangular fuzzy number.
5. *Fifth*: Rank determination using the FAHP approach (hierarchy structure creation) using the following steps:

   5.1. Compose the fuzzy comparison matrix. The fuzzy judgment matrix is formed using triangular fuzzy numbers through pairwise comparison as outlined below:

$$\tilde{A} = \begin{bmatrix} 1 & \tilde{a}_{12} & \cdots & \tilde{a}_{1(n-1)} & \tilde{a}_{1n} \\ \tilde{a}_{21} & 1 & \cdots & \tilde{a}_{2(n-1)} & \tilde{a}_{2n} \\ \vdots & \vdots & \vdots & \vdots & \vdots \\ \tilde{a}_{(n-1)1} & \tilde{a}_{(n-1)2} & \cdots & 1 & \tilde{a}_{(n-1)n} \\ \tilde{a}_{n1} & \tilde{a}_{n2} & \cdots & \tilde{a}_{n(n-1)} & 1 \end{bmatrix} \tag{5}$$

   5.2. Fuzzy-based synthesis to obtain importance rating and λ_{max} (for consistency test).

   5.3. Validation of the consistency of the matrices received from experts.

5.4. Calculation of each option's priority score.

3 Results and Discussion

Based on the AHP survey, a total of 10 experts were asked to evaluate the significance of factors related to the green port. The majority of respondents came from a port operator company which is PT. Pelindo (persero) for about 60%, followed by Academics (30%) such as Semarang Merchant Marine Polytechnic and Semarang Port Authority (10%), with a minimum working experience of 8 years.

3.1 Fuzzy AHP Criteria

The initial step in the AHP model involves creating a hierarchical system comprising multiple levels, including the goal, various types of influencing the main criteria, sub-criteria, and alternatives to determine the most favorable option. The criteria in the same hierarchy must be independent of one another. Referring to several publications of green port, an AHP model for green ports is developed. It consists of three main criteria, eight sub criteria, and three alternatives that will be used. The numerous criteria and sub-criteria of green port practices are adopted based on regulations, guidelines, and previous studies used in the green port performance assessment in Indonesia, such as Ministerial Regulation of Environment and Forestry of the Republic of Indonesia No. 1 of 2021, and APEC Ports Services Network (ASPN). The criteria are summarized in Table 1.

Table 1 Green port criteria

No.	Criteria	Sub-criteria	Alternatives
1	Commitment and willingness (C01)	Development aspirations (S01)	Short-term priority development (A01)
		Green port promotion and training (S02)	
2	Action and implementation (C02)	Green energy structure (S03)	Medium-term priority development (A02)
		Energy saving (S04)	
		Environmental protection (S05)	
		Green management (S06)	
3	Efficiency and effectiveness (C03)	Energy utilization (S07)	Long-term priority development (A03)
		Environmental quality control (S08)	

3.2 The Priority of Green Port Criteria

Table 2 displays the priorities of factors necessary for the successful operation of a green port. The weight assigned to each criterion represents its relative importance in green port development. The table presents the evaluating main criterion, which encompass the most essential factors in achieving the goal of green port is actions and implementation whose importance weight calculated to be 0.564. Then, commitment and willingness are the second important (0.307) and efficiency and effectiveness (0.129) is the last priority. Thus, a low consistency ratio (<10%) indicates that the comparisons were relatively consistent and reliable.

Furthermore, the priority weights of the criteria are shown in Table 3. Based on that table, the priority order is environmental protection with a criteria weight of 0.177 followed by environmental quality control (0.166), green management (0.161), energy saving (0.160), energy utilization (0.092), green development aspiration (0.088), green energy structure (0.083), and green port promotion and training (0.073). The consistency ratio (0.087) in this calculation indicates the comparison result by judgments remains consistent and reliable. That result reveals environmental protection is the highest concern to achieve a green port. In contrast, green promotion and training is less important.

Table 2 Criteria weight priority

Criteria	Synthetic fuzzy value			Min of degree	Weight vector	Weight criteria
	l	*m*	*u*			
Commitment and willingness	0.16	0.303	0.587	0.684	0.316	0.307
Actions and implementations	0.259	0.455	0.717	1	0.581	0.564
Efficiency and effectiveness	0.148	0.242	0.457	0.148	0.134	0.129
Consistency ratio	0.066					

Table 3 Priority ranking of green port sub-criteria

Rank	Sub-criteria	Synthetic fuzzy value			Min of degree	Weight criteria
		l	*m*	*u*		
1	Environmental protection	0.091	0.176	0.307	1	0.177
2	Environmental quality control	0.085	0.162	0.307	0.939	0.166
3	Green management	0.081	0.155	0.307	0.911	0.161
4	Energy saving	0.078	0.155	0.288	0.903	0.160
5	Energy utilization	0.046	0.088	0.186	0.517	0.092
6	Green development aspiration	0.054	0.096	0.170	0.495	0.088
7	Green energy structure	0.050	0.092	0.166	0.471	0.083
8	Green port promotion and training	0.044	0.096	0.160	0.410	0.073
	Consistency ratio	0.087				

After the priority of criteria and sub-criteria have been determined, the analysis comes into alternative levels. In this stage, the researcher analyzed the alternatives priority for each sub-criteria to determine the development priority, as shown in Table 4. The priority weight was obtained with pairwise comparison analysis as in the FAHP method. As a result, the respective sub-criteria for the short-term development priority include green development aspirations, management, environmental protection, and energy utilization. Besides, the sub-criteria which are for middle-term priority are energy saving and environmental quality control. Finally, the long-term development priorities are green energy structure, green promotion, and training.

Table 4 Priority development alternatives of each criterion

Sub criteria	Alternatives development priority	Weight	Rank
Green development aspiration	**Short-term priority**	**0.619**	**1**
	Middle-term priority	0.284	2
	Long-term priority	0.096	3
Green port promotion and training	Short-term priority	0.082	3
	Middle-term priority	0.179	2
	Long-term priority	**0.739**	**1**
Green energy structure	Short-term priority	0.198	3
	Middle-term priority	0.212	2
	Long-term priority	**0.591**	**1**
Energy saving	Short-term priority	0.097	3
	Middle-term priority	**0.558**	**1**
	Long-term priority	0.345	2
Environmental protection	**Short-term priority**	**0.591**	**1**
	Middle-term priority	0.212	2
	Long-term priority	0.198	3
Green management	**Short-term priority**	**0.633**	**1**
	Middle-term priority	0.260	2
	Long-term priority	0.106	3
Energy utilization	**Short-term priority**	**0.441**	**1**
	Middle-term priority	0.393	2
	Long-term priority	0.167	3
Environmental quality control	Short-term priority	0.345	2
	Middle-term priority	**0.558**	**1**
	Long-term priority	0.097	3

The bold values indicate the time span of the selected development priorities that obtained the highest scores based on the pairwise comparison analysis. The definitions of each priority category are as follows: -Short-term priority refers to development aspects planned to be implemented within 1–3 years. -Medium-term priority refers to development aspects planned to be implemented within 3–5 years. -Long-term priority refers to development aspects planned to be implemented within 5–10 years. These time span categories are determined based on the Port Development Guidelines of Tanjung Emas Port

4 Conclusion

The most critical criterion for achieving green port goals is actions and implementation (0.564) followed by commitment and willingness (0.307) and efficiency and effectiveness (0.129). Furthermore, environmental protection (0.177) emerges as the top sub-criterion priority, highlighting the significance of safeguarding the environment within and around the port. Then, environmental quality control (0.166) and green management (0.161) are also key considerations.

Regarding the priority development of each criterion, the short-term development priorities of green port include green development aspirations, management, environmental protection, and energy utilization. Then, mid-term priorities encompass energy saving and environmental quality control, while long-term priorities involve green energy structure and green promotion and training.

Acknowledgments The authors thank to all respondents and parties involved in this research, particularly Port of Tanjung Emas (PT. Pelindo).

References

1. Gupta, K.V., Prakash, G.: Assessment of environmental sustainability issues for south-Asian maritime ports. Aust. J. Marit. Ocean Aff. **1–23** (2022). https://doi.org/10.1080/18366503.2022.2038903
2. UNCTAD: Review of maritime transport 2019. In: Presented at the United Nations Conference on Trade and Development, New York (2019)
3. Chen, J., Zheng, T., Garg, A., Xu, L., Li, S., Fei, Y.: Alternative maritime power application as a green port strategy: barriers in China. J. Clean. Prod. **213**, 825–837 (2019). https://doi.org/10.1016/j.jclepro.2018.12.177
4. Ahmadi, N., Kusumastanto, T., Siahaan, E.I.: Environmentally friendly port development strategy (Greenport) a case study: at the port of Cigading Indonesia. Warta Penelitian **28**, 9 (2018) https://doi.org/10.25104/warlit.v28i1.697
5. van Leeuwen, G., Mohnen, P.: Revisiting the porter hypothesis: an empirical analysis of green innovation for The Netherlands. Econ. Innov. New Technol. **26**, 63–77 (2017). https://doi.org/10.1080/10438599.2016.1202521
6. Nuryakin, N., Maryati, T.: Do green innovation and green competitive advantage mediate the effect of green marketing orientation on SMEs' green marketing performance? Cogent Bus. Manag. **9**, 2065948 (2022). https://doi.org/10.1080/23311975.2022.2065948
7. Pelindo: The port of Tanjung Emas Semarang's environmental assessment report. In: Pelindo III (Persero), Semarang Indonesia (2022)
8. Garg, C.P., Kashav, V., Wang, X.: Evaluating sustainability factors of green ports in China under fuzzy environment. Environ. Dev. Sustain. (2022). https://doi.org/10.1007/s10668-022-02375-7
9. Akbari, A.A., Babaei Samiromi, F., Arjmandi, R., Shojaei, M.: Evaluating ports environmental performance based on the fuzzy analytical hierarchy process (FAHP). IJOH (2022). https://doi.org/10.18502/ijoh.v13i2.8369
10. Emrouznejad, A., Ho, W. (eds.): Fuzzy Analytic Hierarchy Process. CRC Press, Boca Raton (2017)
11. Chiu, R.-H., Lin, L.-H., Ting, S.-C.: Evaluation of green port factors and performance: a fuzzy AHP analysis. Math. Probl. Eng. **2014**, 1–12 (2014). https://doi.org/10.1155/2014/802976
12. Ghodsypour, S.H., O'Brien, C.: A decision support system for supplier selection using an integrated analytic hierarchy process and linear programming. Int. J. Prod. Econ. **56–57**, 199–212 (1998). https://doi.org/10.1016/S0925-5273(97)00009-1

13. Buckley, J.J.: Fuzzy hierarchical analysis. Fuzzy Sets Syst. **17**, 233–247 (1985). https://doi.org/10.1016/0165-0114(85)90090-9
14. Chang, D.-Y.: Applications of the extent analysis method on fuzzy AHP. Eur. J. Oper. Res. **95**, 649–655 (1996). https://doi.org/10.1016/0377-2217(95)00300-2

Alternative Fuels Training for Seafarers to Support Shipping Decarbonization

Mokhamad Aufal Huda, Hilda Emeraldo Ahmad Adiarsa, and Setyo Nugroho

1 Introduction

The marine industry's heavy reliance on fossil fuels poses serious risks to human health and the environment. Sustainable alternatives must be investigated and implemented immediately as international legislations addressing climate change becomes more stringent. Alternative fuel technologies must be implemented for the maritime sector to reach zero emissions by 2050 [1]. The maritime industry will have to transition throughout this period from traditional fuels like heavy fuel oil (HFO) and marine gas oil (MGO) to low- and zero-carbon alternative fuel technologies (LNG/LPG, ammonia, methanol, hydrogen, battery, etc.). Seafarers are essential to marine operations and vessel management. To ensure safe and effective operations, seafarers must learn and acquire new skills and competencies in the decades leading up to 2050 and beyond due to the projected major impact that alternative fuel technologies will have on maritime operations on board ships [2].

1.1 The Background

When LNG/LPG and other alternative fuel technologies become more widely used by 2050, seafarers currently employed on vessels that run on traditional fuel will need additional training. The International Code of Safety for Ships Using Gases or

M. A. Huda (✉) · H. E. A. Adiarsa · S. Nugroho
Institut Teknologi Sepuluh Nopember, Surabaya, Jawa Timur, Indonesia
e-mail: aufal@poltekpel-banten.ac.id

S. Nugroho
e-mail: snugroho@seatrans.its.ac.id

R. S. Widjaja et al. (eds.), *Marine Technology 3*,
SpringerBriefs in Applied Sciences and Technology,
https://doi.org/10.1007/978-3-032-05613-9_15

Other Low-Flashpoint Fuels (IGF Code) previously established seafarers' training requirements on LNG/LPG [3].

It is still unclear, though, if ship operators would adopt other viable alternative fuel technologies (such as methanol, hydrogen, or ammonia) because doing so will necessitate new operational, technological, and safety standards [4]. Furthermore, it is anticipated that decarbonization and the use of alternative fuel sources will result in quick technological improvements, including greater automation and digitization [5]. With the exception of LNG/LPG, no industry-approved training courses cover the other alternative fuel technologies. However, the International Maritime Organization (IMO) is developing guidelines for other alternative fuels that may serve as the foundation for new training standards applicable to ships using specific fuels [6]. When compared to the future demands of training for seafarers, there aren't many approved training courses that give instruction for various alternative fuels, despite the IMO having established Model Courses 7.13 and 7.14 for basic and advanced training, respectively, for ships subject to the IGF code. Though classification societies offer recommendations and restrictions for several of these technologies, there are currently few IMO laws pertaining to alternative fuel technology [7].

1.2 The Questions

The following primary research issues are addressed in this study:

> How might comprehensive programs for seafarers to learn about alternative fuels and technology be developed to help the decarbonization of shipping?

In order to probe more into the research, a number of sub-questions are developed that explore distinct facets of the subject, like:

1. What special difficulties does the marine industry present for alternative fuel training?
2. What are the best procedures and approaches for creating training curricula that address the technical, operational, and safety facets of operating with alternative fuels in accordance with the STCW, IGF code, and IMO model courses?
3. How do shipping firms and educational institutions modify their curriculums to stay up to date with the rapidly changing field of alternative fuel technologies?

2 Methodology

2.1 Conceptual Model

It is assumed in the main question and its sub-questions that comprehensive training programs for sailors constitute the dependent variable in this instance. Conversely, the distinct characteristics and difficulties associated with alternative fuels, the legal

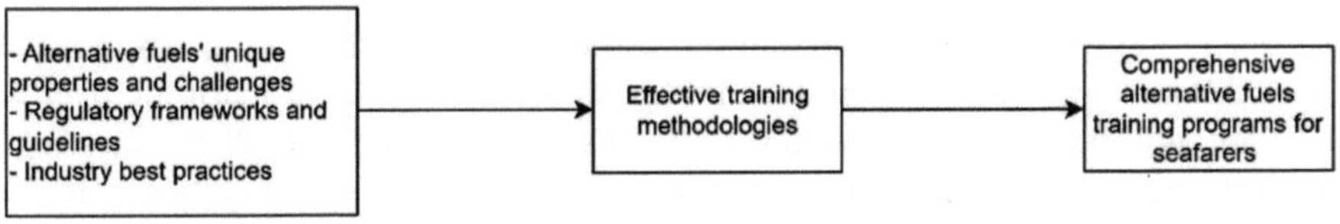

Fig. 1 Conceptual model of the research

frameworks and regulations, and industry best practices derived from accident reports and guidelines for other fuels are independent variables that will influence the outcome of dependent variables. The conceptual model in the manner is depicted in Fig. 1.

Furthermore, it is anticipated that successful training approaches act as mediating variables. An intermediary variable in a causal sequence that links an independent variable to a dependent variable is known as a mediating variable [8]. In this sequence, the independent variable causes the mediating variable, which in turn causes the dependent variable. Comprehensive alternative fuel training programs for seafarers aim to achieve the safe and efficient use of alternative fuels in the marine industry through effective training approaches.

2.2 *Quantitative Approach*

Since the quantitative approach involves many active seafarers as respondents to validate problems, in this case, the surveys and problem validation have been conducted by a marine class authority, Det Norske Veritas (DNV) Holding AS co-sponsored by the Singapore Maritime Foundation (SMF). Representing more than 500 seafarers who work around the world to operate tankers, dry bulk, and container vessels, it is unnecessary and will be redundant to conduct a similar approach. According to a survey conducted among seafarers, 81% said they needed full or partial training to operate the cutting-edge equipment that will be installed on board future ships. Comparably, more than 75% of the participants indicated that they needed either full or partial training on new fuel types, including LNG, batteries, or synthetic fuels. With regard to developing fuels like methanol, hydrogen, and ammonia, this training deficiency increased to 87% of survey participants [5].

The estimated contribution of various alternative fuel technologies (ammonia, methanol, and hydrogen) in the global maritime fleet is based on three scenarios, per the DNV research [5]: (1) The 2021 DNV Maritime Forecast 2050 model uses the IMO 2018 scenario [9]; (2) A 2021 DNV Maritime Forecast 2050 model that represents the DNV Decarbonization by 2050 scenario; and (3) The University Maritime Advisory Services (UMAS) 2019 and Lloyds Register modeled the "Zero Carbon by 2050" scenario [10], the percentages are roughly 21, 43, and 78%, in that order.

Moreover, a noteworthy proportion of seafarers require supplementary training and expertise in alternative fuel technologies such as ammonia, methanol, and hydrogen, and a considerable amount of seafarers possess three scenario models. Three different projected energy technology shares in the global maritime fleet in 2050 are as follows: (1) According to the DNV projected "IMO 2018 scenario," 310,000 seafarers operating on ships use various alternative fuel technologies at their peak; (2) 750,000 is the number from the DNV-modeled decarbonization by 2050 scenario; (3) The LR and UMAS modeled the 1,775,000-person "zero carbon by 2050" scenario. In any case, the investigation reveals that in order to assist the transition of shipping, sailors will require more training in the upcoming years.

2.3 Qualitative Approach

The research data was collected in the Netherlands by interviewing *Samskip*, a Dutch shipping company headquartered in Rotterdam, with some of its fleet already powered by alternative fuels, notably LNG, and the upcoming dual-fuel hydrogen-powered vessels in order. The interview focused on the challenges, the lack, and the needs the company faced while implementing alternative fuels on their fleets, how they dealt with the fuel transition period, improving and upskilling their workforce, conducting training, and keeping up with the evolving landscape of alternative fuels.

3 Results and Discussion

3.1 The Challenges

In response to the first sub-question, the following are some of the particular difficulties that the marine industry faces when it comes to training for various alternative fuels:

1. Insufficient training materials at training facilities: In addition to the few facilities actively offering instruction, there is a dearth of standardized training resources that address the nuances of managing and applying alternative fuels. These documents need to cover a wide range of topics, such as alternative fuel chemistry, fuel transportation and storage, fueling techniques, safety precautions, and emergency response. The training institution's preparedness is in doubt, notwithstanding the shipping industry's trend toward alternative fuels.
2. Complex technical knowledge: In order to use alternative fuels successfully, mariners must acquire new technical know-how. The same qualification and training are required for alternative fuels for sailors and other corporate employees, including superintendents and marine inspectors. It is crucial but presently underappreciated to comprehend the characteristics of fuel, combustion, maintenance techniques, and troubleshooting processes unique to alternative fuels.

3. Because of their special characteristics and handling needs, alternative fuels present particular safety risks. It carries with it various energies and associated risks. To manage and reduce hazards associated with fuel leaks, flammability, toxicity, and potential incompatibilities with current ship systems, seafarers need to receive the necessary training.
4. A complicated web of national, international, and local fuel-use and emissions laws applies to the maritime sector. To guarantee compliance and prevent penalties, seafarers need to be informed about these changing requirements. Shipping firms may face difficulties since, in order to assure compliance, they also need to modernize their safety management system and workforce.
5. In situations involving fuel, mariners must be ready to react quickly and effectively. Training curriculum must incorporate containment tactics, emergency procedures, and appropriate communication protocols. Before boarding a ship, seafarers must receive thorough training. This training must continue after they are on board, with them continuing to learn from any events involving alternative fuels.

3.2 The Training for Seafarers

To answer the second sub-question about the best practices and methodologies for designing training programs, based on the result of the interview, most of the shipping company's current ways to upskill their ship crew are primarily following the existing rules and guidelines based on:

1. *IGF Code*

   The International Code of Safety for Ships using Gases or Other Low-flashpoint Fuels (IGF Code) [3] has emerged as a pivotal regulatory framework in response to the maritime industry's growing interest in alternative fuels. It provides guidelines and safety standards for ships utilizing alternative fuels with low-flashpoint properties, such as liquefied natural gas (LNG). Its primary objective is to ensure the safe design, construction, and operation of vessels powered by these fuels.
2. *STCW Convention V/3, STCW Code A-V/3-1 and 3-2*

   The International Convention on Standards of Training, Certification and Watchkeeping for Seafarers (STCW) Convention V/3 [11] lays down the mandatory minimum requirements for the training and certification of officers and ratings on board vessels equipped with alternative fuel propulsion systems. It explores LNG-specific training programs' content, delivery methods, and assessment criteria.
3. *IMO Model Course 7.13 and 7.14*

   The IMO Model Course 7.13 is dedicated to seafarers' training in handling and using LNG as a ship fuel. The safety representatives of the company explore this course's content, methodology, and relevance, considering the unique safety and operational aspects of LNG-fueled vessels. They also explore the IMO Model

Course 7.14, which extends the training provisions to cover other low-flashpoint fuels, including but not limited to hydrogen and methanol, as they plan to acquire some dual-fuel vessels for their pilot project of feeder vessels.

4. *SOLAS regulation II-1/2.29*

   The SOLAS Regulation II-1/2.29 prescribes the training requirements for seafarers responsible for operating and maintaining ships utilizing alternative fuels. The company explores the specific provisions of the regulation, including the scope of training, qualifications of instructors, and certification requirements. SOLAS Regulation II-1/2.29 is closely intertwined with international codes and guidelines, such as the IGF code.

3.3 Keeping Up with the Evolving Landscape

To answer the third sub-question, based on the result of the interview, the shipping companies are actively adapting to the evolving landscape of alternative fuels and incorporating training programs for seafarers to ensure safety, compliance, and efficiency in their operations. Here are some ways in which they are keeping up with these changes and should be adapted by other companies and training institutions:

1. Investing in Training Programs

   Investment in ongoing education ensures seafarers remain adaptable, competent, and capable of navigating the complexities of evolving energy technologies. It is vital to a safer, more sustainable maritime future. The training programs are regularly scheduled per the company safety management system (SMS) manual.
2. Regulatory Compliance

   Compliance reduces the risk of accidents and helps avoid costly penalties and reputational damage. Moreover, adherence to regulations aligns the industry with global sustainability objectives, demonstrating a commitment to responsible maritime practices. It is the foundation for a secure and sustainable future in the maritime sector.
3. Simulator Training

   Simulators provide a realistic environment where crew members can practice various scenarios, including emergencies, without real-world risks. The third-party trainer offers the use of simulator training. It provides a safe and controlled environment to practice handling new fuels, ensuring proficiency without real-world risks.
4. Supplier and Manufacturer Partnerships

   These partnerships help seafarers become familiar with the specific equipment and technologies used on board. The partnerships depend on the type of fuel onboard the vessels. Collaborating with industry experts ensures access to the latest technologies, equipment, and safety measures. This cooperation enhances training quality, aligns with evolving industry standards, and equips seafarers to handle alternative fuels proficiently and safely.

5. Cross-Training

   It equips seafarers with diverse skills and knowledge, fostering versatility in handling various fuel types. This flexibility enhances safety, operational efficiency, and readiness to navigate the dynamic challenges of alternative fuels, ensuring a competent and adaptable workforce. This versatility ensures seafarers can adapt to vessels and fuel types as needed. The cross-training model can help shipping companies retain their crew since there is a limited crew with knowledge of alternative fuels for the time being.
6. Continuous Learning

   Training programs are often ongoing, with regular updates to inform seafarers about the latest alternative fuels and technology developments. The most important thing from this part is to learn from industry best practices and any previous near misses or incidents regarding alternative fuels. This program usually becomes part of the company safety management system (SMS).
7. Online and Remote Training

   In a dynamic industry like maritime, where seafarers are often at sea, remote training ensures they can stay updated with the latest developments in alternative fuel usage and safety protocols. It also reduces the cost and time associated with traditional in-person training, making it a practical choice for staying aligned with evolving standards.
8. Safety Protocols

   In a rapidly changing field, safety protocols provide a stable framework, fostering confidence and competence among seafarers. Ultimately, these protocols are the linchpin for a sustainable, secure, and efficient maritime industry, enabling the safe transition toward greener alternative fuels.
9. Environmental Awareness

   Environmental awareness ensures compliance with increasingly stringent regulations and aligns the industry with global sustainability goals. It encourages responsible fuel usage, reduces carbon footprints, and minimizes the impact on fragile ecosystems. By fostering a deep respect for the environment, seafarers become stewards of the seas, contributing to a cleaner, safer, and more sustainable maritime future.
10. Certification and Licensing

    Certification verifies seafarer's expertise, assuring employers and regulators of their capability to handle these fuels securely. Furthermore, stringent certification requirements help standardize training and uphold industry standards. They promote a culture of continuous learning and adherence to evolving safety protocols. Ultimately, these credentials enhance the industry's reputation, minimize risks, and ensure seafarers are well-prepared to navigate the complex challenges of alternative fuels in a rapidly changing maritime environment.
11. Feedback Mechanism

    This iterative process ensures that training remains current with the latest industry developments and safety protocols, enhancing the competency of seafarers. It also aids in the early detection and rectification of potential issues, ultimately reducing risks associated with alternative fuels. A robust feedback mechanism is the linchpin for a responsive and resilient training framework in the ever-changing maritime energy landscape.

4 Conclusion

There are still a lot of challenges faced by the shipping industry in the transition to alternative fuels, especially when we look into the unique properties of various alternative fuels. The IGF code can be used as a reference for handling such alternative fuels, but it must be kept in mind that the more alternative fuels are available, the more diverse characteristics and risks will be present. Consequently, the active amendments of the codes are a necessity. The best practices and methodologies for designing training programs according to STCW and its amendments, IGF code and its amendments, and IMO model courses are available but still do not cover all the characteristics of alternative fuels, especially for rapidly developing fuels like hydrogen, which still not matures enough in its development. By staying current with fuel technology advancements and enforcing the ensuing rules, shipping businesses and training facilities must modify their training programs to meet the changing needs of the alternative fuel technologies transition. Future research regarding alternative fuel training for seafarers will need to be more focused on analyzing alternative fuel-related incidents since most shipping-related principles are based on reacting to past events and use the report as the next industry best practice, which is currently quite limited.

Acknowledgments The author sincerely appreciates the research grants by *Samskip* and all the participants' support.

References

1. IMO: Initial IMO GHG Strategy. https://www.imo.org/en/MediaCentre/HotTopics/Pages/Reducing-greenhouse-gas-emissions-from-ships.aspx
2. DNV: The Future of Seafarers 2030: A Decade of Transformation. Det Norske Veritas Holding AS (2023)
3. IMO: IGF Code (MSC.391(95) Code of safety for ships using gases or other low-flashpoint fuels) (2015)
4. Islam Rony, Z., Mofijur, M., Hasan, M.M., Rasul, M.G., Jahirul, M.I., Forruque Ahmed, S., Kalam, M.A., Anjum Badruddin, I., Yunus Khan, T.M., Show, P.-L.: Alternative fuels to reduce greenhouse gas emissions from marine transport and promote UN sustainable development goals. Fuel **338**, 127220 (2023). https://doi.org/10.1016/j.fuel.2022.127220
5. DNV: Insights into Seafarer Training and Skills Needed to Support a Decarbonized Shipping Industry (2022)
6. IMO: Alternative fuel use—regulatory status mapped. https://www.imo.org/en/MediaCentre/Pages/WhatsNew-1841.aspx
7. IMO: Alternative marine fuels: regulatory mapping: GreenVoyage2050. https://greenvoyage2050.imo.org/alternative-marine-fuels-regulatory-mapping/

8. MacKinnon, D.P.: Introduction to Statistical Mediation Analysis (2008)
9. DNV: DNV Maritime Forecast to 2050 (2022)
10. LR: Lloyd's Register Decarbonisation Transition Pathways (2019)
11. IMO Resolution MSC.397 (95), STCW Code Part A Chapter V Section A-V/3 (2015)

MIX
Papier aus verantwortungsvollen Quellen
Paper from responsible sources
FSC® C105338

If you have any concerns about our products,
you can contact us on
ProductSafety@springernature.com

In case Publisher is established outside the EU,
the EU authorized representative is:
Springer Nature Customer Service Center GmbH
Europaplatz 3, 69115 Heidelberg, Germany

Printed by Libri Plureos GmbH
in Hamburg, Germany